Biotechnology in Environmental Management

BIOTECHNOLOGY IN ENVIRONMENTAL MANAGEMENT

Editors

Dr. T.K. Ghosh
Dr. T. Chakrabarti
Dr. G. Tripathi

A.P.H. PUBLISHING CORPORATION
ANSARI ROAD, DARYA GANJ
NEW DELHI-110 002

Published by
S. B. Nangia
A.P.H. Publishing Corporation
Ansari Road, Darya Ganj
New Delhi-110 002
☎ 23274050

ISBN 81-7648-807-0 (for set)
ISBN 81-7648-808-9 (for vol. I)
ISBN 81-7648-809-7 (for vol. II)

2009

Typesetting at
NEW APCON
20, DDA Shopping Complex
Panchsheel Park
New Delhi - 110 017
☎ 26494336, 51750865
Mobile: 31031007

Printed in India at
Chaman Enterprises Patoudi House
Darya Ganj New Delhi-110002
☎ 23271595

Preface

The title of the book *Biotechnology in Environmental Management* is truly descriptive of its contents. Biotechnology is very much an interdisciplinary field linking different branches of science: microbiology, fisheries, biochemistry, engineering etc. Within the past four decades, there has been generous assistance from genetics and molecular biology. The demand for information on this line will gradually increase as a function of Government attention and public awareness. Therefore, an understanding between experts in the various disciplines of the field is essential for the successful application of its basic concepts.

Biotechnology in Environmental Management may serve as a reference with a comprehensive description of the state of the art, as well as a guide to the original literature. The chapters have been selected according to the present need, and each topic provides a comprehensive and effective understanding to satisfy wider audience from students to professionals. Efforts have also been made to incorporate the issues ignored so far. It is specifically directed to environmentalists, microbiologists, aquaculturists, biochemists, bioengineers, chemical engineers, and food and pharmaceutical chemists working in the industries, at universities, or at research institutions. The issues may be of interest to policymakers and lawyers also.

The authors of individual chapters have been chosen because of their recognized expertise and their contributions to the field of biotechnology and environmental sciences. Their willingness to impart this knowledge to their colleagues forms the basis of this book and the editors gratefully acknowledge this. Thanks are due to all those who directly or indirectly cooperated us in compiling the book for the benefit of present and future generations. We are grateful to Mr. S.B.Nangia, Proprietor, APH Publishing Corporation, New Delhi, for publishing *Biotechnology in Environmental Management* with the excellence that readers have become accustomed to expect from this publisher.

Dr. T.K. Ghosh
Dr. Tapan Chakrabarti
Dr. G Tripathi

-Editors

ABOUT THE EDITORS

Dr. T.K. Ghosh, M.Sc., Ph.D. in Zoology from Kalyani University, (born 1951), is working as Scientist Gr.IV (4) in the Environmental Biotechnology Division, NEERI. After joining the Institute in 1979, he has worked in various fields, *viz.* wastewater aquaculture; characterization of sewage, sludge, soil etc.; aquatic and terrestrial biomonitoring; EIA; application of different indices on biological data for evaluating ecological health, trophic levels, trophic pyramids etc. He is the first Indian to start EIA studies around Indian permanent station Maitri in Antarctica in 1994. Presently he is leading the prestigious projects, sponsored by Department of Atomic Energy, Maharashtra Jeevan Pradhikaran, National Vector Borne Diseases Control Programme (GoI), Tehri Hydro Development Corporation, and Department of Biotechnology (GoI). Dr.Ghosh has published 55 papers in national/international journals/symposia proceedings, and 15 chapters in reputed books. Besides, he has contributed in over 100 project reports in NEERI. He is the Fellow of a number of scientific bodies, namely, Academy of Environmental Biology, Inland Fisheries Society of India, Academy of Zoology, Maharashtra Academy of Sciences, Society of Environmental Sciences, and Action for Sustainable Efficacious Development and Awareness. Dr. Ghosh is closely associated with various journals, and he is in the editorial boards of J.Indian Association for Environmental Management, J. Environmental Biology, Indian J. Environment & Ecoplanning, Environment Conservation J., J. Ecotoxicology & Environmental Monitoring, and AQUA TECH. (His contact *e- mail is:* ghoshtk2@rediffmail.com)

Dr. Tapan Chakrabarti did his under graduation post graduation, Doctoral Degree and Law graduation from Nagpur University. Presently, he is working as Director-Grade Scientist & Head, Environmental Biotechnology Division, National Environmental Engineering Research Institute, Nagpur under the Jurisdiction of Council of Scientific Industrial Research (CSIR), New Delhi under the Department of Science & Technology, Government of India. His areas of specialization are Environmental monitoring with special reference to toxic wastewaters and hazardous waste management, biotechnological production of value added chemicals from wastes, treatment and disposal of hazardous wastes genotoxicity of toxic wastewaters and leachates and toxicogenomics of exposed population. Presently he is handling about 30 R&D and Institute Projects. He has more than 30 years of research experience and supervised 14 students for the award of the degree of Doctor of Philosophy in the Faculty of Science, Nagpur University. He has been Principal Investigator of various research projects awarded by different funding agencies including Department of Biotechnology, Ministry of Environmental & Forests, Ministry of Health and Family Welfare etc. and CSIR. Dr. Chakrabarti has over 75 research International and National publications and contributed chapters in books published from renowned publishing companies. He has been honored **"Pitamber Pant National Environment Fellowship Award"** by the Ministry of Environment and Forests, Government of India, New Delhi. He is Life Member and Fellow of many reputed scientific societies and Executive member of various bodies and technical committees. He is member of Monitoring Committee of Supreme Court. His visits abroad include USA, Canada, Denmark, Germany, UK, Kenya, Egypt, Thailand, Korea, Canada and Australia in connection with education, assignment and profession. His name has been incorporated in various International compilations including Who's Who in the World.

Dr. G. Tripathi completed his undergraduate, postgraduate and doctoral degrees from Banaras Hindu University and subsequently worked as CSIR-Research Associate and Associate Professor in

Indo-US Project in this University. Presently, he is working as Associate Professor in the Department of Zoology, J.N. V.University, Jodhpur. His area of specialization centers around Biochemical Physiology and Environmental Biology. He has more than 21 years of research experience and supervised several Ph.D. theses. He has been Principal Investigator of various major research projects of different funding agencies including DBT and ICAR. Dr.Tripathi has over 130 research publications and authored/edited 10 important and interesting books. Maharshi Award from Banaras Hindu University, Varanasi and Zoology Shiromani Award from Indraprasth Shiksha Parisad, New Delhi, have honored him. He is Life Member and Fellow of many reputed scientific societies and executive member of various learned bodies. He is Council Member of The Academy of Environmental Biology, India, and is working on the Editorial Advisory Board of more than half a dozen reputed research journals. He is reviewer of various internationally acclaimed research journals including Com.Biochem.Physiol. (Ireland), Appl.Genom. and Proteom. (Ireland), and Fish Physiol.Biochem.(The Netherlands). He is also member of Eco-ethics International Union, Germany. His name has been incorporated in various international compilations including Who's Who in the World.

Contributors

Dr. S.P.S. Ahlawat
National Bureau of Animal Genetic Resources, P.B. 129. G.T.Road Bye Pass, Karnal- 132001, India

Dr. Sandeep Arora
Department of Biochemistry,
College of Basic Sciences and Humanities
G.B. Pant University of Agriculture & Technology, Pantnagar-263145, India

Dr. Meenakshi Banerjee
Laboratory of Algal Biotechnology, Department of Bioscience,
Barkatullah University, Bhopal, 462026, India

Dr. Anirban Bhattacharya
Environmental Toxicology Laboratory, Department of Zoology, Visva Bharati University, Santiniketan-731235, India

Dr. Shelley Bhattacharya
Centre for Biotechnology, School of Life Sciences, Visva Bharati University, Santiniketan 731235, India

Dr. Tapan Chakrabarti
National Environmental Engineering Research Institute, Nehru Marg, Nagpur-440 020, India

Dr. G.N. Chattopadhyay
Institute of Agriculture, Visva-Bharati
Sriniketan – 731236, India

Dr. Subhendu Datta
Central Institute of Fisheries Education, Kolkata Centre, 32-GN Block,
Sector-V, Salt Lake, Kolkata – 700091, India

Dr. S.P. De
Ex-Deputy Director of Fisheries, West Bengal.
46, S. N. Banerjee Road, Barrackpore, Kolakta – 700 120, India

Dr. T.N. Devaraja
301, Cosmos Apartment, Cosmos Enclave, Kadri Kambla, Mangalore 575 004, India

Dr. S. Saravana Devi
National Environmental Engineering Research Institute, Nagpur-440020, India

Dr. D. M. Dharmadhikari
National Environmental Engineering Research Institute, Nagpur 440020, India

Dr. Abhijit Dutta
Post Graduate Dept. Of Zoology, Ranchi University, India

Dr. Sunita Dutta
Department Of Zoology, Women's College, Ranchi, India

Dr. T.K. Dutta
Central Institute for Research on Goats
Makhdoom, Farah, Mathura – 281 122, India

Dr. N. H. Fuke
Nagpur Veterinary College, MAFSU, Nagpur 440006

Dr. A. K. Gaur
Department of Biochemistry,
College of Basic Sciences and Humanities
G.B. Pant University of Agriculture & Technology, Pantnagar-263145, India

Dr. D. Gayathri
Department of Microbiology, University College, Mangalore-575001, India.

Dr. T.K. Ghosh
National Environmental Engineering Research Institute, Nagpur–440 020, India

Dr. Aruna Golait
National Environmental Engineering Research Institute, Nehru Marg, Nagpur-440 020, India

Dr. Sunita Kanungo
National Environmental Engineering Research Institute, Nehru Marg, Nagpur-440 020, India

Dr. K. Krishnamurthi
National Environmental Engineering Research Institute, Nagpur-440020, India

Dr. Atul Kulkarni
National Environmental Engineering Research Institute, Nagpur-440020, India

Dr. Bharani Kumar
DNA Fingerprinting unit,
National Bureau of Animal Genetic Resources, P.B. 129. G.T.Road Bye Pass, Karnal- 132001, India

Dr. Naveen Kumar
DNA Fingerprinting unit, National Bureau of Animal Genetic Resources, P.B. 129. G.T.Road Bye Pass, Karnal- 132001, India

Dr. M. Suresh Kumar
National Environmental Engineering Research Institute, Nehru Marg, Nagpur 440 020, India

Dr. Ajoy Mandal
Central Institute for Research on Goats
Makhdoom, Farah, Mathura – 281 122, India

Dr. Basudev Mandal
Aquaculture Research Unit, Department of Zoology, Vidyasagar University,
Midnapore–721102, India

Dr. D.P. Mishra
Department of Biochemistry,
College of Basic Sciences and Humanities
G.B. Pant University of Agriculture & Technology, Pantnagar-263145, India

Dr. S.C. Mukherjee,
Director
Central Institute of Fisheries Education, Versova, Mumbai 400061, India

Dr. A.G. Murugesan
M.S. University, Sri Paramakalyani Centre for Environmental Sciences, Alwarkurichi – 627 412, India

Dr. A.K. Pal
Central Institute of Fisheries Education, Seven Bunglows, Versova, Andheri West,
Mumbai – 400061, India

Dr. Dinesh Pandey
National Bureau of Animal Genetic Resources, P.B. 129. G.T.Road Bye Pass, Karnal- 132001, India

Dr. P.N. Pandey
Post Graduate Dept. Of Zoology, Ranchi University, India

Dr. R.A. Pandey
National Environmental Engineering Research Institute, Nehru Marg, Nagpur - 440 020, India

Dr. Bidhan C. Patra
Aquaculture Research Unit, Department of Zoology, Vidyasagar University,
Midnapore–721102, India

Dr. T. Rajyalakshmi
Former Director, CIBA-ICAR, & President SPICAM G-2/1-22-14, Sriramnagar, Kakinada-533 003, India

Dr. Luiz Pereira Ramos
Universidade Federal do Paraná - Departmento de Química, CEPESQ – Centro de Pesquisas em Química Aplicada, Centro Politécnico – CP 19081 – 81531-990 – Curitiba – PR - Brazil

Dr. P.W. Ramteke
Allahabad Agricultural Institute-Deemed University, Allahabad – 211007, India

Dr. Sonali Roy
Environmental Toxicology Laboratory, Department of Zoology, Visva Bharati University,
Santiniketan-731235, India

Dr. John Ruby
M.S. University, Sri Paramakalyani Centre for Environmental Sciences, Alwarkurichi – 627 412, India

Dr. P.K. Sahoo
Central Institute for Research on Goats
Makhdoom, Farah, Mathura – 281 122, India

Dr. N.P. Sahu
Fish Nutrition and Biochemistry Division
Central Institute of Fisheries Education, Versova, Mumbai 400061, India

Dr. Bijaya Ketan Sarangi,
National Environmental Engineering Research Institute, Nehru Marg, Nagpur-440 020, India

Dr. D.B. Satpute
National Environmental Engineering Research Institute, Nagpur 440020, India

Dr. Kestur Gundappa Satyanarayana
Universidade Federal do Paraná - Departmento de Química, CEPESQ – Centro de Pesquisas em Química Aplicada, Centro Politécnico – CP 19081 – 81531-990 – Curitiba – PR - Brazil

Dr. G K. Sehgal
Fisheries Research Complex, Department of Zoology and Fisheries, Punjab Agricultural University, Ludhiana- 141 004, India.

Dr. H.S. Sehgal
Fisheries Research Complex, Department of Zoology and Fisheries, Punjab Agricultural University, Ludhiana- 141 004, India.

Dr. Sutapa Sen
Environmental Toxicology Laboratory, Department of Zoology, Visva Bharati University, Santiniketan-731235, India

Dr. B. A. Shamasundar
Department of Fish Processing Technology
College of Fisheries, Mangalore – 575 002. India

Dr. Rekha Sharma
DNA Fingerprinting unit, National Bureau of Animal Genetic Resources, P.B. 129. G.T.Road Bye Pass, Karnal- 132001, India

Dr. Pratima N. Shastry
Laxminarayan Institute of Technology, Nagpur

Dr. A. R. Sirothia
Nagpur Veterinary College, MAFSU, Nagpur 440006

Dr. K. A. Sirothia
Nagpur Veterinary College, MAFSU, Nagpur 440006

Dr. N. Sukumaran
M.S. University, Sri Paramakalyani Centre for Environmental Sciences, Alwarkurichi – 627 412, India

Dr. G Tripathi
Department of Zoology, J.N.V. University, Jodhpur-342 001, India

Dr. M.K. Tripathi
Department of Biochemistry, College of Basic Sciences and Humanities, G.B. Pant University of Agriculture & Technology, Pantnagar-263145, India

Dr. Fernando Wypych
Universidade Federal do Paraná - Departmento de Química, CEPESQ – Centro de Pesquisas em Química Aplicada, Centro Politécnico – CP 19081 – 81531-990 – Curitiba – PR - Brazil

Dr. Bernhard Zechendorf
European Commission, Research Directorate-General, Square de Meeus 8, B-1049 Brussels, Belgium

CONTENTS

C Aquaculture

D Cleaner Technology

E Bioremediation

F Waste Management

CLEANER TECHNOLOGY

Biotechnological Applications of Cyanobacteria in Development of Clean Technologies for Environmental Management

Meenakshi Banerjee

Introduction

The amazing unbelievable freak – earth was formed some 3.65 billion years ago, since then all life gradually evolved from hydrogen compounds, algae, plant, fishes, amphibia, animal and mammals to man itself. But Man in quest of means to provide itself with adequate food, energy supply and an acceptable personal environment tampered with Earths ecosystem so much that life of man and other living organisms are now under great threat .The social and environmental impacts of modern intensive farming and agriculture, industrial development and energy supply are well documented. Growing population and pollution are the two major concerns of the modern world .The growth of population and increase in civilization has led to an increase in inadequate supply of food in a starved world. Besides this over exploitation has also led to the production of enormous quantities of wastes resulting in serious and alarming environmental problems. All this has focus attention to the urgent need of developing alternate methods of increasing food production by seeking alternatives to chemical fertilizers which are not only costly but also harmful ecologically in the long run and development of environmentally friendly clean technologies to deal with the massive environmental pollution.

The cyanobacteria or blue green algae are a group of photosynthetic, oxygen evolving prokaryotes which have the capacity to fix atmospheric nitrogen .This dual attributes makes them tropically independent and gives them a unique potential to contribute to productivity in variety of agricultural and ecological situations. Besides this cyanobacteria are potentially new and promising sources of drugs, enzymes and bioactive compounds. High value restriction enzymes are already being extracted from some strains of *Anabaena* and *Nostoc* and marketed by chemical companies. Other enzymes like Superoxide dismutase, Phosphatases and a whole range of other enzymes of biotechnological importance have been reported from cyanobacteria (Whitton *et al.,*., 1991, Banerjee *et al.,* 2000 a, b, Chandrashekaran 1997, Cohen 1986).

Out of over 300 cyanobacterial species screened for antiviral activity (Hayashi and Hayashi 1996,Reshef *et al.,* 1997, Archer *et al.,* 1997). Several strains belonging to the Order Chroococcales were found to produce substances that caused significant reduction in cytopathic effect normally associated with viral infection. *Spirulina* is an exceptionally valuable cyanobacterium as human food. .It is a rich source of protein ,vitamin B-12 and iron and as ideal food supplement for vegetarian people .Cyanobacteria are good sources of pigments such as phycocyanin and phycoerythrin which in turn is used for production of health foods ,cosmetics and fluorescent probes. These organisms have potential commercial application as a source of renewable energy especially when the algal growth is coupled to waste water treatment .

Cyanobacteria are also known to produce compounds that stimulate growth of plants. ATP can be conveniently produced photosynthetically by using the thermophillic cyanobacterium *Mastigocladus*. Diverse cyanobacteria are also known to produce hydrogen as a normal metabolic activity by the splitting of H_2O during the photosynthesis. Mankind is already on the threshold of the space age. Satellite and orbiting space stations are becoming increasingly commonplace. As it is expensive to launch and send sufficient supplies of air, water and food for long duration space missions, attractive and low cost alternatives that obviate the need for re-supply from earth are being developed. These recycling life support systems are based on biological processes to regenerate oxygen, water and food from waste materials using cyanobacteria and other microalgae.

A large array of natural products of economic potential may be produced from cyanobacteria which can replace the chemical counter part thereby reducing the toxic effects. Attractive sources of pigments such as the phycobiliproteins and caretonoids, have potential applications as natural dyes and colourants and widely used in fluorescence microscopy, in the food industry as well as for cosmetics (Walsh *et al.*, 1997). Cyanobacteria can also be considered for use as micro particulate tracers in sub surface research (Colwell *et al.*, 1994). Allelochemical from *Fischerella muscicola* have anti-fungal and herbicidal activity (Hagmann and Juttner 1996). Barbamide a chlorinated metabolite with molluscicidal activity from *Lyngbya majuscula* and lignolytic activity from *Anabaena azollae* have been reported (Orjala and Gerwick 1996). Cyanobacterial extracts have been reported to promote somatic embryogenesis and plantlet formation in carrot (Wake *et al.*, 1992) and the colonization of wheat by the nitrogen fixing cyanobacterium has also been reported (Gantar *et al.*, 1995). Curacins Band C , new antimitotic natural products from *Lyngbya* and Raocyclamides A and B, noval cyclic hexapeptides from *Oscillatoria raoi* have been reported (Admi *et al.*, 1996). Extra-cellular leachates of *W. prolifica*. and *Scytonema* were lethal for mosquito larve (Tyagi and Kaushik, 1999). A mosquitocidal cyanobacterium has been developed by introducing the mosquito toxic cryIVD gene from *Bacillus thuringiensis* into cyanobacterium *Agmenellum quadruplicatum* (Murphy and Stevens 1992).

Two serious environmental problems that need to be addressed in the present scenario are the massive environmental pollution due to burning of fossil fuels and other industrial activities, which is a serious threat to the air, soil and water ecosystems and the huge input of chemical fertilizers to make agriculture sustainable. Even in the developed countries the realization is dawning that the use of chemical fertilizers in agriculture is in fact not sustainable in the loosest sense of the word and that massive problems are ahead from diminishing supply, increased cost of inputs and the dreadful residual effects that persist in the environment (Banerjee 2001). The challenge, therefore, is to develop practices, which emphasize on the use of eco-friendly systems rather than depending on traditional practices, which are slowly eating into the core of life quality and destroying it beyond repair. Potential applications of cyanobacteria have already been discussed and there is no doubt that they can serve to lessen the environmental pollution to a great extent because they possess certain distinct advantages over higher plants when grown on solar energy harvesters They grow at a faster rate, are easier to manipulate in open ponds or closed reactors thus ensuring their mass cultivation and generally possess high photosynthetic efficiency. They are also endowed with certain additional traits which make them very attractive as role models for chemical alternatives.

This review, therefore, summarizes potential biotechnological applications of cyanobacteria in development of clean technologies, as a mode of environmental management for a world which is becoming increasingly environmentally unclean. Here the feasibility of using cyanobacteria for 1) production of hydrogen as a source of clean fuel 2) for pollution control and 3) as biofertilizers has been discussed.

Hydrogen production by cyanobacteria

Producing a suitable renewable fuel particularly employing biosystems is one of the priority research interests of the scientific community today throughout the world. Hydrogen is starting to move from a fuel of the future to an energy carrier of the present, promising great reduction in pollution and increased fuel efficiencies. Major goal is, the production of renewable H_2 fuel at affordable costs. Renewable hydrogen production was made less urgent by fossil fuel glut starting in the early 1980s, but is now again of increasing priority as the specter of global warming moves from prediction to reality and as long term supplies and cost of fossil fuels are again of concern. However none of the renewable H_2 production processes currently available, such as photovoltaics-electrolysis and gasification of biomass, can yet deliver H_2 fuel for less than US $20/GJ, a reasonable maximal cost target for renewable H_2 fuel. Thus there is considerable interest in alternative concepts including proposed biological processes and biological H_2 production has been the subject of extensive applied research and development efforts, with major programmes in Japan, Europe, and the USA during the past decade and recently too. Cyanobacteria are also looked upon as a source of biofuels. The biomass of cyanobacteria like other photosynthetic organisms can be used for generation of methane either by phycico-chemical or biological methods (Benemman 2000, Benemman *et al.*, 1986). Cyanobacteria are one among the few life systems capable of producing the energy rich non polluting fuel hydrogen by the photolysis of water with light as the energy source. A number of studies have been carried out since the early 1970s on the utilization of photosynthetic cells to convert solar energy into chemical energy of hydrogen fuel (Bothe 1982, Hall and Rao 1989, Lambert and Smith 1981, Markov *et al.*, 1995a, Hall *et al.*, 1995, Banerjee and Kumar 1989).

Compared to other H_2 producing phototrophic microorganisms the cyanobacteria possess certain advantages which makes them potential candidates for development of photobiological H_2 generating systems. They are O_2 evolving prokaryotes which can grow in air (source of N_2 and CO_2), simple mineral salts and water (as a source of electrons and reductants) with light as the energy source. They adhere easily to polymer surfaces and so the cells can be readily immobilized on solid matrices, which is an advantage in design of bioreactors in corporating these organisms. There are three enzymes involved in hydrogen metabolism by cyanobacteria. H_2 evolution from cyanobacteria is catalyzed mainly by nitrogenase, a metalloenzyme highly sensitive to oxygen. In cyanobacterial fillaments the nitrogenase is located inside specialized thick walled cells called heterocysts. The cell wall of heterocyst is impermeable to oxygen (Peterson and Burris 1978), and is devoid of photosystem II, hence there is no photosynthetic oxygen produced inside these cells. The main function of nitrogenase in cyanobacteria is the reduction of molecular nitrogen to ammonia. However the simultaneous evolution of hydrogen along with nitrogen reduction is an inherent property of the nitrogenase. Under normal conditions the volume of hydrogen evolved is only about a fourth of the ammonia synthesized. Photo production of ammonia and hydrogen are energy consuming reactions. There are two hydrogenases involved in hydrogen metabolism in cyanobacteria. The enzyme designated as reversible hydrogenase is located on the cytoplasmic membrane and catalyzes hydrogen evolution or uptake depending on the redox status of the culture. Cyanobacteria also contain another enzyme whose only function is hydrogen consumption designated as uptake hydrogenase and is located in the thylakoid membrane (Hansel and Lindbald 1998). The diversity of cyanobacterial hydrogenases are also elucidated with molecular approach (Tanagnini *et al.*, 2000). The net hydrogen evolution by cyanobacteria is thus the sum of hydrogen production catalyzed by nitrogenase and hydrogen consumption catalyzed by the uptake hydrogenase. The rate of hydrogen production from cyanobacteria also depends on various parameters such as cell age, nutritional status gas phase and other environmental factors such as microaerobic conditions, partial vaccum, argon or CO_2 enriched atmosphere (Markov *et al.*, 1995 a).

Immobilization of cells by adsorption on solid matrices or by entrapment in gels or polymers like agar, calcium alginate, chitosan, carrageenan, polyvinyl and poly urethan foams increases the functional life time of the cells and influences the hydrogen production. Hall *et al.*, (1995) showed optimization of hydrogen evolution by immobilized cyanobacterial cells in bench top photobioreactors. Park *et al.*, (1991) used a trickling medium column bioreactor for hydrogen production in using *Anabaena azollae*. Hydrogen production from *Anabaena variabilis* cells immobilized on hollow fibre was studied in batch cultures and in column reactors of various dimensions. Rates of hydrogen production varied from 20 to 200 ml $H_2\,g^{-1}$ dw $^{-1}h^{-1}$ (Markov *et al.*, 1993, Benemann 2000). The effect of partial vaccum, temperature and ambient CO_2 levels on photoproduction of H_2 by *Anabaena variabilis* was also studied by Markov *et al.*, (1995 b) in which maximum H_2 evolution rates was observed in the period following photosynthetic CO_2 assimilation.

The practical applications of cyanobacterial photobioreactors for hydrogen production will necessitate efficient systems for harvesting and conversion of solar energy to chemical energy of hydrogen fuel. Otherwise photobiological systems may not be able to compete with already established photovoltaic systems for hydrogen production by water photolysis (Benemann1994, 2000). Light energy conversion is low in cyanobacterial hydrogen production probably because these values are calculated on the basis of radiations incident on the cell rather than the energy absorbed. The two main reasons for this low efficiency probably are photoinhibition of phohotosynthesis and self shading in high density cultures.

What are the prospects to modify cellular hydrogen evolving machinary for higher production of the gas?

Net hydrogen production from cyanobacteria is the sum of hydrogen evolution catalyzed by the nitrogenase and hydrogen consumption by uptake hydrogenase. Hence any modifications which would enhance nitrogenase activity or inhibit uptake hydrogenase would in turn increase hydrogen production. Some of the methods being employed, with positive results are:

1) Increase in number of heterocysts mediated by certain physiological conditions and chemicals which would result in more nitrogenase.

2) Substitution of Mo nitrogenase by V nitrogenase by growing the cyanobacteriain Mo deficient and V enriched medium is found to increase nitrogenase activity and higher hydrogen evolution.

3) Growing cells in Ni deficient medium which in turn would hamper the synthesis of uptake hydrogenase for which Ni is essential would increase hydrogen evolution.

4) Development of uptake hydrogenase deficient mutants (hup⁻) in cyanobacteria would produce more hydrogen.

Thus by the application of biochemical and molecular biology techniques and selection it might be possible to achieve greater levels of hydrogen production which is definately a clean fuel. It was promising that screening of marine cyanobacteria led to the detection of *Synechococcus* sp Miami BG 043511 (Mitsui *et al.*, 1981,Kumazawa and Mitsui 1994) as a strain capable of very high ıates of hydrogen evolution (250µ mol. mg. Chl a $^{-1}$ h^{-1}) with sustained production and little oxidation activity.

Of course, enormous research challenges will be faced in advancing from the present conceptual systems and low efficiency laboratory scale experiments to the large scale, high efficiency sustained processes required in practical applications. Indeed this will require a complete genetic and biochemical make over of the cyanobacterial cell, to redirect essentially the entire metabolism of these organisms from cell growth to cycles of carbohydrate storage and hydrogen release. Although each step in the process is based on known enzymatic capabilities of cyanobacteria, these must still be joined together into one process in a single organism, operating at near maximal efficiency in outdoor systems under full sunlight and subject to contamination and diurnal temperature fluctuations. And even achieving these ambitious goals is not enough: the process must also be contained within a very low capital cost apparatus requiring only minimal operating costs. These dual challenges, of very high efficiencies and very low costs are, of course, also faced by other practical applications of cyanobacteria, of which biohydrogen production is but the most challenging example.

Cyanobacteria in pollution control

Pollution of rivers, lakes and ponds by nitrates and phosphates is a topic of great concern especially when these water bodies are supplied with domestic wastes. Cyanobacteria have been successfully used for tertiary treatment of agriculture and farm waste waters. The use of *Phormidium bohneri* for removal of phosphates and nitrates from the effluents has been studied. Removal of nitrogen and phosphorus from polluted waters using bench top bioreactors incorporating Polyvinyl and Polyurethane immobilized thermophillic *Phormiduim laminosum* has been reported (Garbisue *et al.*, 1992, 1994). A nitrate uptake efficiency of 90% was achieved in the bioreactors.

Significant transitions in the availability, toxicity and subsequent use of metals are predicted to have coincided with the release of di-oxygen in quantity and hence coincided with the first appearance of cyanobacteria. During the rapid adaptive radiation of cyanobacteria there must have been a) selection for the evolution of new metal resistant determinants, and b) greater opportunity for more available metals to be recruited for use in newly evolved proteins (Robinson *et al.,* 2000). Today the availability of metals in many environments fluctuates both spatially and temporarily not only as a result of natural processes but also as a consequence of anthropogenic factors.

Cyanobacteria are known to take up heavy metals in industrial effluents by accumulating the toxic metals within the cells. They thus act as waste traps (Rai *et al.*, 1981, Wilde and Benemann 1993). They can then be harvested to remove the toxic compounds from the water. Zinc, mercury, cadmium, copper, uranuim and lead are some of metal ions that can accumulate as much as several thousand fold in cyanobacteria. Metal accumulation can be either by bioaccumulation or biosorption. Alkali extracted biomass of *Phormidium valderianum* capable of adsorbing Cd^{++} Co^{++} Cu^{++} and Ni^{++} and reusable by desorption and regeneration was reported by Karna *et al.*, (1999). Filamentous cyanobacterial biomass treated with NaOH can be utilized as a novel biosorbent for Cd removal (Nagase *et al.*, 1997). Biofiltration using cyanobacteria has also been attempted. *Oscillatoria annae* cells were mixed with sand and used in a column to treat sewage and oil industry wastes. It could successfully decrease BOD and COD and remove P and N residues. The ability of cyanobacteria to treat various effluents like paper mill effluents, Ossein effluents and domestic sewage by employing a sewage isolate of *Oscillatoria psuedogeminata* was demonstrated by Manoharan and Subramanian (1993).

Cyanobacteria play an important role in streams **polluted with sewage** .The algae use the nitrate, phosphate and other nutrients present in sewage to grow and multiply rapidly .Since they photosynthesize they oxygenate the water and allow the bacteria to thrive and decompose the organic nutrients .The end result of this bacteria - algae cooperation is that the organic wastes are broken down and the water is purified . The cyanobacteria like *Oscillatoria* and *Anacyatis* along with other microalgae play an important role in sewage stabilization ponds. Trickling filters, activated sludge systems, and lagoon system are the three commonly used methods for treatment of municipal sewage. Sewage allowed to remain in an oxidation stabilization pond for 20-30 days can be sufficiently purified through algal –microbial action to be used for irrigation of crop fields. Cyanobacteria also offer an interesting alternative for tertiary waste- water treatment in the Artic and Antarctic zones (Tang *et al.*, 1997). Cyanobacterial biosensors incorporating *Synechococcus* as the biocatalyst has been evaluated as potential biomonitors for **detecting water pollutants**. The biosensor can be used to detect herbicides like dinuron and atrazine that interact with photosynthetic electron transport chain. **Pesticides and other pollutants** can also accumulate in cyanobacterial cells. The algae which have accumulated toxic metals or pollutants may be utilized as biomass for biogas production or in some selected cases as food for animals.

Inevitably, pesticides are commonly used in agriculture and also become a component of industrial effluents and many cyanobacteria are capable of degrading them. Cyanobacteria can degrade methyl parathion as well as phenol from industrial effluents (Megharaj *et al.*, 1994, Shahirekha *et al.*, 1997). .Filamentous cyanobacteria have a natural ability to degrade a highly **chlorinated aliphatic pesticide.** Lindane (Gamma hexa chlorocyclohexane) presents quantitative evidence that this ability can be enhanced by genetic engineering to degrade another cholrinated pollutant, 4-chlorobenzoate (Kuritz and Wolk 1995).

Based on a screening of paddy field heterocystous strains of cyanobacteria, many organophosphorus pesticides, were acted upon and utilized by cyanobacteria (Subramanian *et al.*, 1994) .Studies with *Aulosira fertilissima* ARM68 showed that these pesticides were not only degraded but also used as a source of phosphate probably by phosphatase activity .The effect of **organochlorine** and **carbamate** and **organophosphorus pesticides** by sheathed forms like *Lyngbya major*, *Gloeocapsa atrata* , *Calothrix parietina* and *Scytonema pascheri* showed the ability of cyanobacteria to tolerate them. It has also been shown that the herbicde glyphosate is metabolized into aminomethyl phosphonic acid by *Synechocystis* PCC 6803. Cyanobacteria are capable of degrading various **aromatic hydrocarbons** such as napthalene, n-alkanes and **crude oil spills** by microbial mats of cyanobacteria (Al Hasan *et al.*, 1994.Radwan *et al.*, 2000).

There is increasing evidence that ancient cyanobacteria were among the direct biogenic contributors to oil formation. Evidence indicates that cyanobacterial strains are capable of oxidizing aromatic and aliphatic oil constituents(Al Hasan *et al.*, 1998). Further, in cyanobacteria dominated mats, the cyanobacteria live in natural association with hydrocarbon degrading bacteria and fungi that occur in the cyanobacterial polysaccharide layers. Such mat associations flourish in oil polluted coastal areas. From these studies it is apparent that the cyanobacteria like *Microcoleus* and *Phormidium* provide the oil degraders such as *Rhodococcus, Arthobacter, Pseudomonas,* and *Bacillus* with a number of advantages in mat associations. The cyanobacteria immobilize these oil degraders in their membrane thus protecting them from being washed away into the sea and provide oxygen needed for their initial attack on the hydrocarbon substrate. Oil degradation can also be stimulated by N_2 fixation of cyanobacteria.

Thus, the combined activities of the cyanobacteria and oil degrading organotrophs appears to be crucial and effective in **bioremediating such oil polluted** environments (Radwan *et al.*, 2000). Also, the fact that the cyanobacteria immobilize in their envelope layer oil degrading bacteria and probably have done for geological ages, provides the attractive possibility that genetic interactions could possibly have occurred naturally between the organotrophic and phototrophic partners. Through the coding of this gene for catalyzing enzyme systems, hydrocarbon degradation may have been transferred from the bacteria to the cyanobacteria. Cobley *et al.*, (1993) reported the construction of a shuttle plasmid which could be effectively mobilized from the *E. Coli* to the cyanobacterium *Fremyella diplosiphon*. In other words the presence of cyanobacterial mats appears to be an ideal association for self cleaning of oil polluted environments.

Cyanobacteria form a major component of the biota **of hyper saline environments** including salt lakes, solar salterns, hypersaline lagoons and hypersaline sulphur springs (Des Marais 1995, Oren 2000). The development of massive cyanobacterial mats mostly by fertilization in the saline environments has great biotechnological applications in salt recovery. The colour the cyanobacteria impart to the brine in these ponds are absolutely essential for salt production, since the vapor pressure reaches such low values as to seriously retard evaporation. Increased solar absorption caused by the cyanobacteria leads to elevated temperatures, which helps to offset the effect of low vapor pressures and permits adequate evaporation (Davis 1978, 1993). Also species like *Aphanothece* release massive amounts of slime which has very promising industrial uses. This mucilaginous material helps to seal the porous bottom of evaporation ponds and thus reduces the loss of brine to the ground waters (Borowitzka 1981, De Philippis *et al.*, 1993, Sudo *et al.*, 1993). Management practices that maintain biodiversity and restrained production and organic release by cyanobacteria ,permits these organisms to efficiently strip and oxidize nutrients from the brine thus purifying it (Coleman and White 1993, Davis 1993, Rahaman *et al.*, 1995). Another interesting potential application of halophillic cyanobacteria has been reported by Cohen and Ideses (1989). Extremely high concentrations oxycarotenoids especially canthaxanthin were found in hypersaline cyanobacterial mats. Cultivation of these mats was suggested for the production of these valuable carotenoids. Thus halophilic environments which are generally considered to be harmful and non productive can be converted to a very productive environment both ecologically and biotechnologically by incorporating cyanobacteria which have great adaptability to survive any and everywhere. Pollution of the environment therefore can be controlled to a great extent by the cyanobacteria.

Cyanobacteria as Biofertilizers

The role of cyanobacteria in maintenance of soil fertility is very well documented. As early as De(1939), reported that cyanobacteria were largely responsible for nitrogen recuperation under rice field conditions. The use of cyanobacterial inocula to enhance nitrogen fixation was first tested systematically in Japan. **Algalization** is the term now used for the practice of adding cyanobacteria to soils and is used widely. This was followed by a series of field as well as laboratory studies on biological nitrogen fixation (BNF). BNF through cyanobacteria was not only found to contribute through soil nitrogen but also found to benefit the crop in many ways (Singh 1961, Watanabe 1962, Banerjee and Deb 1996, Banerjee 2001). Biological nitrogen fixation contributes 25-30 Kg N, ha^{-1} per cropping season, by cyanobacterial activity and increases the rice yield by 10-15 % (Kannaiyan 1985). Use of **cyanobacterial inoculum** is already in practice but still not as wide spread as it should be (Sprent and Sprent 1990). Although rice fields provide a favourable environment for the development of a large variety of free living cyanobacteria (Antarikanonda 1984, 1985, Banerjee and Kumar 1992, Banerjee 1991, 2001), it is generally assumed that nutrients fixed by the cyanobacterial associations are made available to the rice plants through exudation autolysis and microbial decomposition.

However under field conditions only a part of this fixed nitrogen is available to the rice plants some being reincorporated by the microflora or volatilized. Under these circumstances it is difficult to control the flow of nitrogen compounds need for the development of the rice plants. The major breakthrough which has come forth as a solution for this problem was to develop strains of cyanobacteria which release ammonium continuously. **An ammonia excreting mutant** (SSI) of the nitrogen fixing rice field cyanobacterium *Anabaena siamensis* was isolated by Thomas *et al.*, (1991), Boussiba *et al.*, (1991).

Fertilization of the rice plants under laboratory conditions by application of another MSX resistant mutant of *Anabaena variabilis* SA1 was found very successful. Similar mutants have been obtained since then for mny cyanobacteria like *Nostoc muscorum* and *Anabaena variabilis* (Singh *et al.*, 1983, Spiller *et al.*, 1986, Kamuru *et al.*, 1998). Since then, isolation and scaling up of the technology has taken place from the laboratory to the rice fields in order to evaluate their biofertilizer potential. Work of Kamuru *et al.*, (1998) showed that inoculum of nitrogenase depressed mutants of *Anabaena variabilis* (Strain SA1) to rice and wheat plants greatly increased the yield. The contribution of the mutant cyanobacterium to growth and yield of the rice plant was equivalent to the application of 71-73 Kg N, ha^{-1}. Efforts have also been made to study the accumulation of ^{15}N in rice inoculated with the parent and a nitrogenase depressed mutant of *Anabaena variabilis*. Significantly high levels of ^{15}N accumulated in plants inoculated with the mutant strain indicating that more fixed nitrogen was readily available for root uptake and assimilation (Kamuru *et al.*, 1997). Progress has also been achieved in producing some rice field isolates in open raceway reactors (Fontes *et al.*, 1987, Boussiba 1988, 1991, 1993, Tredici *et al.*, 1991).

This has great significance for cyanobacteria to be used as biofertilizer because the prevalent techniques being used has its limitations mainly due to its inability to efficiently produce good quality inocula and the difficulty in some cases in reestablishing the inoculated strains in the field (Grant *et al.*, 1986, Roger and Watanabe 1986). The tubular reactor systems have emerged as a reliable means of supplying desired microorganisms for different purposes and when used to produce nitrogen biofertilizers it has other specific advantages such as direct application of these dense cultures into the field thus reducing the cost of production, improved cell viability, prevented dilution of the culture media by rain water and constant gave a rich source of protein for food and fodder.

Ammonia excretion by *Anabaena azollae* was demonstrated by Kannaiyan (1993). This organism and *Anabaena variabilis* when immobilized on poly urethane and poly vinyl foams and calcium alginate beads released ammonia extracellularly (Brouers *et al.*, 1988, Aruna *et al.*, 1998, Kannaiyan *et al.*, 1997). Cyanobacterial cultures under immobilized conditions with PU, are well protected inside the pores of PU and also increases the nitrogen use efficiency (NUE) of rice owing to its continuous photoproduction of ammonia in rice fields. The effect of inoculation of immobilized cyanobacteria in the transplanted rice was also significant (Kannaiyan *et al.*, 1994, 1997).

Extensive studies have also been done on the factors affecting nitrogen fixation of cyanobacterial communities in rice fields which is a pre-requisite to be able to exploit these organisms for maximum benefit (Banerjee and Kumar 1987, 1988, Banerjee 1991, Banerjee and Hader 1996, Banerjee *et al.*, 1998). Attempts have also been made to study the relationship between abundance of nitrogen fixing cyanobacteria and environmental factors of rice fields which showed that the ratio of dissolved inorganic nitrogen to soluable reactive phosphate determined cyanobacterial distribution (Quesada and Fernandez, 1996, Quesada *et al.*, 1997, Banerjee and Kumar 1986, Banerjee 2001). Short term *in situ* acetylene reduction assays conducted on mats of *Aulosira*, one of the strongest nitrogen fixers of Indian paddy fields showed a diurnal pattern with higher activity at day times and reduced activity in the dark. One outstanding observation was that this organism could fix nitrogen considerably the night hours (Banerjee and Kumar 1992), which adds to its importance as a biofertilizer even more.

Although most of the studies using cyanobacteria as biofertilizers have been conducted on rice, its effect on other crops have been studied and found to be very stimulatory (Banerjee and Abraham 2000a). The protein rich non nitrogen fixer *Spirulina* was also found to improve productivity in rice and wheat (Banerjee and Deb 1996). Cultivation of *Azolla* with its associated nitrogen fixing *Anabaena*, provides another means of adding combined nitrogen to ecosystems. *Azolla* is mostly inter- cropped with rice in flooded fields contributing combined nitrogen, green manure and sometimes acting as food for fish in ditches within rice fields. The surface floating mats of *Azolla* plants can bring about a marked reduction in weed density including some serious rice field nuisance. The *Anabaena-Azolla* association has an advantage over most free-living cyanobacteria in being less sensitive to inhibition by combined nitrogen so less caution is needed if a nitrogen fixer is added. Isolated *Anabaena azollae* and *Anabaena cycadae* were also found to have a profound effect on the growth and yield of rice and wheat under field conditions and pot cultivation (Banerjee and Abraham 2000a, b,). Another novel approach has been the use of flyash, an environmental pollutant and cyanobacteria as a manure for crops like rice wheat mustard with significant stimulatory results (Banerjee 1992, Banerjee and Deb 1996, Banerjee *et al.*, 1999). Cyanobacterial biomass harvested from lakes have also been used as a biofertilizer for rice wheat and mustard (Banerjee and Sharma 1994, Banerjee *et al.*, 1997).

Organic phosphorus is a major component of the global phosphorus cycle and dissolved organic phosphate constitutes more than half of the filterable total phosphorus in waters of rice fields. Phosphatase activity is found in microorganisms capable of hydrolysing phosphorus compounds and is wide spread among cyanobacteria. Phosphatase therefore not only plays a very significant role in hydrolyzing organic phosphate for cyanobacterial growth but might be also providing the rice plant with inorganic phosphate. It appears from studies that cyanobacterial phosphatase activity plays a significant role in rice fields where the concentration of phosphate may vary considerably, not effecting phosphatase induction or repression, and under such conditions cyanobacteria are able to hydrolyze organic phosphates to support their own growth as well as the growth of rice plants (Banerjee 2003). From the above studies there seems little doubt that increasing the cyanobacterial biomass in rice fields will enhance nitrogen and possibly phosphate supply for the rice plants and subsequent crops thereby acting as a dual fertilizer. The possible ways, this might be achieved in addition to use of cyanobacterial inocula, are as follows:

1) Any extension of the period when the field is wet or flooded to the fallow season is likely to increase cyanobacterial growth, though it may encourage weed growth and the need for extra ploughing which can be undertaken.

2) Addition of phosphate fertilizers and in some cases liming increases nitrogen fixation of the cyanobacteria present.

3) Deep placement of nitrogen fertilizers overcomes the inhibitory effect on cyanobacterial growth and nitrogen fixation naturally.

4) Controlling the level of grazing but not reducing it to such low levels so as to hinder nitrogen cycling.Suppression of ostrocod grazing by neem tripled the cyanobacterial biomass and increased nitrogen fixation rates ten fold.

5) Attempts to develop new symbiotic associations: the most direct way to enhance nitrogen fixation in crops would be either to develop new symbiotic associations or to introduce a functional nitrogenase system into the cells of the crop. Cyanobacteria are especially well suited for this purpose because of the great variety of symbiotic associations that occur between them and other organisms and because most of the cyanobacteria involved are nitrogen fixers. The well documented cases of symbiotic nitrogen fixers possess heterocysts which are presumably the site where the fixation occurs. However several possible non- heterocystous symbiotic nitrogen fixers have been reported and a relatively wide range of free living non heterocystous cyanobacteria have been studied including a few records of unicellular forms. These unicells are especially suited for incorporation into eukaryotic cells.

Thus, undoubtedly cyanobacterial biofertilizers is one of the most potent clean technologies that hold promise of replacing chemical nitrogenous and probably phosphorus fertilizers in the near future and then we can hope to be freed of the malefic effects of traditional fertilizer applications which is causing nothing but a polluted environment.

Conclusion and future prospects

In this review an attempt has been made to discuss the potential and applications of cyanobacterial biotechnology in the management of the environment by adopting clean technologies using cyanobacteria as substitutes for generation of fuel, fertilizer and remedy to curtail pollution. It would be most desirable to have processes that could be developed in the near term, without requiring the very high efficiencies and low costs of production. One plausible approach would be to combine cyanobacterial harvesting based on waste-water treatment. As cyanobacterial role in waste- water treatment processes is well documented, the combination of these technologies into an integrated system for cyanobacterial biomass production would greatly increase hydrogen production, depollution of the water and also biofertilizers. In such waste- water treatment processes CO_2 is often the limiting factor, suggesting that CO_2 by recycled internally and hydrogen become the major fuel output and the biomass on harvesting serve as the biofertilizer. This type of management process would greatly reduce the economic constrains and performance criteria such as photosynthetic efficiency. With the development of PCR based methods for rapidly analyzing DNA sequences from organisms in different environments the technology of molecular genetics is increasingly being used to address ecological, environmental and evolutionary questions related to pollution and toxicity what ever be the cause. In this context, to utilize the study of repetitive DNA and of genes involved in pollution and toxicity in cyanobacteria, the use of environmental knowledge is essential to evaluate the extent of damage the pollutants are causing. The complete sequencing of the cyanobacterial genome has revealed a substantial number of functionally unknown FUN genes. Others show sequence similarity to known genes but their functions are only tentatively assigned. Therefore researchers engaged in the quest of gene function are increasingly eager to understand the challenge imposed by natural environment.

Therefore till molecular genetics can come forth with answers to certain environmental problems it is essential to utilize natural resources like the cyanobacteria to fight this massive deterioration of the environment. The important features of cyanobacterial systems are versatility and adaptability. Intensive research is thus warranted to understand many of the basic aspects pertaining to production of metabolites and/or usage through intermediary pathways with the concurrent evolution of applied research towards the large scale production of the products or cyanobacteria. Mass culture of cyanobacteria for biofertilizer, biofuel and bioremediation represents a novel biotechnology which, naturally prompts questions concerning the future of this endeavour. Two distinct modes of production system have emerged, one which relates to industrial production based on modern technology, that is fully controlled sophisticated system with selected strains engineered to produce desirable products and the second is based on local resources and simple techniques. Both these technologies can be applied to cyanobacteria to reap immense benefits.

Used as **biofertilizer, biofuel** and in **bioremediation,** the cyanobacteria offer yet another way to help feed the people through soil renewal, providing economic opportunity without resorting to the vicious cycle of chemical fertilizers, soil exhaustion, pollution of air, water and soil and dependence on imports.

The future, thus holds a promise for continuous development towards a pollution free world, if alternate clean technologies are implemented. In this effort, the cultivation and exploitation of cyanobacteria is a potential natural wealth of immense practical value in every aspect of life conceivable.

Summary

Cyanobacteria are primitive phototropic prokaryotes and their ubiquity on the planet and dominance in every conceivable ecosystem is a measure of their ecophysiological resilience and adaptability. They have been studied extensively as primary producers in all kinds of ecological niches including extra terrestrial habitats such as Mars. In addition to their dual capacity to fix carbon and nitrogen they can tolerate high and low extremes of temperature, screen excessive solar radiation (PAR, and UV-B), and also retain capacity for shade adaptations which makes them eminently suitable to colonize diverse habitats. Their role in maintaining soil fertility and as biofertilizers is very well documented. Apart from this they are instrumental in reducing eutrophication of water ecosystems and play significant role in ameliorating all sorts of toxicity in air, water and soil. Cyanobacteria not only play key role in global cycles of several elements but they are becoming increasingly important economically in both positive and harmful ways. Effective progress in cyanobacterial biotechnology requires shifts in attitude to research including a more critical approach to choice of strains, how these strains are maintained, suitability of culture media and caution in interpreting the results. There are already many strains available, which should be fairly representative of the environment from which they were taken so a direct utilization of such strains from the niche should be used if the work is directed to answering questions about the environment and its management.

This review therefore is an attempt to discuss the potentials of cyanobacteria to develop clean technologies to replace the traditional ones, which are not only toxic but also non biodegradable and therefore totally undesirable. The role of cyanobacteria to act as biofuel, biofertilizer and in depolluting the environment has been discussed. It is time now to consider alternate methods to keep the environment clean if disaster in future has to be prevented, and cyanobacterial biotechnology may play a critical role in this direction.

References

Admi, V., Afek, U. and Carmeli, S.(1996) Raocyclamide A and B,noval cyclic hexapeptides isolated from the cyanobacterium *Oscillatoria raoi.* J. Natural Products (USA). 59 : 396-399.

Al-Hasan, R. H., Sarkhoh, N. A., Al-Bader, D. and Radwan, S.S. (1994) Utilization of hydrocarbons by cyanobacteria from microbial mats on oily coasts of the gulf. Appl. Microbial. Biotechnol.41: 615-619.

Al-Hassan,R. H., Al Barder, D. A., Sorkhoh, N.A. and Radwan, S.S.(1998) Evidence for n-alkane consumption and oxidation by filamentous cyanobacteria from oil contaminated coasts of the Arabian Gulf Marine Biology 130:521-527.

Antarikanonda, P. (1984) Production of extracellular free aminoacids by cyanobacterium *Anabaena siamensis*. Current Microbiology 11(4): 191-196.

Antarikanonda, P. (1985) A new species of the genus *Anabaena : Anabaena siamensis* sp nove (cyanophyceae) from Thailand. Nova Hedwigia 41:343-352.

Archer,S.D., Mc Donald, K. A. and Jackman, A.P. (1997) Effect of light irradiance on the production of sulfolipids from *Anabaena* 7120 in a field batch photobioreactor. Appl. Biochem. Biotechnol 67:139-152.

Aruna, S.J., Balachander, D. and Kannaiyan, S. (1998) Effect of immobilization of nitrogen fixing cyanobacteria in polyurethane foam and sugarcane waste on ammonia production; In: Cyanobacterial biofertilizer for rice crop ed S. Kannaiyan; TN Agri Univ Coimbatore, TN.

Banerjee, M. (1991) Blue green algal ecology of paddy fields. Bionature 11(1): 45-49

Banerjee, M. (1992) Possible utilization of flyash for green manure production. Chem . Environ. Res 1(2) : 89-91

Banerjee, M. (2003) Biotechnological implications of phosphatase activity by two non hair forming strains of Calothrix on phosphorus dynamics of rice fields. J. Ecophysiol Occup Hlth 3:117-126.

Banerjee, M and Kumar, H. D. (1986) Effect of algalization on growth of rice variety Jaya. Proc. Int. Sym. Soc Greeen Veg Res 2: 109-115.

Banerjee, M. and Kumar, H.D. (1987) Effect of light quality, anaerobiosis and glucose on nitrogen fixation by the rice field cyanobacterium *Aulosira fertilissima* Ghosh . Microbiologia Hungarica 34(1): 57-63.

Banerjee, M. and Kumar, H.D.(1988).Effect of carbon sources on growth and nitrogen fixation of *Aulosira fertilissima* ghosh. Biochem. Physiol. Pfanzen 183:51-58.

Banerjee, Meenakshi and Kumar, H. D. (1989) Factors regulating nitrogenase activity and hydrogen evolution in *Azolla Anabaena* symbiosis .Int .J. Hydrogen Energy 14(12) 871-879.

Banerjee ,M .and Kumar,H.D. (1992) Diurnal N_2 fixation by *Aulosira fertilissima* . Use of short term *in situ* assay .Naturalia 17:51-58

Banerjee, M. and Sharma, A. (1994) Agronomic significance of utilizing algal biomass from lakes as a green manure for cultivation of *Triticum vulgare* .Cientifica 22(2) :195-203

Banerjee, M. and Deb, Mallika. (1996) Potential of flyash and *Spirulina* combination as a slow release fertilizer for rice field . Cientifica 24(1) : 55-62.

Banerjee, M. and Hader, Donat P.(1996) Effect of UV radiation on the rice field cyanobacterium *Aulosira fertilissima* .Environ. and Exp. Bot.36(3): 281-291.

Banerjee, M. Sharma, A. and Thomas, T. (1997) Potential use of Lake cyanobacteria biomass as fertilizer for cultivation of food crops .World J. Micobiol. Biotechnology 13(15) :595-596

Banerjee, M., Sinha, Rajeshwar, P.and Hader, Donat P.(1998) Biochemical and Spectroscopic changes in Phycobiliproteins of the cyanobacterium, *Aulosira fertilissima,* induced by UV-B radiation. Acta Protozoologica 37:45-148.

Banerjee, M. Thomas, T. and Abraham, Sheena.(1999) Use of environmentally hazardous flyash in combination with cyanobacteria as a green manure for cultivation of *Brassica juncea.* Ecol Env and Cons 5(2) : 107-110

Banerjee, M., and Abraham, S. (2000a) Biofertilizer potential of isolated cyanobionts on productivity of rice. J. Environ Sci (In Press)

Banerjee, M. and Abraham, S.(2000b) Comparative evalution of biofertilizer potential of isolated and symbiotic and free living cyanobacteria on *Triticum aestivum* World J. of Microbiol and Biotechnology (In Press).

Banerjee, M., Whitton, B.A. and Wynn-Williams, D.D. (2000a). Phosphatase activities of endolithic communities in rocks of the Antarctic Dry Valley. Microbial Ecology. 39 : 80-91.

Banerjee, M., Whitton, Brian., A. and Wynn- Williams, D.D. (2000b). Surface phosphomonoesterase activity of a natural immobilized system : *Chroococcidiopsis* in an Antarctic desert rock. J. Applied Phycology. 12 :549-552.

Banerjee, M. (2001) Biological N_2 fixation for sustainable agriculture, p 312-320. In :Algal Biotechnology ; ed P.C. Trivedi ;Pointer Publishers, Jaipur.

Benemann, J.R.(1994) Feasibility analysis of photobiological hydrogen production ,p.931-940,.In : Hydrogen Energy Progress X. Proc of the 10th World Hydrogen Energy Conference Florida ;ed, D.L. Block and T.N. Veziroglu.

Benemann, J.R.(2000) Hydrogen production by microalgae. J. Applied Phycology 12 : 291-300.

Benemann, J.R., Weissman, J.C., Goebel, R.P. and Augenspein, D (1986) In : Algal Biomass Technologies p176-191 ed W.R. Bardley and Mc Intosh Berlin, :J.Gramer.

Borowitzka, L. J. (1981) The microflora : Adaptations to life in extremely saline lakes .Hydrobiologia 81:36-46

Bothe ,H. (1982) Hydrogen production by algae .Experientia 38:59-64

Boussiba ,S. (1991) N_2 - fixing cyanobacteria - potential uses .Plant Soil 137; 177-180.

Boussiba , S. (1988) N_2 fixing cyanobacteria as nitrogen biofertilizer .A study with the isolate *Anabaena azolle* .Symbiosis 6: 129-136

Boussiba , S. (1993) Production of the nitrogen fixing cyanobacterium *Anabaena siameisis* in a closed tubular reactor for rice farming. Microbiol. Releases 2: 35-39.

Brouers, M., Shi, D.J. and Hall D.O.(1988) Immobilization methods for cyanobacteria in solid matrices. Methods in Enzymology 167:629-636.

Chandrasekaran, M.(1997) Industrial enzymes from marine microorganisms. The Indian Scenario. J. Marine Biotechnology 5: 86-89.

Cobley, J.G., Zerweek, E., Reyes, R., Mody, A., Seludo-Unson, J. R., Jaeger, H., Weerasuriya, S. and Navankasattusas, S. (1993) Construction of shuttlie plasmids which can be efficiently mobilized from *Escherichia coli* into the chromatically adapting cyanobacterium *Fremyella diplosiphon* . Plasmids 30:90-105

Cohen, Z. (1986) Products from microlagae ,p 421-454., In : Handbook of Microalgal Mass Culture : ed, A. Richmond : Boca Raton FL , USA

Cohen, Y. and Ideses, R. (1989) Hyper saline cyanobacterial mats:environmental manipulation for productivity of specific carotenoids and specific polysaccharides, p.135-138; In: Current topics in Marine Biotechnology; ed, S, Miyachi, I. Karube, and Y. Ishida; Japanese Society for Marine Biotechnology.

Coleman, M.U. and White, M.A. (1993) The role of biological disturbances in the production of solar salt: p-623-631, Vol I, In : Seventh symposium on salt .Elsevier Science Publishers, Amsterdam.

Colwell, F, S., Pryfogle, P.A., Lee, B.D. and Bishop, C.L.(1994) Use of cyanobacteria as a particulate tracer for terrestrial subsurface applications. J. Microbiol. Methods 20:93-101.

Davis, J. S. (1993) Biological management for problem solving and biological concepts for a new generation of solar salt works; p 611-616 . In : Seventh symposium on salt Vol-1. Elsevier Science Publishers , Amsterdam.

Davis, J.S. (1978) Biological communities in a nutrient enriched saline. Aquatic Botany 4:23-42

De, P.K.(1939) The role of blue green algae in nitrogen fixation in the fields. Proc. Roy Soc . Lond .B: 127;121-139

De Philippis, R., Margheri, M.C. Pelosi, E. and Ventura, S. (1993) Ecopolysaccharide production by a unicellular cyanobacterium isolated from a hypersaline habitat. J. Applied. Phycology 5:387-394.

Des Marais, D. J. (1995) The biogeochemistry of hypersaline microbial mats; p251-274.Vol-14 In: Advance in Microbial Ecology, ed J. G. Jones , Plenum Press, New York.

Fontes, A.G., Vargas M.A., Morino, J., Guerrco M.G. And Losade H. (1987) Factors affecting the production of biomass by a nitrogen fixing blue green algae in outdoor culture. Biomass. 13: 33-34.

Ganter, M., Kerby N.W., Rowell, P., Obreht, Z. and Scrimgeour, C.(1995) Colonization of wheat (*triticum vulgare* L) by N_2 fixing cyanobacteria: dark nitrogenase activity and effect of cyanobacteria on natural ^{15}N abundance in plants. New Phytol 129:337-343.

Garbisu, C., Hall, D.O. and Serra J.L. (1992) Nitrate and nitrite uptake by free living and immobilized N-starved cells of *Phormidium laminosum*. J. Chem. Tech. Biotechnol 57:181-189.

Garbisu, C., Gill J.M., Bazin, M.J., Hall, D.O. and Serra, J.L.(1994) Inorganic nitrate and phosphate removal from water by freé living and poly vinyl immobilized *Phormrdium laminosum* in batch and continuous flow bioreactors. Enzyme and Microbial Technology.16:395-410.

Grant, I.F., Roger, P.A. and Watanabe, I (1986) Ecosystem manipulation for increasing biological N_2-fixation by blue green algae (cyanobacteria) in low land rice fields. J.Biol. Agric. Hortic 3: 299-315

Hagmann, .L. and Juttner, F.(1996) FischerellinA a novel photosystem II allochemica of the cyanobacterium *Fischerella muscicola* with antifungal and herbicidal activity. Tetrahedron Lett (UK) 37:6539-6542.

Hall, D.O. and Rao, K.K. (1989) Immobilized photosynthetic membrane and cells for the production of fuels and chemicals. Chemica Oggi .7(3) 41-47.

Hall, D.O., Markov, S.A. , Watanabe, Y.and Rao, K.K.(1995) The potential applications of cyanobacterial photosynthesis for clean technologies .Photosynthetic Research 46:159-167.

Hansel, A. and Lindbald, P(1998) Appl. Microbiol. Biotechnology50:153-160

Hayashi, T. and Hayashi, K.(1996) Calcium spirulan, an inhibitor of enveloped virus replication from a blue green algae *Spirulina* . J. Nat. Prod 59:83-87

Karna, R.R., Uma, L., Subramaniam, G. and Mohan, P. M. (1999) World .J. Microbiol and Biotechnol 15:729-732.

Kamuru, F., Albrecht, S.L., Allen , L.H. and Shanmugam ,K.T. (1997) Growth and accumulation of ^{15}N in rice innoculated with the parent and nitrogenase depressed mutant strain of *Anabaena variabilis* .Applied Soil Ecol 5:189-195

Kamuru, F., Stephan ,L., Albrecht, L.H., Allen, Jr. and Shanmugan, K.J.(1998) Dry matter and nitrogen accumulation in rice innoculated with a nitrogenase derepressed mutant of *Anabaena variabilis.* Agron. J. 90:529-532

Kannaiyan, S.(1985) Advances in Applied Phycology; p95-102; ed A.C. Shukal and S.N. Pandey , Narendra Publishing House ,New Delhi.

Kannaiyan, S.(1993) Algal biofertilizer technology for rice . Indian Phycol.Rev. 2:73-82.

Kannaiyan, S., Rao K.K. and Hall D.O.(1994) Immobilization of *Anabaena azollae* from *Anabaena filicoides* in poly vinyl foam for ammonia production in photo bioreactor system. World J. Microl. Biotechnol 10:55-58.

Kannaiyan, S., Aruna, S. J., Kumari, S.M.P. and Hall D.O.(1997) Immobilized cyanobacteria as a biofertilizer for rice crops. J. Applied. Phycology 9: 167-174

Kumazawa, S. and Mitsui, A.(1994) Biotechnol.Bioeng 44:854-858

Kuritz, T. and Wolk, C. P. (1995) Use of filamentous cyanobacteria for biodegradation of organic pollutants. Appl. & Environ. Microbiol 61:234-238

Lambert,G.R. and Smith, G.D.(1981) The hydrogen metabolism of cyanobacteria .Biol. Rev 56:589-660.

Megharaj, M., Madhavi., D.R., Sreenivasulu, C., Umamaheswari Subramaniam, G. Sekar, S. and Sampoornam, S. (1994) Biodegradation and utilization of organophosphorous pesticides by cyanobacteria. Int Biodeterioration and Biodegradation (UK) 33:129-143.

Mitsui, A., Murray, Jr. R.L., Ententman, B., Miyazawa, K.and Polk, E.(1981) p 216-225, In: Biosaline research –A look to the future . Environmental Science Research, Vol :23 , Plenum Press, New York.

Manoharan, C. and Subramaniam, G. (1993) Ind. J. Environ. Health 35:88-96

Markov, S. A., Lichtl, R., Rao, K. K. and Hall, D.O.(1993) A hallow fibre photobioreactor for continuous production of hydrogen by immobilized cyanobacteria under partial vacuum . Int. J. Hydrogen Energy 18:901-906

Markov, S. A., Bazin , M. J. and Hall ,D.O. (1995a) The potential of using cyanobacteria in photobioreactors for hydrogen production .Advance in Biochem .Eng. Biotech 52:61-86

Markov, S.A. Lichtl, R., Bazin , H.J. and Hall, D.O. (1995b) Hydrogen production and carbon dioxide uptake by immobilized *Anabaena variabilis* in a hollow fibre photobioreactor. Enzyme and Microbial .Biotech .17:306-310

Murphy, R.C. and Stevens S.E. Jr (1992) Cloning and expression of the cryIVD gene of *Bacillus thuringiensis* sub sp Israelensis in the cynobacterium *Tolypothrix tenuis*. J. Fermentation and Bioeng 84:151-154.

Nagase, H., Inthron, D., Isaji, Y., Oda, A., Hirata, K. and Miyamoto, K.(1997) Selective cadmium removal from hard water using NaOH –treated cells of the cyanobacteria *Tolypothrix tenuis* . J. Fermentation & Bioeng 84:151-154.

Oren, A. (2000) Salt and Brines; p 281-306. In : In : The ecology of cyanobacteria, ed Brian .A. Whitton and Malcolm Potts, Kluwer Academic Publishers, Netherlands.

Orjala, J. and Gerwick, W.H.(1996) Barbamide, a chlorinated metabolite with molluscidal activity from the Caribbean cyanobacterium *Lyngbya majuscula*. J. Natural Products(USA) 59: 427-430.

Park, I. H., Rao, K. K. and Hall, D.O. (1991) Photoproduction of hydrogen, hydrogen peroxide and ammonia using immobilized cyanobacteria. .Int. J. Hydrogen Energy 16:313-318.

Peterson, R.B. and Burris R.H. (1978) Hydrogen metabolism in isolated heterocysts of *Anabaena* 7120. Arch. Microbiol 116: 125-132

Quesada, A., Leganes, F. and Fernandez, V.E. (1997) Environmental factors controling N_2-fixation in Mediterranean rice fields. Micro.Ecol.34:39-48.

Quesada,A. and Fernandez V.E. (1996) Relationship between abundance of N_2 –fixing cyanobacteria and environmental factors of Spanish rice fields. Micro. Ecol 32:59-71

Radwan, S., S.Al-Hasan, Redha H.(2000) Oil pollution and cynobacteria ;p 307-319. In : The ecology of cyanobacteria ,ed Brian .A. Whitton and Malcolm Potts, Kluwer Academic Publishers, Netherlands.

Rahaman, A. A., Ambika Devi , M. and Sosamma-Esso (1993) Biological management of Indian Solar salt works ; p-633-643,Vol I. In: Seventh symposium on salt . Elsevier Science publishers .Amsterdam.

Reshef, V., Mizrachi, E., Maretzki, T., Silberstein , C., Loya, S., Hizi, A. and Carmeli, S.(1997) New acylated sulfoglycolipids and related known glycolipids from cyanobacteria with a potential to inhibit the reverse transcriptase of HIV -1. J. Nat . Proc 60: 1251 –1260

Robinson, N .J., Rutherford, J. C. ,Mathew ,R. P. and Jennifer S.C.(2000) Metal metabolism and Toxicity : Repetitive DNA .p 443-463 In : The ecology of cyanobacteria ,ed Brian .A. Whitton and Malcolm Potts, Kluwer Academic Publishers, Netherlands.

Roger ,P. A. and Watanabe, I.(1986) Techniques for utilizing biological N_2-fixation in wetland rice : potentialities ,current usage and limiting factors .Fert. Res 9:39-47

Shashirekha, S. Uma. L. and Subramaniam, G.(1997) Phenol degradation by the marine cyanobacterium *Phormiduim valderianum* BDU 30501. J. Indus. Microbiol. & Biotech. 19:130-133

Singh, R. N., Singh ,R. K. and Sharma , R. (1983) An L-Methionine DL-Sulfoximine resistant mutant of the cyanobacterium *Nostoc muscorum* showing inhibitor resistance glutamyl transferase defective ,glutamine synthetase and producing extracellular ammonia during N_2 fixation . FEBS Lett 154:10-14

Singh, R.N.(1961) Role of blue green algae in nitrogen economy of Indian agriculture .Indian Council of Agricultural Research , New Delhi

Spiller, H., Laltore, C., Hasan, M.E. and Shanmugam, K.T.(1986) Isolation and characterization of nitrogen derepressed mutant strains of cyanobacterium *Anabaena variabilis* J. Bacteriol 165:412-419

Sprent, J. I. and Sprent P. (1990) Nitrogen fixing organisms: pure and applied aspects, Chapman and Hall, London, UK.

Subramaniun, G., Sekar, S. and Sampoornam, S.(1994) Biodegradation and utilization of organophosphorus pesticides by cyanobacteria. Int. Biodeterioration and biodegradation (UK) 33 : 129-143.

Sudo, H., Burgess, J.G., Takemasa, H., Nakamura, N. and Matsunaga, T.(1995) Sulphated exopolysaccharide production by the halophilic cyanobacterium *Aphanocapsa halophytica*. Curr. Microbiol 30: 219-222.

Tanagnini, P., Costa, J. L., Almeida, L., Oluverira, M.J., Salema, R.and Lindbald, P. (2000) Curr. Microbiol 40:356-361.

Tang, E. P.Y., Vincent , W.F., Proulox, D., Lessard, P. and Noune, J dela (1997) Polar cyanobacteria versus green algae for tertiary waste water treatment in cool climates .J. Appl.Phycol.9: 371-381

Tredici, M. R., Chini, P., Lihelli, G. and Materassi, R.(1991) A vertical alveolaar panel for outdoor mass cultivation of microalgae and cyanobacteria. Bioresources Technol.38:153-159

Thomas, S.P., Zaritsky, A. and Boussiba, S.(1991) Ammonia excretion by a mutant of the nitrogen fixing cynoabacterium *Anabaena siamensis* . Bioresources Technol. 38:(2-3) 161-166

Tyagi, R. and Kaushik, B.D. (1999) Cyanobacterial toxins and their effect on certain microorganisms and Anopheles mosquito larve.; In : Toxicology and Environmental Health, ed S.B. Vohra and V.P. Agarwal. Asiatech Publ.Inc. New Delhi.

Wake, H., Akasaka, A., Umetsu, H., Ozeki, Y., Shimomura, K., and Matsunga, T. (1992) Promotion of plantlet formation from somatic embryos of carrot with a high molecular weight extract from a marine cyanobacterium. Plant Cell Reports (Germany) 11 : 62-65.

Walsh, K., Jones, G.J. and Dunstan, R. H. (1997) Effect of irradiance on fatty acid, carotenoid, total protein composition and growth of *Microcystis aeruginosa*. Phytochemistry (UK) 44 : 817-824.

Watanabe, A.(1962) Effect of nitrogen fixation blue green algae *Tolypothrix tenuis* on the nitrogen fertility of paddy soil and on the crop field of rice plant. J. Gen. Appl. Microbiol 8(2):85-91

Whitton, B.A., Grainger, S. L. J., Hawley, G. R.W. and Singh J. N.(1991) Cell bound and extra celular phosphatase activities of cyanobacterial isolate. Microbial Ecology 21:85-98

Wilde, E.W. and Benemann, J.R. (1993) Bio removal of heavy metals by the use of microalgae. Biotech. Adv 11: 781-812

Microbial Desulphurization of Coal Containing Organic Sulphur - An Overview

R.A. Pandey

IntroductionSulphur, released into the atmosphere through combustion of coal, is in the oxidized form as sulphur dioxide (SO_2). Through heterogeneous oxidation on particulate matter, SO_2 is further oxidized to SO_3 which, being highly soluble in water, results in the formation of acid rains. It is well recognized that acid rain is an important factor responsible for environmental damage to plant life in lakes, fish kill and damage to other aquatic fauna and flora (Kilbane, 1989). Acid rain also often threatens to damage buildings and historic monuments, a notable example being Taj-Mahal (Monticello *et al.*, 1985). It has been reported that an unprecedented increase in forest damage (more than 60 percent in some regions) in Germany has been registered as result of acid rains. Because of its global effects of acid deposition, regulatory standards of many countries have been made stringent to reduce the level of oxides of sulphur and nitrogen in the atmosphere. Clean Air Acts of many countries have forced to adopt control measures to minimize emissions of hydrogen sulphide and sulphur dioxide into ambient air; and it is stipulated that their concentrations should not exceed 8 μg and 60 $\mu g/Nm^3$ respectively as an annual arithmetic mean (Kargi, 1984). Three different standards of SO_2 in ambient air have been laid down. These are 120 $\mu g/Nm^3$ for industrial use, 80 $\mu g/Nm^3$ for residential and mixed use, and 60 $\mu g/Nm^3$ for sensitive areas (Kargi, 1984). Reduction in sulphur content of coal in association with subsequent improvement in quality needs to be achieved through development of novel precombustion desulphurization technologies. One of such precombustion technologies is microbial desulphurization of coal. The present paper outlines a critical review on the status of development in the field of precombustion desulphurization process developed for removal of organic sulphur from the coal.

Organic sulphur constituents in coal

There are two major types of sulphur compounds in coal (Dugan, 1986). Inorganic sulphur present mostly in the form of sulphate and sulphides while organic sulphur constituents exist as integral part of the coal molecule (Kargi, 1992). The structural existence of organic sulphur compounds along with inorganic sulphides (pyrite) can be presented in postulated structure of coal molecule as follows (Spiro *et al.*, 1984).

Organic sulphur compounds present in coal are bound to the other aromatic rings of molecular structure of coal. The various types of possible sulphur compounds, viz., aliphatic, aromatic, polyaromatic and heterocyclic sulphurous compounds associated with coal molecules are presented in Table 1. However, the exact forms of these sulphur compounds in coal are not clearly known (Kargi, 1982). One of the major problems in desulphurization of coal containing organic sulphur is the removal of these organic sulphur compounds from coal matrices. Removal of these organic sulphur compounds from the coal matrix is associated with loss of carbon atoms from the bulk of the coal thereby reducing the calorific value of the coal.

Table 1 : Organic sulphur constituents in coal

Organic Constituents	Molecular formulae / structure	Reference
Aliphatic sulphur compounds		
Alkyl monosulphides	R_1-S-R_2 (General)	Spiro *et al.*, 1984
Methyl sulphide	CH_3-S-CH_3, CH_3-S-C_2H_5	
Ethyl methyl sulphide	C_2H_5-S-C_2H_5, $(C_3H_5)_2S$	
Ethyl sulphide	$(C_2H_3)_2S$	
Allyl sulphide Vinyl sulphide		
Alkyl disulphide	R_1-S-S-R_2 (General)	
Methyl disulphide	$(CH_3)_2S_2$	
Diethyl disulphide	C_2H_5-S-S-C_2H_5	
Allyl disulphide	$(C_3H_5)_2S_2$	
Ethyl sulphoxide & sulphone	C_2H_5-SO-C_2H_5, C_2H_5-SO_2	
Alkyl mercaptans	RSH (General)	Boudou, *et al.*, 1987; Nishioka, 1988; Attar, 1979
Ethyl mercaptan	C_2H_5SH	
n-propyl mercaptan	C_3H_7SH	
n-butyl mercaptan	C_4H_9SH	
Unspecified 27 different mercaptans and other thiols compounds from coal and lignite		
Aromatic Sulphur Compounds		Boudou, *et al.*, 1987; Attar, 1979; Nishioka *et al.*, 1985
Aryl mono sulphides	ϕ-S-ϕ_2 (ϕ & ϕ are aryl groups)	
Benzyl mono-sulphide	C_6H_5-S-C_6H_5	
Methyl benzyl mono-sulphide	$C_6H_4.CH_3$-S-$C_6H_4.CH_3$	
Aryl di-sulphide	ϕ_1-S-S-ϕ_2 (General formula)	
Benzyl disulphide	C_6H_5-S-S-C_6H_5	
Methyl benzyl disulphide	$C_6H_4.CH_3$-S-S-$C_6H_4CH_3$	
Bis (4-hydroxy phenyl) disulphide Unspecified other disulphide		
Aryl Mercaptans	SH (General formula)	
Benzyl thiol	C_6H_5SH	
Methyl benzylthiol	$C_6H_4CH_3SH$	
Unspecified other Thiols		

Various physico-chemical methods have been cited in the literature of removal of organic sulphur from the coal (Spiro *et al.*, 1984) which are summarised in Table 2 (Spiro *et al.*, 1984; Wise, 1971; Reid, 1960; Masciantonio, 1965; Stambaugh *et al.*, 1975; Angelova *et al.*, 1972; Haber *et al.*, 1940; Block *et al.*, 1975). These processes operate at high temperature and pressures and are observed to be energy intensive. The chemical desulphurization processes are also associated with a loss in the valuable coking property of coal which is of the prime importance from the commercial point of view. Thus the microbial process from removal of organic sulphur from coal with minimal destruction of coking property appears to be promising subject to the availability of suitable strains possessing high organic sulphur removal capability.

Organic Constituents	Molecular formulae / structure	Reference
Thiophene & its substitute derivative		13, 14, 15 Willey, et al., 1981; Nishioka, et al., 1986; Ignasiak et al., 1978
Thiophene		
Thiophene carboxylic acid	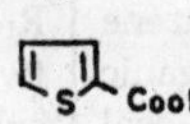	
Benzothiophene	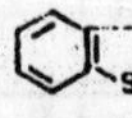	
Dibenzothiophene	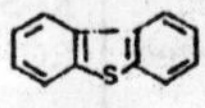	
Naphtho [2,1-b] benzothiophene	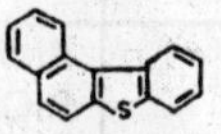	
Naphtho [2,3-b] benzothiophene	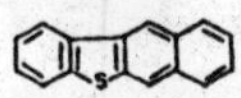	
Phenanthro [4,5-b] thiophene	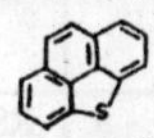	
Naphtho [2,3-b] phenanthro [1,2-d] thiophene	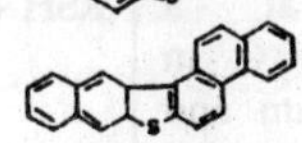	
Sulphidic derivatives of benzene	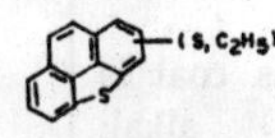	
Heterocyclic sulphur containing constituents (approx. 123 compounds)	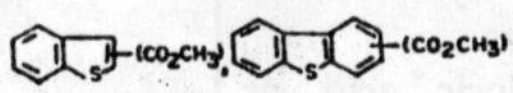 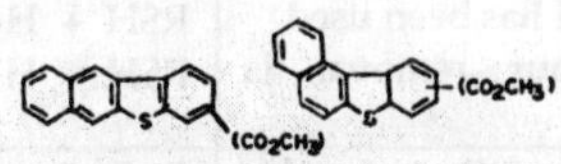	

Microorganisms for removal of organic sulphur from coal

The availability of organic sulphur compounds for biodegradation in the coal environment is a function of its solubility, mobility, sorption and desorption. These parameters may prove to be critical limiting factors for controlling biodegradation rates of many organic sulphur compounds as these organic sulphur compounds present in coal have low solubility in the aqueous phase.

However, microorganisms belonging to various groups having capability of removal of organic sulphur from coal are cited in the literature. The list of these microorganisms is presented in Table 3. A thermophilic *Thiobacillus* like strain (TH1) has been reported to remove both inorganic (pyrite) and organic sulphur from the Turkish lignite (Eliot, 1978). Turkish lignite containing 1.5 percent sulphur was subjected to sequential removal of inorganic and organic sulphur. The coal samples having sizes of 270-80 mesh along with TH1 strain, was inoculated in specific medium maintained at a temperature of 50°C. The results indicated 90-95 percent removal of pyrite within 25 days (Eliot, 1978). The coal was treated with TH1 strains grown in medium containing yeast extract at neutral pH. After pyrite removal, 50 to 57 percent removal efficiency for organic sulphur could be achieved within 25 days (Eliot, 1978). Thus, the total time period required for treatment was about 50 days. However, the products of organic sulphur solubilization remained unidentified and mass balance for substrate and product also could not be established (Spiro *et al.*, 1984). Thermophilic *Sulfolobus acidacaldarius* had also been used for desulphurization of coal containing organic sulphur. Pretreated coal with 10 percent sulphuric acid was inoculated with consortium of *S.acidocaldarius* at 70°C. The *S.acidocaldrius* could remove 44 percent of organic sulphur from coal within 30 days (Couch, 1988). However, other workers were not able to reproduce the activities of *S.acidocaldarius* for removal of organic sulphur from coal (Gockay *et al.*, 1983). Thermophilic cultures require expensive arrangement for maintenance of temperature for the metabolic activity. Therefore, any reactor operation for desulphurization of coal with use of thermophilic cultures will be energy intensive and costly process for large scale operation (Kargi *et al.*, 1982).

Table 2 : Physico-chemical methods for desulphurization of coal containing organic sulphur

Sr. No.	Methods	Details	General chemical reactions
1	Solvent extraction - Phenol P-Cresol - Nitrobenzene - Amines	- Phenols and nitrobenzene selectively dissolved organic sulphur compounds from coal at temperature of 180-120°C and pressure of 500-1500 psi	R_1-Sx-R_2 + X → Sx.X (where R_1 & R_2 are alkyl groups, X is solvent)
2	Reduction methods - Hot air (Air stream method)	- 20 and 30% removal of organic sulphur from coal is achieved with hot air	R_1-Sx-R_2 → R_1R_2_Sx RCH_2CH_2SH → $RCH{=}CH_2$ + H_2S
3	Oxidation methods - Hypochlorite - Nitric acid - Hydrogen peroxide - Ozone	- Loss of carbon - Deteriorate the coal quality	R_1-Sx-R_2 → R_1-SO_3H_R_2-SO_3H → R_1OH + R_2OH + $2H_2SO_4$
4	Displacement methods - Alkyl based displacement method - Acid based displacement method	- Method operates at a temperature of 150-400°C in molten salt of potassium and sodium hydroxides - Coal yield is poor as coal is contaminated by alkali aluminium silicates - Bioling hydrochloric acid or hydroiodic acid has been used - Organic sulphur removal is very poor	RSH + OH → RS + H_2O RSH + HCL → RCL + H_2S RSH + HI → RI + H_2S
5	Gas solid process (in stationary phase) - Hydrogen (Hydrosulphurization) - Steam - Air	- Volatilization followed by reduction as results of pyrolisis cause removal of organic sulphur from coal - Process operates at a temperature 600-900°C - Result in loss of coal (coal yield is poor) with organic sulphur removal efficiency of 70-80%	R_1-Sx-R_2 + 4H → R_1H + R_2H + H_2Sx R_1-Sx-R_2 + $2R_3H$ → H_2Sx

A *Psedomonas* species, designated as CB1, was used for desulphurization of coal enriched in organic sulphur CB1, and was capable of removing 30 percent organic sulphur from coal within a period of 9 hours (Isbister, *et al.*, 1985). However, the laboratory where CB1 was originally isolated, no longer possessed a culture with desulphurization activity (Isbister, *et al.*, 1985). Chandra and co-worker (1979) used a mixed thermophilic culture isolated from soil and enriched with debenzothiophene for removal of organic sulphur from coal. Nearly 20 percent reduction in organic sulphur content could be achieved within 10 days (Chandra 1979). Presterilized coal gave better removal efficiency of sulphur (Chandra, 1979). Several microorganisms belonging to *Arthrobacter* and *Pseudomonas* genera were isolated from different natural sources by Hedric (Hedric, 1982). These microorganisms were tested in a mixture of removal of organic sulphur from coal. Hedric was able to achieve 26 percent removal of organic sulphur from coal within 15 days (Hedric, 192).

Table 3 : Different Microorganisms used for removal of organic sulphur from coal

Microorganisms used	Organic sulphur removal	Time period required	Reference
Mixed heterotrophic culture	20%	20 days	Chandra et.al. (1979)
Thermophilic *Thiobacillus* like strain (THI)	50%	25 days	Gokay & Yurteri (1983)
Sulfolobus acidocaldarius	44%	28 days	Kargi & Ribionson (1982)
Mutagen IGTS7	91%	212 days	Kilbane (1989)
Pseudomonas CBI	30%	9 hrs	Isbister & Kobylinski (1982)
Arthrobacter & *Pseudomonas*	26%	Not mentioned	Hedrick (1982)
Rhodococcus rhodochrous (pregrown cells)	21-72%	48 hrs	Kilbane & Jackowski (1992)
Rhodococcus rhodochrous (growing cells)	40-67%	3-6 weeks	Kilbane & Jackowski (1992)

A consortium of mixed cultures was evaluated for desulphurization of coal containing organic sulphur by Kilbane and Bielage (Kilbane, 1988; Kilbane *et al.*, 1990). Mixed cultures possessed matabolic activity of removal of specific sulphur atom from the organic constituents without affecting the carbon skeleton of the ring. The consortium was of *Bacillus* and *Rhodococcus* genera (Kilbane, *et al.*, 1990). These cultures were preserved on an agar slant enriched with dibenzothiophene. The consorium of mixed cultures indicated an efficiency of 26 percent removal of sulphur (Kilbane, *et al.*, 1990). Further, Olson and Brincleman (1986) studied the factors affecting biological removal of organic cultures (Olson, *et al.*, 1986). However, the extent of organic sulphur removal was found to be only 25 percent and treated coal also lost some of the valuable properties including calorific value (Kilbane, 1990). Further, Kilbane (1990) suggested to use sulphur limiting growth medium for the mixed cultures for achieving better sulphur removal efficiency. A scheme for estimation of sulphur by adopting bioavailability assay and protein estimation was also reported (Kilbane, 1992).

An isolated culture namely *Rhodococcus rhodochrous* IGTS8 has exhibited the ability to use sulphur as a sole source of energy and also the tendency to grow on various sulphur compounds including dibenzothiophene, without affecting the carbon skeleton of these compounds. IGTS8 pregrown cells were shown to be capable of removing organic sulphur from coal with better retention of carbon and calorific value at a shorter treatment time (Galagher, *et al.*, 1993; Warhurst, *et al.*, 1994). The pre-grown cells of IGTS8 require 200-fold smaller reactor volume as compared to normal growing cell system. The pre-grown cells indicated 21-72 percent removal of organic sulphur from water soluble coal derived material in 48 hours where as growing cells showed removal of 40-60 percent sulphur within 3-6 weeks (Gallagher, *et al.*, 1993; Warhurst, *et al.*, 1994). A mixed bacterial culture of IGTS7, selected after mutagenesis, was used for desulphurization of coal. The results indicated 91 percent removal of sulphur from coal within 212 days (Gallagher, *et al.*, 1993; Warhurst, *et al.*, 1994).

Recent developments for acquisition and amplification of specific gene in a selected group of microorganisms through recombinant DNA technology have occurred of enhancing the capabilities of microorganisms for desulphurization of coal containing organic sulphur. Further, Isbister and Kobylinski (Isbister, *et al.*, 1985) developed a mutant CBI from dibenzothiophene oxidizing microorganism and evaluated the same for removal of organic sulphur from coal. The

CBI mutant culture could remove upto 47 percent of the organic sulphur from coal (Isbister, 1985).

A noval approach has been reported for enhancing the efficiency of microbial desulphurization of coal containing organic sulphur. The surface properties of coal are transformed by incorporating a pretreatment step prior to microbial activities. The use of either surface active substance or solvent as a pretreatment step for coal has been cited. The surfactants are known to enhance the aqueous dispersion of organic sulphur compounds from coal having lesser solubility and hence use of surfactant helps in overcoming the limitation of solubility in microbial desulphurization of coal containing organic sulphur. The isolate P12 produces biosurfactant during its growth on dibenzothiopene derivatives which are present as a major fraction of organic sulphur compounds in coal. Hence, crude extract of the biosurfactant produced by P12 was adopted for exploiting its capability to desulphurize coal. About 45-50 percent reduction in organic sulphur from coal could be achieved after treatment with culture P12. Similarly pretreated coal with methanol, followed by microbial treatment using P12 culture, resulted in reduction of about 62-65 percent of organic sulphur from coal (Dhillon, *et al.*, 1994).

Some of the reports cited in the literature indicated the use of anaerobic microorganisms for removal organic sulphur from coal (NEERI, 1993). The anaerobic sludge sediments collected from various sources were evaluated for desulphurization of coal containing organic sulphur. As there was no formation of hydrogen sulphide at the end of experimental period, the results were not encouraging (NEERI, 1993; Report on Environment, 1992).

Several other molds are reported to decompose some of the sulphur compounds. Sulphonated aromatic compounds have been metabolized to sulphate by fungal cultures (Starkey, 1984; Kari, 1984). However, there are no reports in the literature indicating use of molds of removal of organic sulphur from coal. The isolation and identification of molds capable of oxidizing organic sulphur compounds present in coal is a challenging but promising since these are reported to be more effective for biological desulphurization of coal (Kargi, 1984).

Poria placenta (ATCC 11538), obtained from American type culture collection, was grown on yeast extract and peptone medium in submerged culture. Coal was pretreated with myculial mass, obtained from this fungus and then subjected to selected treatment with *S.brierleyi* bacteria grown. There was no significant increase in the removal efficiency of organic sulphur from the coal (Kargi, *et al.*, 1982; Krawise, 1988).

Conventional microbial desulphurization of coal containing organic sulphur was carried out in aqueous solution. The organic sulphur compounds, present in coal and being insoluble in water, will not be readily available for microbial action. Therefore, the activity of the microorganism particularly for removal of organic sulphur results in poor efficiency. Significant improvement relating to biological experiments has been made by keeping experimental controls that were often lacking in earlier work. The absence of these controls sometimes led to erroneous conclusions and overspeculation about the efficiency of microbiological processes. However, the use of microorganisms for desulphurization of coal containing organic sulphur can be made successful by adopting the following measures:

- selection of conditions for enhancing the solubility of nonpolar organic sulphur compounds of coal

- shifting of thermodynamic equilibria to favour the synthesis of enzymes, required for solubilization of organic sulphur compounds
- maintenance of required quantity and specific characteristics of microorganisms for desulphurization in a suitable reactor system

Pretreated coal with various agents, viz. surface active agents, solvents, microorganisms may also help in the enhancement of activities of microorganisms for desulphurization. Therefore, to explore the possibilities for the use of microorganisms for desulphurization of a pretreated coal different strategies have to be worked out. This will help in changing the surface properties of coal thereby increasing the attachment of microorganism on the surface of coal for enhancing microbial activity.

Pathways for Microbial Desulphurization of Coal Containing Organic Sulphur

Coal is a fossil fuel which is generated through the activities of the microorganisms which were present in ancient time. These microorganisms had also affinity for the sulphur constituents. Metabolism of present day microorganisms is likely to be similar to that of microorganisms during fossil deposition, providing the basis for metabolic alteration of the sulphur found in fossil fuels. The chemical composition of the coal indicates that most of the organic sulphur compounds present in coal belong to either alkyl of aryl sulphides, thiols and thiophene derivatives. However, among these the thiophene and its derivatives predominate in the coal. Therefore, the pathways adopted by microorganisms or metabolization of thiophene and its derivatives have been studied in detail. However, the metabolism, other than sulphur bearing, organic constituents, has not been given due consideration.

The microorganisms for metabolization of sulphur constituents followed two routes namely aerobic and anaerobic pathways.

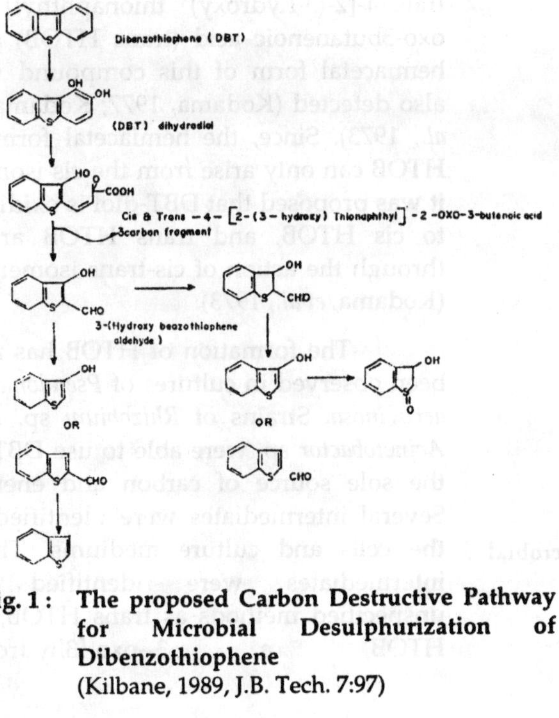

Fig. 1 : The proposed Carbon Destructive Pathway for Microbial Desulphurization of Dibenzothiophene
(Kilbane, 1989, J.B. Tech. 7:97)

Aerobic Pathways for Metabolization of Organic Sulphur Compounds

Microbiological enzymatic reactions responsible for metabolization of organic sulphur compounds can be categorized as hydrolases, Kinases and oxidases. The hydrolase enzymes generated by microorganisms are specific and hydrolyse the sulphur compounds present in the coal which results in solubilization of organic sulphur constituents from coal to soluble forms (Dodgson, *et al.*, 1975). The enzymes that catalyze the hydrolysis of organic sulphur compuonds are well documented in the literature (Gokay *et al.*, 1983; Jakoby *et al.*, 1987 Gleen, 1994). Some of the important ones are aryl sulphohydrolases, alkyl sulphohydrolases, carbohydrate sulphohydrolases, sulphatophosphatase and sulphamidases (Casagrande, *et al.*, 1977). The kinase enzymes act for cleaving of C-S-C bond that releases the sulphur element from the rings (Gleen, 1949).

Metabolic Pathways for Metabolization of Dibenzothiophene (DBT) and its Substituted Derivatives Present in Coal: The relative abundance of aromatic dibenzothiophene (DBT) linkage in coal has led to the acceptance of this molecule as a general model for organic sulphur compounds of the coal. Two distinct pathways have been reported for metabolization of dibenzothiophene. These are carbon degradative and 4S pathways for metabolization of DBT. The induction of DBT oxidation in *Pseudomonas jianii* has been demonstrated by Kodama (Kondama, *et al.*, 1970). Several metabolic intermediates of DBT oxidation have been isolated from cultures of *Pseudomonas* and *Beijerinckia* after incubation with DBT (Kondama, *et al.*, 1970, 1973; Kondama, 1977). These microorganisms appear to have a common pathway for DBT oxidation which is shown in Fig. 1. The first two oxidation products, (+)-cis -1, 2-dihydroxy-1, 2-dibnzothiophene-5-oxide (DBT-5-oxide) were isolated from culture medium after cells of a mutant strain (B8/36) of *Beijerinckia* which has been incubated with DBT. Both compounds were identified on the basis of infrared, ultraviolet and mass spectra (Kodama, *et al.*, 1973). A cis relative stereochemistry was assigned to DBT dihydrodiol from evidence obtained by characterization of its acetonide derivatives. Cell extracts of wild type *Beijerinkia* species incubated with DBT dihydrodiol and NAD+ under anaerobic conditions formed a new product 1,2-dihydroxydibenzothiophene (DBT diol) (Kodama, *et al.*, 1973). The addition of air to the same compound resulted in the formation of a compound which was identified as 3-hydroxy-2-formylbenzothiophene (HFBT) on the basis of its ultraviolet absorption spectrum (Kodama, *et al.*, 1973). The isolation of DBT oxidation products from cultures of *P.jianii* also reveals the presence of a 3-hydroxy-2-formylbenzothiophene-5-oxide, and trans-4-[2-(3-hydroxy) thionaphthyl] 2-oxo-3butanenoic acid (trans HTOB) and hemiacetal form of this compound was also detected (Kodama, 1977; Kodama, *et al.*, 1973). Since, the hemiacetal form of HTOB can only arise from the cis isomer, it was proposed that DBT-diol is oxidized to cis HTOB, and trans HTOB arises through the action of cis-trans isomerage (Kodama, *et al.*, 1973).

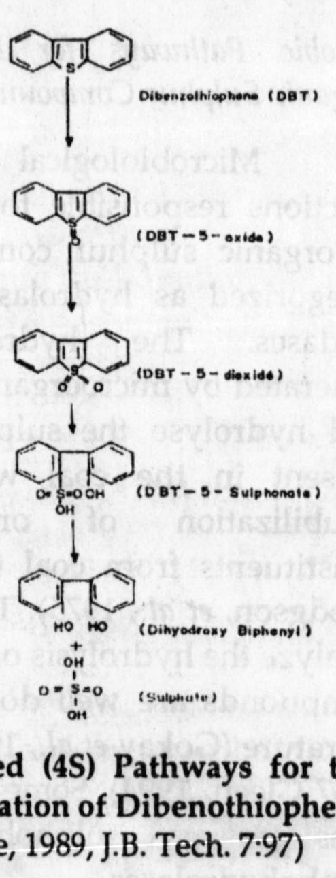

Fig. 2: Proposed (4S) Pathways for the Microbial Degradation of Dibenothiophene
(Kilbane, 1989, J.B. Tech. 7:97)

The formation of HTOB has also been observed in cultures of *Pseudomonas aeruginosa*. Strains of *Rhizobium* sp. and *Acinetobactor* sp. were able to use DBT as the sole source of carbon and energy. Several intermediates were identified in the cells and culture mediums. These intermediates were identified by unspecified methods as trans HTOB, cis HTOB, a 3--oxo-[3hydroxythionaphthenyl] 1-2-methylene-dihidrothionaphthene and an unidentified compound with absorption maxima at 450 nm (Hou, *et al.*, 1976).

A different mechanism of aerobic DBT metabolism, involving specific release of sulphur from thiophene ring has also been observed using mixotrophic microorganism of species *S.acidocaldarius*. The sulphur released from thiophene ring gots converted into sulpho-oxide, sulphone, sulphonate and sulphate (Fig.2). Furthermore, several other *Pseudomonas* strains, isolated from the soil and stripmine sites have also been found to mineralize DBT to liberate sulphate (Faght, *et al.*, 1988). Faght and Westlake (1988) have cited the *Pseudomonas* sp. (HL>6) which metabolized DBT-5-oxide to unknown water soluble products. More recently Stoner and Cowerkers have investigated the pathway for metabolization of DBT using ten isolates separately as well as in mixture.

Recently, another pathway of metabolization of DBT has been reported by using *Rhodococcus rhodochrous* IGTS8 which is able to use DBT as a sole sulphur source but not as carbon source. The proposed pathway for DBT metabilization by *Rhodococcus rhodochrous* is presented in Fig. 3. Growing cells of the culture metabolized DBT to 2,2-dihydroxybiphenyl whereas stationary phase cells transformed it to 2-hydroxybiphenyl and these two metabolites were not metabolized further (Gallagher, *et al.*, 1993).

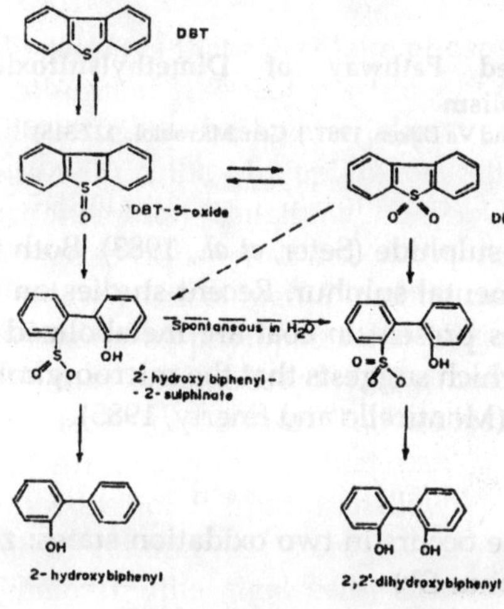

Fig. 3 : Desulphurization of DBT by *Rhodococcus rhodochrous* IGTS8

There is a single report in the literature concerning the microbial degradation of benzothiophene by *Pseudomonas* sp. isolated by enrichment from oil contaminated soil and supplemented with benzothiophene as a carbon source.

Another important thiophene derivative which is likely to be presented in coal is thiophene carboxylate. This compound is metabolized to ketoglutarate and sulphate by several groups of microorganisms has been reported by Cripps (1973). The pathway follows hydrolysis and opening of the ring as C-S linkage and conversion of sulphur to sulphate (Fig.4).

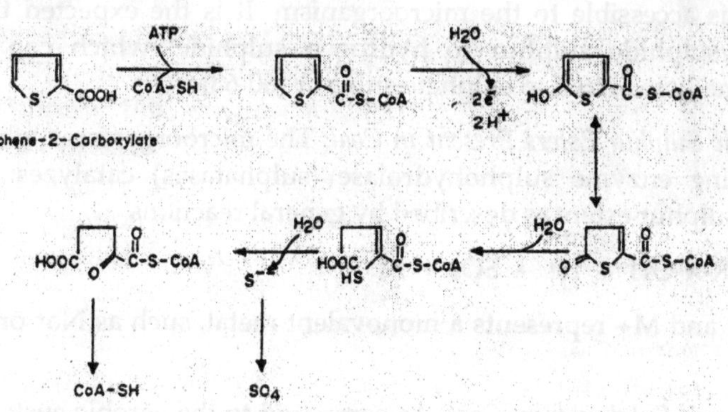

Fig. 4 : Proposed Pathway for Thiophene-2-Carboxylate
(Cripps, 1973, J. Biochem., 134:353)

Aerobic Pathway for Removal of Alkyl and Aryl Sulphides and Alkyl Sulphoxides from Coal: The hypothermophilic sulphur metabolizing archaebaterium is found to transform dimethyl sulphoxide, dimethyl sulphide and mercaptans. These compounds are observed to be transformed to aldehyde, formic acid and finally to carbon dioxide. The microorganisms belonging to Thiobacillus group have capability to generate hydrogen sulphide but this mechanism is not clearly defined. The hydrogen sulphide generated is oxidized to sulphate (Fig.5) (Fiala, *et al.*, 1986; Seter, *et al.*, 1993). Work has focused on two microorganisms *Pyrodictium brockii* and *Pyrococcus furiosus*. *P.brockii* has an autotroph (uses CO_2 as major carbon source) and oxidizes hydrogen sulphide (Seter, *et al.*, 1983). Both the microorganisms are described as being able to reduce elemental sulphur. Recent studies on the subject have suggested that some of the organic sulphides present in coal are metabolized by *P.furiosus* to methyl mercaptans from methyl trisulphide, which suggests that the microorganism recognizes the central sulphur atom in zero oxidation state (Monticello and finerty, 1985).

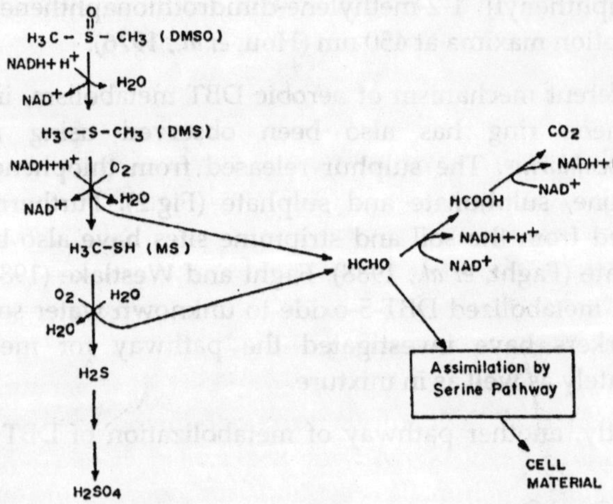

Fig. 5: Proposed Pathway of Dimethylsulfoxide Metabolism
(deBont and Va Dijken, 1981, J. Gen. Microbiol., 127:315)

$$CH_3\text{-}S\text{-}S\text{-}S\text{-}CH_3 \rightarrow CH_3S^- + S^{2-}$$

(Sulphur in organic trisulphide and inorganic polysulphide occurs in two oxidation states: zero termed polysulphide or excess sulphur and minus two sulphur S^{2-})

Studies on this subject suggest that the polysulphide form of organic sulphur can be recognised as the position where sulphur is attacked by *P.furiosus*. Thus, the polysulphide organic sulphur, if present in coal, is accessible to the microorganism. It is the expected that *P.furiosus* reduces the organic polysulphide sulphur to hydrogen sulphide, which can be precipitated as metal sulphide and finally oxidized to sulphate sulphur (60-63).

Aerobic Pathway for Removal of Organic Sulphur Esters Present in Coal: The microorganisms which are associated with peat and having enzyme sulphohydrolase (sulphatases) catalyzes the microbial transformation of organic sulphur esters as described by general reactions

$$R\text{-}X\text{-}O\text{-}SO_3\text{-}M \rightarrow R\text{-}X\text{-}OH + H^+ + SO_4 + M^+$$

Where R presents an organic moiety and M+ represents a monovalent metal, such as Na^+ or K^+ (Cripps, 1973).

Anaerobic Pathway for Removal of Organic Sulphur from Coal: As compared to the aerobic systems, the anaerobic degradation of model organic sulphur compounds has not been studied extensively. Most of the anaerobic microorganisms utilize sulphate as a terminal electron acceptor in place of oxygen during anaerobic respiration or through the use of completely or partially reduced sulphur compound as energy source (sulphide, elemantal sulphur, tetrathinate, thiosulphate). Relatively large amount of sulphur is utilized by cells with capability for dissimilatory metabolism. In comparison, all cells, including the dissimilatory sulphur metabolizers, must assimilate relatively small amount of sulphur into structural and functional components of the cell, such as proteins, co-enzyme A, vitamin B, sulpholipid, etc. however, the organic sulphur compounds, because of their poor solubility remain intact in coal molecule. The anaerobic microorganisms are not able to modify these components from coal (Dugan, 1972; Peck, 1962; Peck, 1968; Peck, 1961).

Anaerobic Pathway for Degradation of Organic Sulphoxide Sulphides/Sulphones: A variety of microorganisms have been reported, which can reduce dimethyl sulphoxide to dimethyl sulphide. Dimethyl sulphoxide reduction has been demonstrated in a wide range of prokaryotes, viz., *Aspergillus niger* and *Saccharomyces cerevisiae* (Zinder, *et al.*, 1978).

Reduction of these compounds have also been cited in the cell free extracts of *E.Coli* grown in anaerobic conditions. Bamforth and Anness (1979) found that dimethyl sulphoxide reduction by cell free extract of *Saccharomyces cerevisiae* required NADPH. Further, these authors have also concluded that dimethyl sulphoxide may be reduced by the same enzyme, present in the yeast. Dimethyl sulphide has been reported to be metabolized anaerobically to methane, carbon dioxide and hydrogen sulphide by using anaerobic lake sediments (Zinder, *et al.*, 1978). These products were formed from the terminal intermediate methanethiol as reaction indicates.

$$4CH_3SH + 3H_2O \rightarrow 3CH_4 + HCO_3 + 4H_2S + H^+$$

Methane and carbon dioxide are also reported to be produced in the ratio of 3:1 from methane thiol by using methanol fermenting microorganism *Methanosarcina barkeri* (Kluyver, *et al.*, 1974). Thus, the anaerobic metabolic pathway for the organic alkyl sulphide and alkyl sulphoxide has been cited, however, the application of these microorganisms and their metabolism for the coal containing organic sulphur compound is limited.

Reactor System for Microbial Desulphurization of Coal Containing Organic Sulphur

The possible reactor systems for microbial desulphurization of coal containing organic sulphur may constitute the following unit operations and the possible flow sheets for such systems are given in Fig.6.

- Pulverization of coal of required size (200-300 mesh)

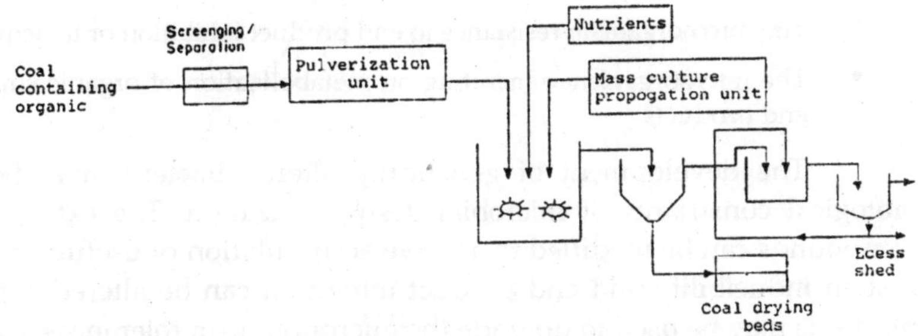

Fig. 6 : Schematics of Flowsheet for Precombustion Microbial Desulphurization of Coal Containing Organic Sulphur

- Units for mass culturing of specific culture
- First stage reactor unit with mixing arrangement along with aeration devices
- Settlers
- Coal drying bed for desulphurized coal
- Second stage reactor unit for stabilization of solubilized products of the organic sulphur constituent of the coal

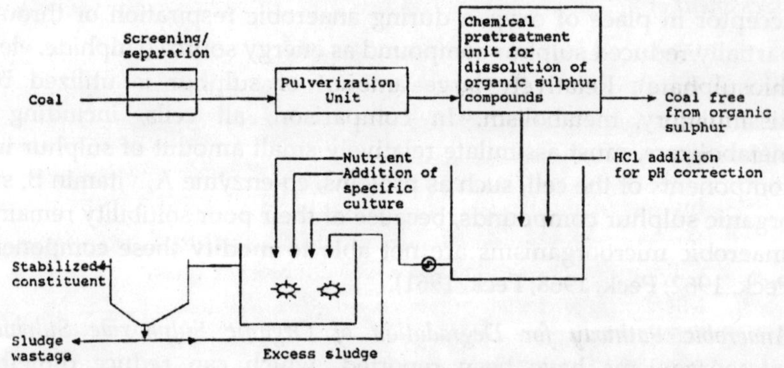

Fig. 7: Schematics of Flowsheet for Precombustion Microbial Desulphurization of Pretreated Coal Containing Organic Sulphur

The second flowsheet is based on the removal of organic sulphur compounds from coal derived liquid after treatment with alkali (sodium hydroxide) and ammonia under controlled conditions. The coal derived liquid contains excess of organic sulphur which can further be treated by biological process by incorporating suitable and efficient microorganisms in a completely mixed activated sludge process (Fig.7).

Further Research on Microbial Desulphurization of Coal Containing Organic Sulphur

Needs for Development of a Suitable Microorganism

The contribution of microbiological activity to desulphurize the coal containing organic sulphur has been unequivocally demonstrated in the forgoing sections. However, there is a need to isolate and develop dominant microorganisms for microbial desulphurization of coal containing organic sulphur having following characteristics

- Microorganism having substrate specificity for organic sulphur moiety present in the coal
- The microorganism of robust properties having more tolerance of the relatively harsh nature of the coal
- The microorganism resistance to end product inhibition or toxicity
- The microorganism generating on metabolization of organic sulphur from coal to usable or safe end products

The development of genetically altered bacteria may help to alleviate many of the biological constraints of microbial desulphurization. The extent of the oxidation of DBT-like compounds can be modified to achieve accumulation of useful (or least toxid) byproducts, while system inducibility and end product inhibition can be altered. Similarly, heavy metal resistant plasmids may be used to upgrade the microorganism tolerances. Genetic modification of enzyme specificity may be used to increase the system's ability to discriminate between sulphur and non sulphur aromatic hydrocarbons.

Assessment of Pathways for Solubilization and Stabilization of Organic Sulphur Compounds

Various mechanisms for biotransformation of organic sulphur compounds have been cited in the literature as discussed earlier. However, no supportive evidences are available for these pathways since these are the postulated pathways. It is essential to investigate the following aspects

- Investigations on metabolic pathways of model organic sulphur compounds present in coal
- Investigations on organic sulphur compounds interaction studies during microbial transformation using multisubstrate system which are present in coal
- Microorganism-microorganism interaction studies as specific mixed sulphur compounds while stabilization of solubilized products are achieved in the other system
- Polysulphides are identified to be significant intermediates during microbial desulphurization of coal containing organic sulphur. Are polysulphides significant intermediates in the transformation of either organic or inorganic coal sulphure? This question is to be answered after proper investigation.

These aspects of investigations will help in selection of proper microorganisms for development of techno-economically feasible process for desulphurization of coal containing organic sulphur.

Method for Determination of Biomass in the Biological Reactor for Microbial Desulphurization of Coal

Microorganisms are involved in carrying out different microbial reactions, the rates of which vary in magnitude in various coal environments. Assessment of microbial activity in mixed biomass system was carried out by indirect means through the measurement of the amount of viable (or functioning) material that was present and not the activity as such. Because of the difficulty of accurately determining the number of viable microorganisms present, the most ideal methods for estimating cell concentration are mass measurement, such as dry weight, total suspended solids and volatile suspended solids. However, these methods measure the total mass of material present, both functional and non-functional in coal based biological reactor system. It is difficult to differentiate coal particles and biomass in these methods.

Some of the methods for assessment of biomass in the microbial desulphurization units incorporate determination of nitrogen content, since protein content of DNA content is used as a indirect measurement of biomass concentration. However, these methods require more critical and efficient instruments for measurement of nitrogen concentration with reference to coal slurry reactors for biooxidation of organic sulphur compounds present in the coal. Other methods for measurement of biomass in terms of Adenosine Triphosphate (ATP) and reducing activity are the need for assessment and evaluation as controlling parameters for coal based bioreactor system.

Preservation and Maintenance of Microorganisms for Desulphurization of Coal Containing Organic Sulphur

A suitable method for preservation and maintenance of microorganisms in desulphurization of coal needs to be developed, since the specific microorganisms, maintained other substrates for organic sulphur removal, are reported to loose their activity after a period of time. If these microorganisms are preserved on coal it self, their activity is likely to be prolonged for desulphurization.

Enhancement of Rate for Microbial Desulphurization of Coal Containing Organic Sulphur

The rate of removal of organic sulphur from coal is slow and requires high residence time for activities of the microorganisms. Higher rate of removal of sulphur may be achieved through

- Incorporating active microorganisms with high potential for solubilization and stabilization of organic sulphur compounds from coal
- Involving microorganisms of defined characteristics and communities having potential for desulphurization
- Optimization of kinetic parameters form maximum removal rate without affecting the coal quality, viz. organic sulphur content of coal/microorganism ratio, nutritional requirement, oxygen requirement, suitable reactor system for accommodating upto 40 percent of coal solid concentration with optimal flow sheet, development of suitable down stream processing scheme for removal/stabilization/utilization of solublilized organic sulphur compounds from coal, modeling, scale up and possibilities of automation for developed scheme.

Development of Enzyme Based Catalysed Process for Removal of Organic Sulphur from Coal

Enzymatic catalysis in non-aqueous media can be considered for use in solubilization of organic sulphur compounds from coal. The use of horseradish peroxidase and several other oxidative enzymes in organic media and application of these enzymes for desulphurization of coal can be explored in controlled conditions. Specially, the horseradish peroxide, laccase, and chloroperoxidase were found to be capable of oxidizing dibenzothiophene in suitable solvents. Thus, such enzymatic reactions in conjunction with further microbial steps can be used to remove organic sulphur from coal.

Development of Biological Process Coupled with Physico-Chemical Process for Removal of Organic Sulphur from Coal

Hydrogen sulphide is generated in conventional hydrosulphurization system for desulphurization of coal. Hydrogen sulphide is produced from organic sulphur compounds in the coal. National Environmental Engineering Research Institute, Nagpur, India has developed a desulphurization process for removal and recovery of sulphur from hydrogen sulphide in different gaseous emissions. The same process can be adopted as a part of the hydrodesulphurization system for removal of organic sulphur from coal. Schematic of the proposed system is presented in Fig.8. The incorporation of the process for downstream processing of emissions generated in the hydrodesulphurization process will make the conventional hydrodesulphurization process most cost effective as elemental sulphur of high purity can be recovered from the process emissions. Cost effective process can be achieved through other additional advantages of the process, viz. no waste generation and the process operates at controlled condition resulting in better efficiency.

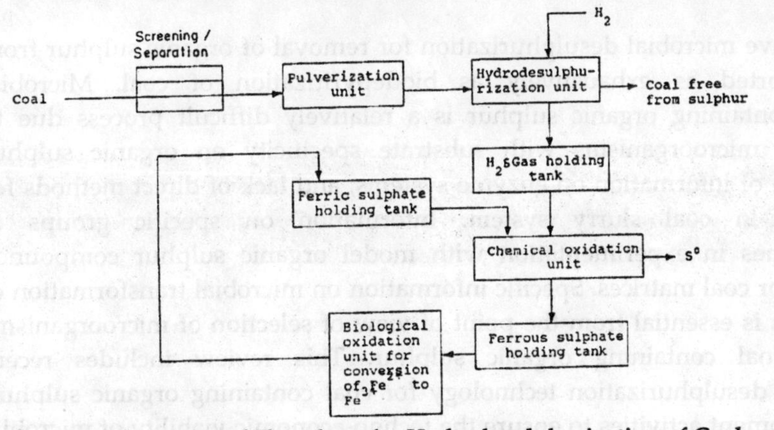

Fig. 8 : Schematics of Flowsheet for Hydrodesulphurization of Coal with Recourse to Recovery of Elemental Sulphur

Appropriate Method for Measurement of Organic Sulphur in Coal

There is a need for developing and modifying the existing procedures for analyzing the organic sulphur present in coal. The most common method used is the American society method no. D2492, in which organic sulphur is determined indirectly by chemical method by substracting pyritic sulphur and inorganic sulphur contents from total sulphur content of coal. The ASTM organizaton is currently revising and reviewing this method for modification, as organic sulphur determined by this procedure is not accurate and reproducible (Straszhien, *et al.*, 1983).

Another method for determining the sulphur content of coal is elemental analysis (Peck, 1965). The elemental analyzer estimates the total sulphur or coal after removal of inorganic/pyritic sulphur is subjected to further analysis. The biologically treated coal does not give the accurate and reproducible results as detector, incorporated in analyzer, is a thermal conductivity detector (TCD). The sensitivity for responsing the sulphur at lower concentration is very poor (Marckuszewski, 1981).

The Extended X-ray Absorption of Fine Structure (EXAFS), Electron Probe Micro Analysis (EPB) and Scanning Electron Microscopy coupled with Energy Dispersive X-ray Spectrometry (SEM & EDS) are the other techniques used for determination of organic sulphur in coal (Huffman, *et al.*, 1978; Salomon, 1977). However, these sophisticated methods cannot be adopted for routine analysis of coal and monitoring of microbial desulphurization process. A need, therefore, exists for development of a simple, reliable, accurate and reproducible method for monitoring of organic sulphur, present in coal.

From foregoing discussions, it can be observed that considerable advances have been made in exploring the possibilities of use of microbial process for desulphurization of coal containing organic sulphur over last ten years. However, techno-economical aspects of the microbial processes imparted constraints in formulation of a suitable efficient scheme for desulphurization of coal containing organic sulphur. A need, therefore, exists for investigations on the issues, highlighted in this review specially on aspects for making the process techno-economically viable.

Summary

Studies on preamptive microbial desulphurization for removal of organic sulphur from coal have not been reported as exhaustively as biodepyritization of coal. Microbial desulphurization of coal containing organic sulphur is a relatively difficult process due to nonavailability of efficient microorganisms with substrate specificity on organic sulphur constituents of coal, absence of information on enzyme systems, and lack of direct methods for measurement of biomass in coal slurry system. Information on specific groups of microorganisms and enzymes in experimentation with model organic sulphur compounds cannot be directly utilized for coal matrices. Specific information on microbial transformation of organic sulphur compounds is essential from the point of view of selection of microorganisms for desulphurization of coal containing organic sulphur. This review includes recent developments on microbial desulphurization technology for coal containing organic sulphur. Future research and development activities to ensure the techno-economic viability of microbial processes are also discussed.

Acknowledgement

Authors are thankful to Director, NEERI, Nagpur for his kind permission to publish this paper

References

Attar, A.J.R.H. (1982) Sulphur Groups in Coal, In : C.Karr (Edito), Anal. Methods for Coal and Coal Products Ch 56, Vol. III, P.256

Attar, A. (1979), Hydrocarbon Process, 1, 175

Angelova, G. and Kovacheva (1972), Chem. Abstract 72.23245J

Boudou, J.P., Boulegue, J., Malachan, L., Nip, M., deLecuw, J.W. and Boon, J.J. (1987), Fuel, 66, 1558

Block, S., Sharp, J.B. and Darlage, L. (1975), Fuel, 54, 113

Bamforth, C.W. and Anness, B.J. (1979) Soc. Gen. Microbiol. Quart., 6, 160

Couch, G.R. (1988) Resources, Conservation and Recycling, 1, 207

Chandra, D. (1979), Fuel, 58, 549

Casagrande, D. And Siefert (1977) Science, 195, 75

Cripps R.E. (1973) Biochem. J. 134, 353

Dugan (1986) Biotechnol. Bioengg. Symp. No. 16, 196

Dhillon, J.K., Junagade, S.V., Raman, V.K., Pandey, R.A. and Bal, A.S. (1994), Proceeding of Symposium on AMT at Ludhiana, 9-11 Feb.

Department of Biotechnology, New Delhi/National Environmental Engineering Research Institute, Nagpur (1993) Report on Microbial Desulphurization of Fossil Fuels (Coal, Oil & Gas)

Dodgson, K.S. and Rose, F.A. (1975) Sulphohydrolses in D.M. Greenberg (Editor), Metabolic pathways page 359, 3rd ed. vol. 7 Academic Press New York

Dugan, P.R. (1972), Biochemical Ecology of Water Pollution Plenium. New York, Page 123

Eliot, R.C. (1978), Coal Desulphurization Prior to Combustion, New Jersey : Noyes Data Corporation, 6, 33

Faght, J.M. and Westlake, D.W.S. (1988) Can. J. Microbiol., 34, 1135

Fiala, G. and Stetter, K.O. (1986). Arch. Microbiol., 145, 56

F.F., Nack, H. and Oxley, J.H. (1975), Hydrocarbon Processing, 115

Gockay, G.F. and Yurteri, R.N. (1983), Fuel, 63, 1223

Gokay, C.F. and Yurteri, R.N. (1983) Fuel, 62, 1223

Gallagher, J.R., Olson, E.S. and Stanely, D.C. (1993) FEMS Microbiol. Lett. 107:31

Gleen, H. (1949) J. Gen. Microbiol. 3, Vol 13

Haber, F. and Bruner, Z. (1940), Electrochem., 10, 697

Hedric, H.G. (1982), Proceeding of Symposium on Biological and Chemical Removal of Sulphur and Trace Elements in Coal and Lignite, Louisiana Tech. University, Ruston Louisiana USA, Nov. 1-12, 73

Hou, C.T. and Laskin, A.I. (1976), Ind. Microbiol., 17, 351

Huffman, G.P., Huggins, F.E., Shah, N. (1978) In Proceeding Utilization of High Sulphur Coal II, Ed. Chugh, J.P. and Candle, R.D., Elsevier Publ., New York, Page 3

Ignasiak, B.S., Fryer, J.F. and Jadernak, P. (1978) Fuel, 57, 578

Isbister, J.D. and Kobilinski, E.A. (1985) Coal Science and Technology Series No. 9 (Attia, J.A. ed) Elsevier, 627

Isbister, J.D. and Kobylinski, E.A. (1985) in Processing and Utilization of High Sulphur Coals (Ed. Y.A. Attia) Elsevier, Amsterdam Page 627

Jakoby, W.B. and Griffith O.W. (Editors), (1987) Methods in Ezymology Vol 143, page 582, Academic Press New York

Kilbane John (1989) TIBTCH 7, 97

Kargi, F. (1984) Advances in Biotechnological Processes, 3, 241

Kargi, F. (1982) Enzyme Microbiol. Technol., 4, 13

Kargi, F. and Robinson, J.G. (1982), Biotechnol. Bioengg., 24, 2115

Kilbane, J.J. (1988) Am. Chem. Soc. Div. Fuel Chem. 33, 586

Kilbane, J.J. and Bielage, B.A. (1990) Chemtech, 34, 747

Kilbane, J.J. (1990), Kiban, J.J. (1990) Bioprocessing and Biotreatment of Coal D.L. Wise (Ed), Marcel-Dekker New York Page 487

Kilbane, J.J. (1992) Biotechnol. Bioengg. 40:1107

Kargi, F. (1984), Advance in Biotechnological Process, 3, 241

Kargi, F. and Robinson, J.M. (1982) Appl. Envi. Microbiol., 44, 878

Krawise, S. (1988) In Biotechnology Applied to Fosil Fuels, D.L. Wise (Ed.), Page 136 Marcel Dekkar, New York

Kodama, K., Natakani, S., and Umehara, K. (1970) Agric. Biol. Chem. 34, 1320

Kodama, K., (1977) Agri Biol. Chem., 41, 1193

Kodama, K., Umehara, K. and Shinizu, K. (1973) Agric. Biol. Chem. 37, 45

Kluyver,, A.J. and Schnellen C.G.T.P. (1947) Arch. Biochem., 14, 57

Malik, K.A. nd Claus, D. (1976) Fifth Int. Ferment. Sump (Berlin) 23, 421

Marckuszewski, R. (1981) Am. Chem. Soc. Ser. No. 169, 401

Masciantonio, P.X. (1965), US Patent 3, 166, 483

Masciantonio, P.X. (1965), Fuel, 44, 269

Monticello and Finnerty (1985) Ann. Rev. Microbiol., 39,371

Naryan, R., Kullerud, G. and Wood, K.V. (1988). A New Perspective on the Nature of Organic Sulphur in Coal in: Division of Fuel Chemistry Pre Prints, 195th Meeting of the American Chemical Society, Toronto, Canada

Nishioka, M. (1988) Energy and Fuels, 2, 214

Nishioka, M., Lee, M.L., Kudo, H., Muchirim D., Baldwin, L.J., Parkray, S., Stuart, J.G. and Castle, R.N. (1985) Anal. Chem. 57, 1327

Nishioka, M., Lee, M.L. and Castle, R.N. (1986), Fuel, 65, 390

Olson, G.J. and Brincleman (1986), Fuel, 65, 1638

Peck, H.D. (1962). Bact. Rev., 26, 67

Peck, H.D. (1968) Ann. Rev. Microbiol., 22, 489

Peck, H.D. (1961) J. Bacteriol., 82, 933

Reid, E.E. (1960), Organic Chemistry of Bivalent Sulphur, Chemical Publishing Co. Inc., New York

Report on Environment, New Delhi (1992) Evaluation of S-Reducing Microorganisms for Organic Desulphurization of Coal, Central Fuel Research Institute, Dhanbad, India

Spiro, C.L., Wong, J., Lytle, F.W., Greegor, R. Maylotte, D.H. and Lamson, S.H. (1984), Science, 226, 48

Starkey, R.L. (1984), Ind. Eng. Chem. 48, 1420

Stoner, D.L. and Wey, J.E. (1990) Appl. Env. Microbiol., 56, 2667

Schicho, R.N., Olson, S.H., Parks, E.J. and Kelly, R.M. (1989), Fuel, 68, 1368

Sagardia, F., Rigou, J.J., Martinez Lahoz, A., Fuentes, F., Lopez C. and Flores W. (1975) Appl. Microbiol. 29, 575

Stetter, K.O., Konig, H. and Stackebrandt, E. (1983), Appl. Microbiol. 4, 535

Stambaugh, E.P., Miller, J.E., Tam, S.S., Chauhan, S.P., Feldmann, H.D., Carleton, H.E., Foster, F.F., Nackl, H. and Oxley, J.H. (1975), Hydrocarbon Processing, 115, 228

Straszhein, W.C., Green, R.T. and Markuszewski, R. (1983) Fuel, 63, 1070

Salomon, R.R. and Monizone, A.V. (1977), Fuel, 56, 393

Wise, W.S. (1971), Solvent Treatment of Coal, Mills and Boon Ltd., London

Warhurst, A.M. and Fewson, C.A. (1994) Crit. Rev. Biotechnol 14:29

Zinder, S.H. and Brock, T.D. (1978) J. Gen. Microbiol., 105, 335

Zinder, S.H. and Brock, T.D. (1978) Appl. Environ. Microbiol., 35, 344

24
Microbial Detoxification of Chlorinated Aromatic Compounds

M. Suresh Kumar

Introduction

Public concern over the possible effects of chemicals on humans and their environment has largely focused on a few classes of compounds. Of these compounds chlorinated aromatics are the most spectacular. Over the last few decades, significant quantities of chloroaromatic compounds have been used for crop protection and industrial application. Chlorinated organic compounds constitute one of the largest groups of pollutants for chemical synthesis (Alexander, 1981). Chlorinated aromatic hydrocarbons include a wide range of compounds, viz. chlorophenols, chlorobenzenes, chloronitro benzenes, chloroaniline, polychlorinated biphenyls (PCBs) and many pesticides. Aromatic chlorinated compounds are released as by product after manufacturing commercial product in the industry. The pulp and paper mill is a major industrial sector utilizing huge amounts of lignocellulosic materials and water during manufacturing process, and releases chlorinated phenols and chlorinated hydrocarbons in the wastewater (Witteh *et al.*, 1992).

Sources of Waste Containing Chloroaromatics

Most chlorinated benzenes are used mainly as intermediates in the synthesis of fine chemicals, and enter the environment as losses and wastes from production sites. As losses are normally 1-2% of raw materials, their total quantities do not exceed a few hundred tones per annum. The exceptionable aspects are the dispersive uses of mono-and dichlorobenzenes and waste disposal.

PCBs were first produced in 1929 in the United States. Total cumulative world production of PCBs in 1980s has been estimated at about 7,50,000 tonnes, of this total, approximately 60% have gone into closed electric uses (e.g., transformers, capacitors), 15% to nominally closed uses (hydraulic and heat-transfer fluids), and 25% to dispersive uses like plasticizers, paint and printing ink components, adhesives, and additives for cutting oils, textile auxiliaries, and pesticides (Reineke & Knackmuss, 1988).

Two types of pollution may result from releases of chemicals into the environment. In point source pollution the concentrations of the chemicals is high this may occur in landfills, waste dumps, and industrial effluents or at sites of accidents associated with transportation and application of chemicals. In dispersed pollution the concentrations are low and result from losses from production sites via volatilization or from agricultural uses.

Different strategies are necessary to stem the impact of low concentrations of dispersed chemicals occurring worldwide and the high concentrations of point source pollutants that occur in smaller areas. If dispersed chemicals such as an agrochemicals tends to persist, its use and introduction into the environment can only be regulated legislatively. Technological solutions, however, can be applied for more highly concentrated pollutants, e.g., in industrial wastewaters and dumps.

Role of Microorganisms in the Degradation of Chloroaromatics

Microorganisms are key agents for recycling organic matter in the biosphere. Many chlorinated organic compounds are relatively resistant to microbial attack and thus accumulate in the environment. The mechanisms bacteria use to detoxify chlorinated compounds are being studied extensively. Understanding dechlorination mechanisms will pay important dividends in efforts to develop more efficient biological processes for treating hazardous wastes.

Four fundamental mechanisms of biodehalogenation have been identified. These are 1) hydrolytic, 2) oxidative, 3) reductive and 4) eliminative reactions. Most dehalogenating enzymes studied to date catalyze net hydrolytic carbon-halogen bond cleavage. Oxidative dechlorination is often mediated by monooxygenases or dioxygenases. These reactions are usually cooxidative. That is, the chlorinated compound undergoing oxidation is not the primary substrate of the enzyme and it does not support growth of the organisms producing the oxygenase.

Reductive dechlorination may be very important environmentally. Anaerobic dechlorination is typically very slow, but highly chlorinated molecules undergo dechlorination via this mechanism almost exclusively. Some biologically relevant reductive dechlorination reactions may be mediated by transition metal co-enzymes found in anaerobic bacteria. The reduced cofactors shown to be reactive in vitro are hematin, vitamin B_{12} and coenzyme F_{430}. An understanding of these and other mechanisms of biodehalogenation will allow the use of innovative biotechnololgical approaches for enhancing bioremediation.

Persistency and Toxicity of Chloroaromatics

Halogenated organic compounds comprise the largest group of Environmental Protection Agency Priority Pollutants (Lawrance, 1991). These compounds have attracted widespread attention due to this toxicity and their general persistence in the environment. For example, chlorinated pesticides were expected to last for prolonged periods in soils and waters. Also poly-chlorinated biphenyls (PCBs) were used as transformer fluids because of their great stability. A flurry of patents were issued for newly synthesized chlorinated organic compounds during the period from 1950 to 1970 (Hutzinger & Veerkamp, 1981). However, the often toxic and/or carcinogenic potential of halogenated compounds has more recently become apparent. In this context, the environmental longevity of halogenated organic molecules is now perceived to be a curse rather than a blessing. New applications for chlorinated compounds have waned. Further more, many older compositions have been banned. Despite this, chlorinated compounds are still prevalent in the environment as a result of the persistence of old applications and contamination from continued used of less toxic materials.

In general, chlorinated organic compounds exert both toxic and carcinogenic effects in mammals as a result of metabolic activation which can produce reactive intermediates (Henschler, 1985). The first metabolic step is often catalyzed by cytochrome P-450 monooxygenase. Although this enzyme sometimes serves to detoxify xenobiotic compounds, the monooxygenation of chlorinated compounds often produces very unstable, highly reactive products. The reaction of these metabolites with critical cellular molecules, most significantly DNA, can mediate cellular toxicity and potentially carcinogenicity.

Aerobic Degradation of Chlorinated Aromatic Compounds

One requirement for microbial degradation of any compound is the need to induce enzyme production. Not all halogenated aromatic compounds will induce enzyme production even when their nonhalogenated form can (Reineke, 1984). For example, some organisms that obtain energy from diphenylmethane cannot grow on diphenylmethane when chlorine atoms are substituted in the para position of both benzene rings. It the nonhalogenated diphenylmethane is added as primary substrate, the chlorinated substitute form is degraded by co-metabolism.

Most halogenated aromatic compounds that are degraded under aerobic conditions are probably acted upon through co-metabolism. It is also possible that a halogenated aromatic compound is transformed to a toxic product that prevents further aerobic degradation. Complete aerobic mineralization of chlorinated aromatic compounds has been reported in the past (John & Cookson, 1995). However, the persistence of these compounds reflects the inability of many microorganisms to degrade these compounds. The nature and number of chlorine substitutions, and the substitution positions influence the extent of degradation. Degradation of chlorine-substituted aromatic compounds frequently does not follow the reaction pathways of the unsubstituted parent compounds (Suthan, 1997).

Many aerobic organisms are reported to co-metabolize halogenated aromatic compounds. The relatively non-specific nature of the enzymes that transform benzoate to catechol accounts for this. However, in some cases this degradation is not complete in that chlorinated benzoates and catechols are the end result of oxidative degradation (Reineke, 1984). Once a halogenated catechol or phenol is formed, there appear to be more organisms that can degrade these under aerobic conditions. An extensive list of microorganisms have been identified that can transform various chlorinated phenol compounds. The most common transformation pathway is the elimination of the halide after ortho cleavage of the halocatechols. An example is the degradation of 3-methyl-5-chlorocatechol.

The complete aerobic mineralization of halogenated aromatic compounds has been confirmed for several chemicals and microbial consortium. However, the persistence of these compounds illustrates the ineffectiveness of many microorganisms. The nature, the number of substitutions, and their substitution position affect the ease and degree of degradation. The degradation of these halogen-substituted aromatic compounds frequently does not follow the reaction pathways of the unsubstituted parent compounds.

Chlorinated aromatic compounds can be used as sole energy sources. The use of 3-chlorobenzoate as a sole carbon source for *Pseudomonas* sp. is one example. Once the chlorine has been removed, the compound is mineralized as typical of oxygenated aromatic compounds. The aerobic replacement of the halide with the hydroxyl group is typical of many halogenated aromatic transformations. However, if the substituted chlorine atom is changed to from 2-chloro- or 4-chlorobenzoate, the microorganisms cannot use the compound as a sole energy source. Other microorganisms such as *Nocardia* sp. and *Arthrobacter globiformis* can utilize the 4-chlorobengoate as a carbon and energy source. Present evidence indicates that the aerobic enzymes that provide energy and carbon source by elimination of halogens from aromatic compounds are highly specific. For most compounds, the aerobic rate of transformation decreases as the chlorine substitutions increase.

The response of the microbial community toward organic compounds does not depend on total concentration, but mainly on the water-soluble concentration. Bioavailability of a compound is of extreme importance, because if frequently accounts for the persistence of compounds that are biodegradable and that might otherwise be assumed to be readily degraded.

When two or more different sequential microbial populations are required for complete degradation, longer than normal acclimatization times may be involved. For difficult to degrade compounds, this is rather a rule than an exception. Different kinetics of the various degradation steps and this acclimatization time are the reasons why the overall disappearance seems to follow a cyclic pattern. It can be assumed that nonlinear responses in reality are rather a rule than exception. Other factors that may impact the bioavailability, and thus the kinetics of biodegradation are weathering, sequestering, and complexation of substrate. Weathering of a contaminant results in easily biodegradable compounds being degraded early and formation of an aged residue. Sequestering of a compound occurs when a compound becomes less available or essentially wholly unavailable for biodegradation when it enters or is deposited in a micropore that is inaccessible to microorganisms. Complexation of a compound affects biodegradation when the contaminant forms insoluble complexes in association with inorganic or organic substances present in the environment.

Pentachlorophenol readily degrades in the environment by both chemical and biological processes. Bacteria and fungi are active degraders of PCP. Microbial species include the bacteria of *Pseudomonas* sp. *Flavobacterium* sp. and fungi of *Phanerochaete chrysosporium, P. ordida, Trichoderma* sp. and *Trametes versicolar* (Kaufman, 1977; Lamar and Dietrich, 1990; Glaser, 1988). Pentachlorophenol is rapidly mineralized to CO_2 by *P. chrysosporium,* although the transformation product of 1,4-tetra chlorobenoquinone is also produced. Successful field bioremediation of PCP has been conducted with both bacteria and fungi in solid-phase and slurry-phase reactors.

Most of the chlorinated phenoyalkanoates are degraded by soil microorganisms. However, little degradation is experienced at temperatures below 10°C and at low moisture (Smith and Lafond, 1990). The common soil bacteria *Arthrobacter* sp. Cleaves the ether linkage of 2,4-D (Fig. 1). This oxidative mechanism yield 2,4-dichlorophenol and alanine. As with the aerobic oxidation of the benzene ring, metabolism of 2,4-D yield a catechol structure as an intermediate. Both 4-chlorocatechol and 3,5-dichlorocatechol are reported as oxidation products. These chlorocatechols are degraded by several pathways that result in one or more forms of

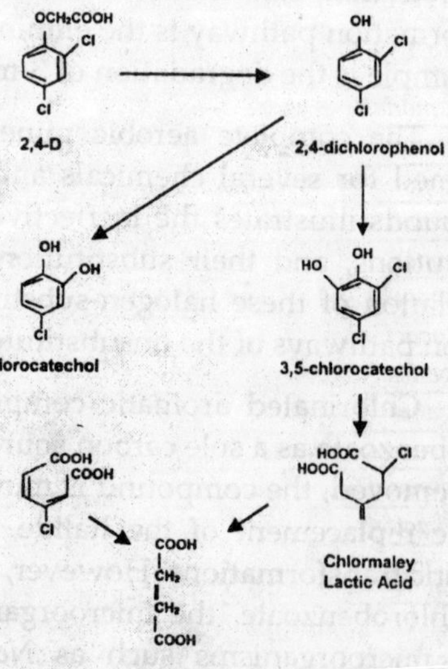

Figure 1 : Aerobic degradation of 2, 4-D

dichloromuconic acid and chloromaleylacetic acid and degrade to products such as succinic acid.

PCBs chlorinated with four or more chlorine atoms are resistant under aerobic conditions. No single organism is responsible for the degradation of multiple chlorinate PCBs. Degradation is achieved through a consortium of selected organisms. The co-metabolic transformation steps of 4,4′-dichlorobiphenyl (4,4-DCBP) to 4-chlorobenzoate by two species of *Acinetobacter* sp. has been well established (Adriaens *et al.*, 1989). The proposed degradation pathway is illustrated in **Figure 2.** The first step of the pathway is a hydrolytic ring dehalogenation. The aerobic biodegradation of PCBs appears to be limited to congeners with five or fewer chlorine atoms and two adjacent unsubstituted carbon atoms (Quensen *et al.*, 1990). Analog substrate enrichments have produced varied results for PCB degradation. The addition of biphenyl as an analog substrate had significant effect on the degradation of Aroclor 1242 (Brunner *et al.*, 1985).

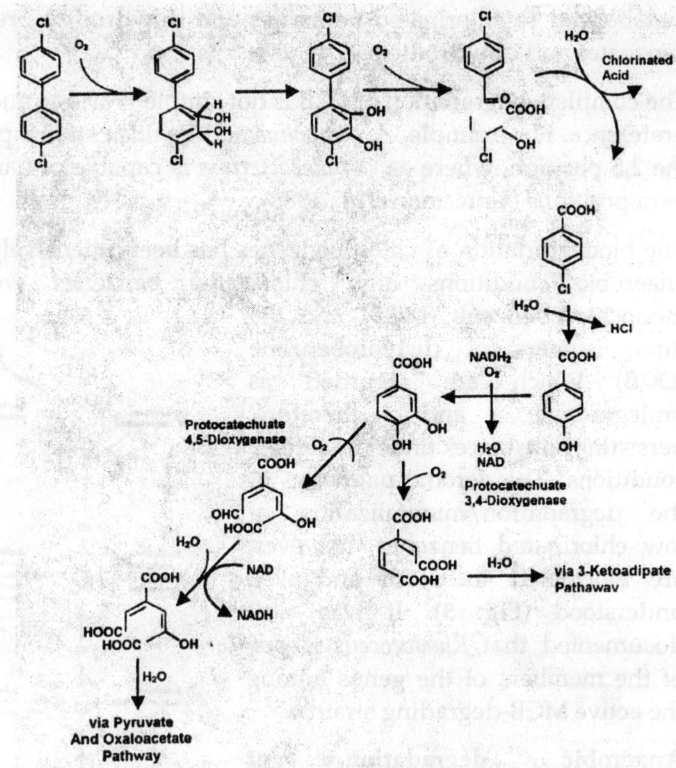

Figure 2 : Degradation of 4,4′ –dichlorobiphenyl (4,4′DCBP)

Table 1 : PCB Degradation Organisms

Pseudomonas putida
Pseudomonas cepacia
Pseudomonas testosteroni
Pseudomonas sp.
Pseudomonas aeruginosa
Pseudomonas sp.
Acinetobacter sp.
Alcaligenes odorans
Alcaligenes faecalis
Alcaligenes eutrophus
Alcaligenes denitrificans
Arthrobacter sp.
Corynebacterium sp.

Source: Unterman, 1988; Hernandez, 1991; Kohler, 1988; Kohler, 1988a; Clark, 1979

Although analog enrichment was the largest single factor for increasing biodegradation of PCBs, this was not the only factor of significance. Natural soil microflora may not provide the necessary microflora to degrade PCB′s. The addition of *Acinetobacter* with the biphenyl substrate reduced the adaptation period compared with uninoculated soil. Since *Acinetobacter* can not dehalogenate any of the products that it forms from cometabolism of PCBs, further degradation of these compounds is carried out by other soil microflora. A diverse group of organisms has been identified as participating in PCBs degradation (**Table 1**). PCBs are co-metabolized to chlorinated benzoates and dihydrodiols, resulting in a mixture of halogenated benzoates and dihydrodiols.

The complete degradation of PCB is not simple. Various microbial strains have distinct congener preference. For example, *A. eutrophus* and *P.putida* strains prefer congeners with substitutions at the 2,5 position, where as *Corynebacterium* is capable of handling congeners substituted in both para positions (Unterman *et al.*, 1988).

The biodegradation of chlorobenzenes has been intensively studied in the last decades. Under anaerobic conditions high chlorinated benzenes were reductively dechlorinated to monochlorobenzene (MCB) and the three isomers of dichlorobenzene (DCB) which are regarded as undegradable and therefore, persisting substances under anaerobic conditions. The aerobic pathways of the degradation/mineralization of low chlorinated benzenes, however, are elucidated in detail and well understood (**Fig. 3**). It was well documented that *Rhodococcus* is one of the members of the genus among the active MCB-degrading strains.

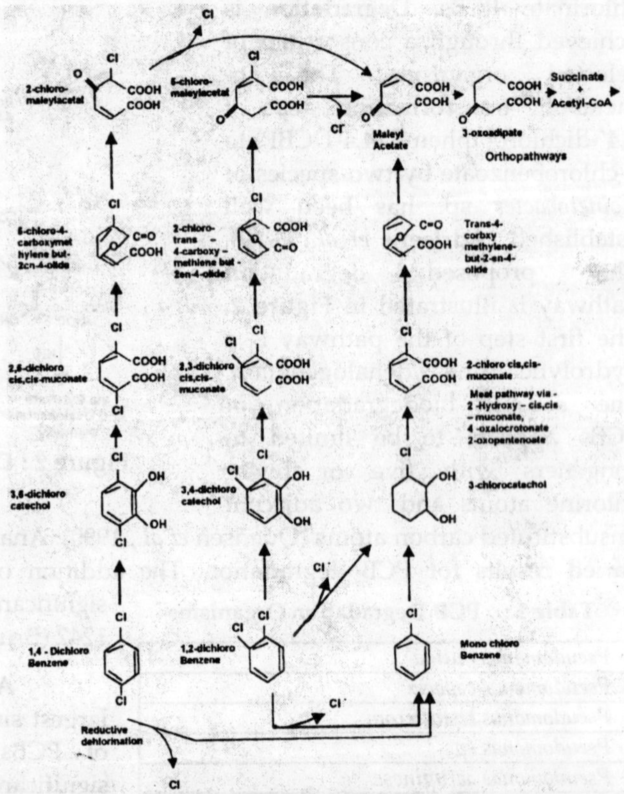

Figure 3 : Pathways for the degradation of the main chloroaromatics (MCB, 1,4 – and 1,2 –DCB)

Anaerobic degradation of chlorinated aromatic compounds

Chlorinated aromatic hydrocarbons that are recalcitrant under aerobic conditions are some times degraded by one or more reductive dechlorinations under anaerobic conditions. Chlorinated organic compounds serve as the electron acceptors, and the primary substrate supplies the electron due to oxidation. Since reaction rates are influenced by the availability of electron acceptors and donors, the rate of dehalogenation is linked to the rate of substrate oxidation.

Microbial populations can mediate the anaerobic dehalogenation of halogenated compounds. Halogens at the ortho and para positions are more resistant to dehalogenation than those at the meta position. Once the halogen is removed, ring fission leads to methane and carbon dioxide (Young, 1984). The main mechanism for transformation of halogenated aromatic compounds is by reductive dehalogenation. This reaction results in the replacement of the halogen by hydrolysis. Although the chlorine is usually replaced with hydrogen, it can be replaced with an amino, chloryl, hydroxyl, or methyl substitute (Dolfing, 1992). Reductive dehalogenation has been confirmed on a number of halogenated aromatic acids, mono- to hexa-chlorobenzenes, chlorophenols, chlorophenoxyacetate, herbicides, DDT, and PCBs. In reductive dehalogenation the halogenated organic compound serves as the electron acceptor (**Figs. 4 and 5**). Most anaerobic dehalogenation is probably a result of co-metabolism. However, theoretical evidence from calculation of Gibbs free energy available from dechlorination indicates that microorganisms can benefit from the use of halogenated aromatic compounds as electron acceptors under anaerobic conditions. Experimental evidence comes from studies on the dechlorination of 3-chlorobenzoate to benzoate by Desulfomonile tiedjei (Tiedje and Stevvens, 1987). Several chlorinated benzoic acids are used as sole sources of carbon and energy by bacteria. Metabolism of 3- and 3,5-dichlorobenzoate proceeds by direct deoxygenation, followed by decarboxylation of the aromatic ring to give 3- and 3,5chlorocatechol, respectively, prior to ring fission and subsequent dehalogenation. In contrast, 4-chlorobenzoate is dehalogenated hydrolytically to 4-hydroxybenzoate prior to ring fission.

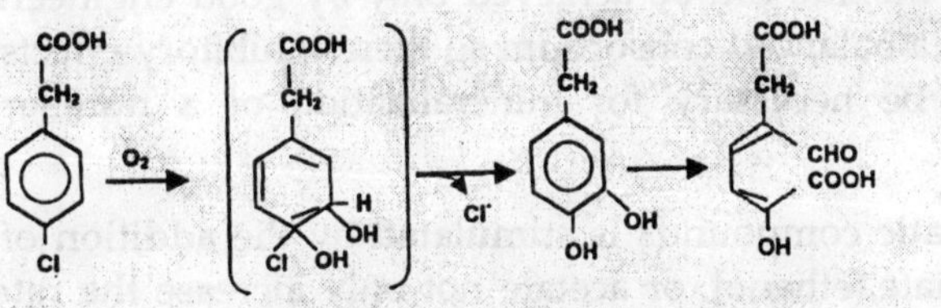

Figure 4 : Reaction Sequence in the Degradation of 4-Chlorophenyl Actate

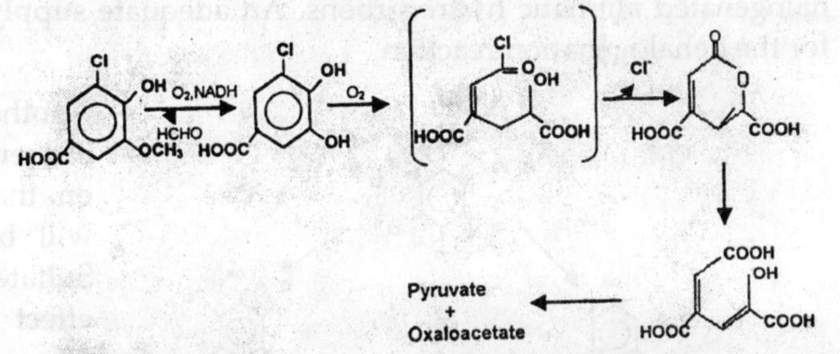

Figure 5 : Reaction Sequence Proposed for the Degradation of 5-chlorovanillate

Methanogenic metabolism has successfully dehalogenated several chlorinated aromatic compounds. These includes 2,4,5-trichlorophenoxyacetate, 3-chlorobenzoate, 2,4-dichlorophenol, 4-chlorophenol, 2,3,6-trichlorobenzoate, and 2,4-, 2,5-, 2,6-, 3,4- and 3,5-dichlorobenzoates (Zhang and Wiegel, 1990; Gibson and Suflita, 1990) (**Fig. 6**). Methanogenic cultures show preferential removal of ortho-chlorines, with meta- or para-chlorines removed at slower rates (Haggblan *et al.,* 1991). Pseudomonas putida is reported as one of the potential degraders of 4-chlorophenol through co-

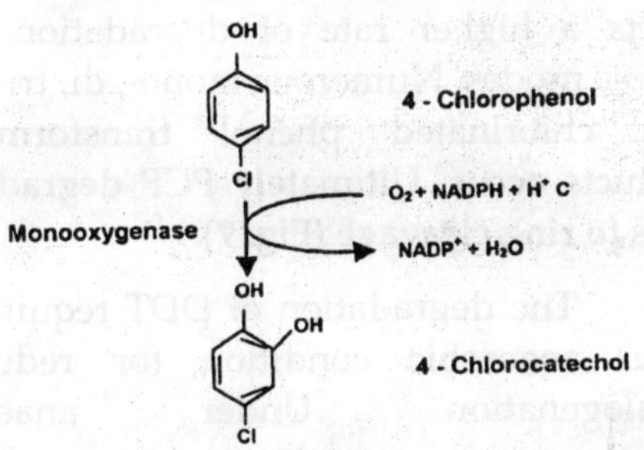

Figure 6 : Proposed degradation of 4-Chlorophenol by Pseudomonas putida

metabolism (Saez and Rittman, 1991). The major problem is the effect of phenol, an intermediate on the degradation pathway. Phenol buildup can prevent the dechlorination of 4-chlorophenol. The interactions of these microbial systems and the sensitivity of the overall mineralization of 2,4-dichlorophenol to methane and carbon dioxide can be achieved only by good engineering. The process must be controlled to maintain a balanced consortium so these inhibitory effects do not commence. Similar interactions may be necessary for mineralization of a number of halogenated aromatic compounds.

Anaerobic dehalogenation of aromatic compounds is stimulated by the addition of an electro donor. Addition of butyrate, propinate, ethanol, or acetate not only increase the rate of dehalogenation but increases the extent of degradation (Gibson, 1990). The addition of p-cresol and propionate enhances the methanogenic degradation of 2,4-dichlorophenol (Haggblom, 1991). The enhancement of dehalogenation by volatile fatty acids has also been found for halogenated aliphatic hydrocarbons. An adequate supply of these electron donors is necessary for the dehalogenation reaction.

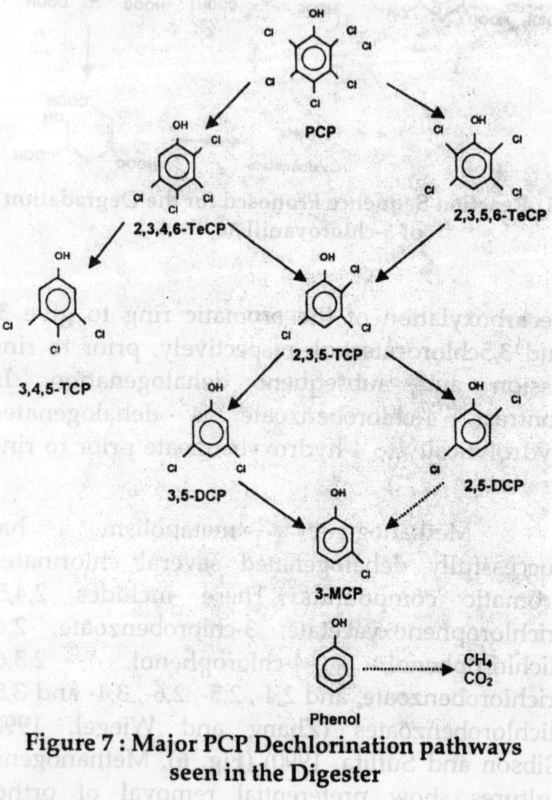

Figure 7 : Major PCP Dechlorination pathways seen in the Digester

Another important consideration is the influence that various electron acceptors have on the reaction. Certain electron acceptors will block the desired reduction reaction. Sulfate has been reported to have an adverse effect on the dehalogenation of some aromatic compounds.

Bacterial degradation of PCP is more rapid in soils containing higher organic content and moisture. Anaerobic metabolism affects a higher rate of degradation than aerobic modes. Numerous mono-, di, tri-, and tetra chlorinated phenol transformation products occur. Ultimately PCP degradation leads to ring cleavage (Fig. 7).

The degradation of DDT requires an initial anaerobic condition for reductive dehalogenation. Under anaerobic methanogenic and sulfur-reducing conditions, some persistent pesticides can be detoxified. By combining metabolism under both anaerobic and aerobic conditions, it is possible to completely mineralize DDT (Fig. 8). DDT is metabolized under anaerobic conditions to yield p-dichlorodiphenyl-methane (Slater, 1984).

Chlorobenzenes have also been shown to dechlorinate to lower chlorinated benzenes. Chlorobenzene dechloronation is also position specific. Chlorine with two adjacent chlorines are

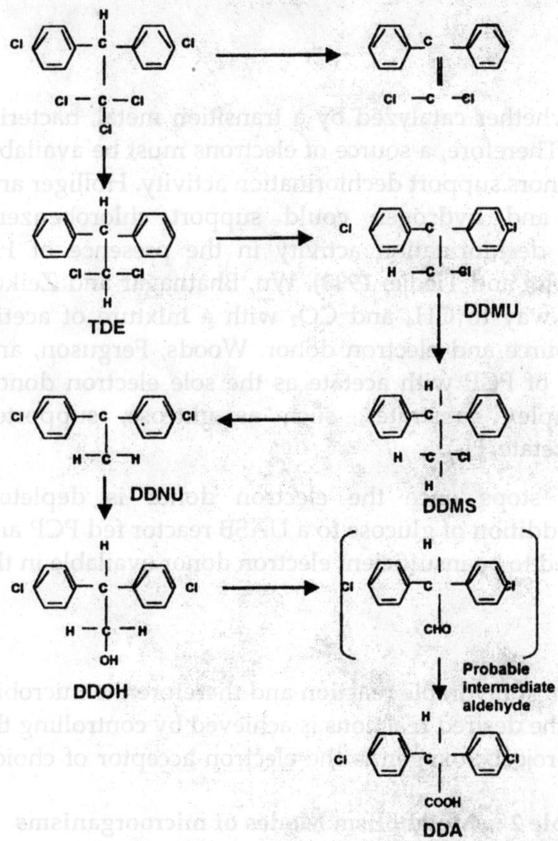

Figure 8 : Biotransformation of DDT

removed more easily than chlorine with only one adjacent chlorine. Chlorine removal with no adjacent chlorines is not commonly observed. Fathepure, and his coworkers (1988) observed haxachlorobenzene (HCB) dechlorination to trichlorobenzenes (TCB) and dichlorobenzenes (DCB) in anaerobic biosolids in three weeks. Using nitrate and sulfate as alternative electron acceptors a significant reduction of the MCB concentration in the culture suspensions was measured under different conditions. Anaerobic degradation processes were only observed when mixed cultures were used as inoculum. Although denitrifying and sulfate-reducing taxa could be detected within these mixed cultures by analyzing their 16s rDNA, pure cultures capable of degradation of MCB could be isolated not yet. Both denitrifying and sulfate-reducing enrichment culture grow with nitrate and sulfate, respectively, as electron acceptors on MCB as sole growth substrate. The expected metabolites CO_2, nitrites (in denitrifying cultures) and sulfide (in sulfate-reducing bacteria) could be detected during the cultivation processes, but up to now it was impossible to establish complete, reliable mass balances. Although these experiments were performed with the common laboratory equipment for anaerobic work, it can not be excluded absolutely that traces of oxygen got admitted to the culture suspensions (e.g., by diffusion through tubes and plugs, as contaminant of the gas phases and solutions etc.). These traces of oxygen might be necessary to initiate the degradation process. After that, further degradation might take place via the typical pathways of denitrification and sulfate reduction.

Advantages of Anaerobic Degradation

Anaerobic metabolic modes have a significant advantage over aerobic modes for PCBs. Dechlorination of the highly chlorinated Aroclor 1260 even occurs to a significant extent under anaerobic conditions. Anaerobic reductive dehalogenation is the only known biodegradation process for the more highly chlorinated PCB mixtures (Morris *et al.*, 1992). The dehalogenation is usually associated with syntropic communities. Only one obligate anaerobic that is capable of aryl reductive dehalogenation has been isolated (*Desulfomonile tiedjei*). Anaerobic metabolic modes have a significant advantage over aerobic modes for PCBs, thermodynamics would favor anaerobic dehalogenation of PCBs over aerobic systems. Even dechlorination of the more highly chlorinated Aroclor 1260 occurs to a significant extent under anaerobic conditions (Quensen *et al.*, 1990).

Role of Electron Donors in Dechlorination

Reductive dehalogenation reactions, whether catalyzed by a transition metal, bacterial cofactor, or an enzyme, require two electrons. Therefore, a source of electrons must be available for the reaction to take place. Many electron donors support dechlorination activity. Holliger and his co-workers (1992) found that lactate and hydrogen could support chlorobenzene degradation. Desulfomonile tiedjei exhibited dechlorination activity in the presence of H_2, formate, or pyruvate but not with acetate (Mohn and Tiedje, 1990). Wu, Bhatnagar and Zeikus (1993) were able to dechlorinate PCP all the way to CH_4 and CO_2 with a mixture of acetic, propionic, and butyric acids as the carbon source and electron donor. Woods, Ferguson, and Benjamin (1989) demonstrated dechlorination of PCP with acetate as the sole electron donor. Rasmussen *et al.*, (1993) found that complex substrates, such as glucose, supported dechlorination better than simple substrates (acetate, H_2).

Evidence exists that dechlorination stops once the electron donor is depleted. Dehalogenation of PCP was enhanced by the addition of glucose to a UASB reactor fed PCP and phenol (Hendriksen *et al.*, 1992). There appeared to be insufficient electron donor available in the reactor fed only phenol.

Role of Electron Acceptors in Dechlorination

The electron acceptor establishes the most favorable reaction and therefore the microbial metabolism mode for degradation. Obtaining the desired reactions is achieved by controlling the electron acceptor. For many bioremediation projects, oxygen is the electron acceptor of choice. However, some compounds degrade more favorably under one of the anaerobic modes. Controlling one or more of these specific electron acceptors is particularly important for the dehalogenation of chlorinated organic compounds. The metabolism modes of microorganisms are presented in Table 2.

Table 2 : Metabolism Modes of microorganisms

Type	Electron Acceptors
Aerobic respiration	Oxygen (O_2)
Anaerobic respiration	
Denitrification	Nitrate (NO_3^-)
Nitrate reduction	Nitrate (NO_3^-)
Sulfate reduction	Sulfate (SO_4^{-2})
Ferric iron reduction	Iron (Fe^{+3})
Obligate proton reduction	Protons (H^+)
Sulfur reduction	Sulfur (S)
Organic respiration	Fumarate
Organic reduction	Trimethylamine oxide
Fermentation	Organic compound
Methane fermentation (Methanogenesis)	Carbon dioxide (CO_2)

Common soil minerals such as manganese (IV) and iron (III) oxides also may serve as alternative electron acceptors. The importance of such alternative electron acceptors depends on their availability. An indication of the available electron acceptors within a site can be estimated from redox measurements. The redox potential expresses the electron availability as it affects oxidation states. One may have the option of using either an aerobic or an anaerobic mode of metabolism in designing a bioreamediation system. Some organic compounds can be degraded under either aerobic or anaerobic conditions. For these contaminants, selection of the microbial metabolism may well depend upon the existing redox potential. The practical range of redox potential in the natural environment is from +800 mV (high O_2, with no O_2 depleting processes) to about –400 mV (systems high in H_2). However, it should be noted that the concentrations of particular oxidants or reductants will affect microbial metabolic activity regardless of the redox potential. The common reductants and oxidants in bacterial respiration is presented in Table 3.

Table 3 : Reductants and Oxidants in Bacterial Respiration

Reductant Electron Donor	Oxidant Electron Acceptor	Product	Example Organisms or Group
Aerobic Respiration			
H_2	O_2	H_2O	Hydrogen Bacteria
Organic	O_2	$CO_2 + H_2O$	Many Bacteria
NH_3	O_2	$NO_2 + H_2O$	Nitrifying Bacteria *Nitrosomonas*
NO^{2-}	O_2	$NO_3 + H_2O$	Nitrifying Bacteria *Nitrobacter*
Fe^{+2}	O_2	Fe^{+3}	*Ferrobacillus*
S^{-2}	O_2	$SO_4^{-2} + H_2O$	*Thiobacillus*
Anaerobic Respiration			
Organic	NO_3^-	$N_2 + CO_2$	Denitrifying Bacteria
Organic	SO_4^{-2}	$H_2O + S^{-2}$	Sulfate Reducers
H_2	SO_4^{-2}	$H_2O + S^{-2}$	Sulfate Reducer *Desulfovibrio*
H_2	CO_2	CH_4, H_2O	Methanogens
Organic	NADP	CH_4, CO_2	Methanogens
Organic	H^+	H_2	Acetogens
Organic	Organic	Organic	Many Fermentative Bacteria

Source: John T. Cookson, Jr. 1995. Bioremediation Engineering Design and Application. McGraw-Hill. Inc. NY

Bioremediation of Chloroaromatics Contaminated Sites

In situ bioremediation is the application of biological treatment to the cleanup of hazardous chemicals present in the subsurface. The optimization and control of microbial transformations of organic contaminants require the integration of many scientific and engineering disciplines.

Hazardous compounds persists in the subsurface because environmental conditions are not appropriate for the microbial activity that results in biochemical degradation. The optimization of environmental conditions is achieved by understanding the biological principles under which these compounds are degraded, and the effect of environmental conditions on both the responsible microorganisms and their metabolic reactions.

During the process of in situ bioremediation, microorganisms use the organic contaminants for their growth. In addition, compounds providing the major nutrients such as nitrogen, phosphorus, and minor nutrients such as sulfur and trace elements are also required for their growth. Two categories of transformations exist. In the first, biodegradation provides carbon and energy to support growth, and the process, therefore, is growth-linked. In the second, biodegradation is not linked to multiplication, but to obtaining the carbon for respiration in order for the cells to maintain their viability. This maintenance metabolism may take place only when the organic carbon concentrations are very low. Co-metabolic transformations also fall into the second category.

Microbial transformations of organic compounds are frequently described by the terms, degradation, mineralization, detoxification, and activation. Degradation means that the initial substrate no longer exists. Mineralization refers to the complete conversion of the organic structure to inorganic forms such as CO_2, H_2O, and Cl^-. Detoxification is the transformation of the compound to some intermediate form that is nontoxic or less toxic. The process of forming toxic end products or intermediate products is known as activation.

Microorganisms are capable of catalyzing a variety of reactors; dechlorination, hydrolysis, cleavage, oxidation, reduction, dehydrogenation, dehydrohalogenation, and substitution.

- **Dechlorination**: the chlorinated compound becomes an electron acceptor; in this process, a chlorine atom is removed and is replaced with a hydrogen atom.
- **Hydrolysis**: frequently conducted outside the microbial cell by exoenzymes. Hydrolysis is simply a cleavage of an organic molecule with the addition of water.
- **Cleavage**: cleaving of a carbon-carbon bond is another important reaction. An organic compound is split or a terminal carbon is cleaved off an organic chain.
- **Oxidation**: breakdown of organic compounds using an electrophilic form of oxygen.
- **Reduction**: breakdown of organic compounds by a nucleophilic form of hydrogen or by direct electron delivery.
- **Dehydrogenation**: an oxidation-reduction reaction that results in the loss of two electrons and two protons, resulting in the loss of two hydrogen atoms.
- **Dehydrohalogenation**: similar to dechlorination, results in the loss of a hydrogen and chlorine atom from the organic compound.
- **Substitution**: these reactions involve replacing one atom with another.

Examples of these reactions are shown in **Table. 4.**

Table 4 : Microbially Catalyzed Reactions

Reaction	Example
• Dehalogenation	$Cl_2C = CHCl + H^+ \longrightarrow ClHC = CHCl + Cl^-$
• Hydrolysis	$RCO - OR' + H_2O \longrightarrow RCOOH + R'OH$
• Cleavage	$RCOOH \longrightarrow RH + CO_2$
• Oxidation	$CH_3CHCl_2 + H_2O \longrightarrow CH_3CCl_2OH + 2H^+ + 2e^-$
• Reduction	$CCl4 + H^+ + 2e^- \longrightarrow CHCl_3 + Cl^-$
• Dehydrohalogenation	$CCl_3CH_3 \longrightarrow CCl_2CH_2 + HCl$
• Substitution	$CH_3CH_2Br + HS^- \longrightarrow CH_3CH_2SH + Br^-$

Source: McCarty, P.L and Semprini, L. 1994. Groundwater treatment for chlorinated solvents, in Handbook of Bioremediation, Norris, R. D., et al., Eds. Lewis Publishers, Boca Raton, FL.

Uses of Bacteria in Bioremediation

Bioremediation has been defined as a "biological response to environmental abuse" (Hamer, 1993). This definition serves to distinguish between the use of microorganisms to remediate contaminated sites and their application in biotreatment/biorecycle processes designed to reduce organic and inorganic contaminant emissions at source (David, 1997). Bioremediation is concerned with the biological restoration of historically contaminated sites and with the cleanup of areas contaminated in more recent times, either accidentally or incidentally, as a result of the production, storage, transport, and use of organic chemicals. By contrast, bioremediation may be viewed as a more effective and environmentally benign cleanup technology since it results in the partial or complete bioconversion of organic pollutants to microbial biomass and stable, nontoxic end products.

Bioremediation Technologies

Bioremediation treatment technologies have been broadly divided into two categories based on whether biodegradation is stimulated *in situ* or carried out *ex situ* in compost heaps or bioreactors (Baker and Herson, 1994; Blackburn and Hafker, 1993). *In situ* techniques involve enhancement of the biodegradation rate of organic contaminants within affected soil, sediment, surface water, or ground water environments. *Ex situ* techniques require physical removal of the contaminated material (usually soils or sediments) followed by treatment under contained conditions in bioreactors, biopiles, compost heaps, ponds, or lagoons. Although *in situ* bioremediation, by definition, assumes treatment of the contaminated material in place, "pump and treat" technologies are usually included in this category, despite the fact that they involve the removal, treatment, and return of associated water from contaminated soil zones. In addition, air venting and soil vapor extraction techniques that include biological treatment of the extracted, contaminated air, prior to discharge to the atmosphere, are usually defined as *in situ* techniques.

In situ Bioremediation Techniques

Bioremediation of organic contaminants *in situ* requires stimulation of the degradative activities of endogenous microbial populations by the provision of nutrients and/or external electron acceptors. Alternatively, it may involve the addition of competent, exogenous microbial inocula, with or without nutrient and electron acceptor supplementation.

The addition of nutrients to contaminated soils has also been investigated with varying degrees of success (Morgan and Watkinson, 1989; Compeau *et al.*, 1991; St. John and Sykes, 1988). In the majority of soil bioremediation applications, nutrient addition has been combined with oxygen provision via percolation of oxygen saturated water, air sparging, or bioventing (Hinchee *et al.*, 1995). Hydrogen peroxide has also been widely used as an alternative oxygen source in percolation trials (Pardick *et al.*, 1992; Lu, 1994). Air-sparging involves injection of air into the water-saturated zone of soils and sediments. Bioventing is one of the most widely used methods of remediating vadose-zone soils contaminated by petroleum hydrocarbons. In practice, soil vapor extraction and bioventing techniques are often combined in order to speed up the remediation of highly contaminated sites.

In situ bioremediation under anaerobic conditions may also be enhanced by the provision of electron acceptors, such as nitrate or sulfate (Sherwood *et al.*, 1995; Freedman *et al.*, 1995). Stimulation of in situ methanogenic bioremediation has not, so far, been investigated to any great extent, although there is general recognition that degradation of organic pollutants in anaerobic microniches in soil, sediment, and ground water environments contributes significantly to overall *in situ* bioremediation rates (Fathepure *et al.*, 1995; Field *et al.*, 1995).

Ex situ Bioremediation Techniques

Ex situ bioremediation techniques are usually aerobic and involve treatment of contaminated soils or sediments using solid or slurry-phase systems. Solid phase systems include land forming (Soil-treatment units), compost heaps, and engineered biopiles (Litchfield, 1991). Land farming should more correctly be regarded as an *in situ* treatment system since it involves plowing, tilling, and raking of the contaminated soil on site, together with the application of water, nutrients, and microbial inocula, if required. Containment is essential in order to prevent contamination of ground water by leachate from the tilled zone. Compost heaps consist of contaminated soil or sediment supplemented with composting material (straw, wood chips and so on) in order to enhance water-and air-holding capacity and improve physical handling properties. Periodic mixing or turning is applied in order to ensure adequate aeration. Nutrients and microbial inocula are added, if required.

Slurry – phase bioremediation is a batch – treatment technique in which excavated soils or sediments are mixed with water and treated in reactor vessels or in contained ponds or lagoons. Effective bioremediation has been obtained with slurry – phase systems for soils and sediments contaminated with a wide range of organic compounds, including pesticides, petroleum hydrocarbons, pentachlorophenol, polychlorinated biphenyls (PCBs), creosote, coal tars, wood – preserving wastes (Yare, 1991).

Above – ground reactors are also used for the treatment of extracted ground water in "pump and treat" systems and for bioremediation of volatile organics removed from contaminated sites by vapor extraction or bioventing techniques. "pump and treat" technologies employ both *in situ* and *ex situ* treatment for aquifer bioremediation. Alternatively, the contaminated ground water may be treated in above – ground bioreactors in which nutrients supplementation, pH, aeration, temperature, and cosubstrate addition can be optimized. Soil vapor extracts may be treated by physico-chemical means or biologically by passage through biofilters (Mercer *et al.*, 1990).

Environmental Factors Influencing *in situ* Bioremediation Rates

Both the rate and the extent of microbial remediation of organic contaminants *in situ* are affected by a variety of environmental factors, some of which may be manipulated whereas others are difficult to modify within the contaminated site. The major important environmental factors are;

- pH
- Temperature
- Water content and Geological characteristics
- Nutrient availability
- External electron availability
- Bioavailability of pollutants
- Cometabolism etc.,

Optimization of conditions are critical for successful implementation of both *in situ* and *ex situ* bioremediation technologies. Many variables governing the efficacy of bioremediation processes are a function of environmental conditions, type of contaminants, contaminated media, effectiveness of primary substrates, supplemental nutrients, electron acceptors, and their mode of delivery.

Bioaugmentation

In instances in which a competent degradative population either does not exist or in which the time needed for adaptation/acclimation may be too long, microbial inoculants may be added to promote and/or enhance bioremediation rates (Singleton, 1994; Bouwer and Zehnder, 1993; Liu and Suflita, 1993). This technique is referred to as bioaugmentation and many involve the addition of natural isolates or genetically engineered microorganisms (GEMs). Concern about the potential negative impacts of release of GEMs to the environment has resulted in strict international control, with GEMs augmentation studies consequently being confined to contained laboratory-based, bioreactor applications.

Bioaugmentation trails using natural isolates have met with varying degrees of success. Degradation of pentachlorophenol in soil was reported by various authors to be enhanced by the inoculation of *Rhodococcus chlorophenolicus* (Middledrop *et al.*, 1990), *Flavobacterium* sp. (Crawford and Mohn, 1985) and *Arthrobacter* sp. (Edgchill and Finn, 1983).

Summary

Exploitation of the degradative capacities of microorganisms is the fundamental basis of organic pollutent bioremediation. Consequently, knowledge of the physiological characteristics, biochemical capabilities, ecology and genetic plasticity of the involved microbial species or consortia is an essential prerequisite for successful bioremediation in the field, however, requires a multidisciplinary approach involving hydrologists, chemical, environmental and agricultural engineers, oceanographers, chemists as well as microbiologists, biochemists, ecologists, and molecular biologists. The development and application of cost-effective, efficient, and environmentally sound bioremediation technologies depends on the integration of these many different disciplines, as well as on advances in our understanding of the physico-chemical, biological, and ecological factors that govern microbial degradation of organic contaminants both *in situ* and *ex situ*.

References

Adriaens, P, Kohler, H.P.E., Kohler-Staub, D. and Focht, D.D. (1989). Bacterial dehalogenation of chlorobenzoates and coculture biodegradation of 4,4-dichlorobiphenyl. Appl. Environ. Microbiol. Vol. 55: 887-892

Alexander, M. (1981). Biodegradation of chemicals of environmental concern. Science. 211: 132-138

Alexander, M.(1994). Biodegradation and bioremediation, Academic press, New York

Baker, K.H and Herlon, D.S. (1994). Microbiology and biodegradation, in bioremediation, McGraw-Hill, New York. Pp, 9-60

Blackburn. J.W and Hafker, W.R. (1993). The impact of biochemistry, bioavailability and bioactivity on the selection of bioremediation techniques. Trends in Biotechnol. Vol. 11: 328-333

Bouwer, E.J and Zehnder, A.J.B. (1993). Bioremediation of organic compounds putting microbial metabolism to work. Trends in Biotechnol. 11: 360-367

Brunner, W., Sutherland, F.H and Focht, D.D. (1985). Enhanced biodegradation of polychlorinated biphenyls in soil by analog enrichment and bacterial inoculation. J. Environ. Qual, 14: 324-328

Crawford, R.L and Mohan, W.W. (1985). Microbiological removal of pentachlorophenol from soil using a *Flavobacterium* Sp. Enz. Microb. Technol. 7: 617-620

Clark, R.R., Chain, E.S.K and Griffin, R.A. (1979). Degradation of polychlorinated biphenyls by mixed microbial cultures. Appl. Environ. Microbiol. 37: 680-685

Compeau, G.C., Mahaffery, W.D., and Patras, L. (1991). Full-scale bioremediation of contaminated soil and water, in environmental biotechnology for waste treatment. Vol. 25. (Sayler, G.S., Fox. R, and Blackburn, J.W. eds). Plenum, New York. Pp. 91-109

David Sheehan. (1997). Bioremediation protocols. Humana Press. Totowa, New Jersy. Pp. 3-22

Dolfing, J. (1992). Gibbs free energy of formation of halogenated aromatic compounds and their potential role as electron acceptors in anaerobic environments. Environ. Sci. Technol. 26: 2213-2218

Edgchill, R.U and Finn, R.K. (1983). Microbial treatment of soil to remove pentachlorophenol. Appl. Environ. Microbiol. 45: 1122-1125

Fathepure, B.Z., J.M. Tiedje, and S.A. Boyd. (1988). Reductive dechlorination of hexachlorobenzene to tri- and dichlorobenzene in sewage sludge. Appl. Environ. Microbiol. 54: 327-330

Fathepure, B.Z., Youngers, G.A., Richter, D.L and Downs, C.E. (1995). In situ bioremediation of chlorinated hydrocarbons under field aerobic – anaerobic environments, in bioremediation (Baker, K.H and Herson, D.S., eds) McGraw-Hill, New York, Pp. 169-186

Freedman, D.L., Lasecki, M., Hashsham, S and Scholze, R. (1995). Accelerated biotransformation of carbon tetrachloride and chloroform by sulfate-reducing enrichment cultures, In bioremediation. (Baker, K.H and Herson, D.S., eds.). McGraw-Hill, New York. Pp. 123-128

Field, J.A., Stams, A.J.M., Kato, M and Schraa, G. (1995). Enhanced biodegradation of aromatic pollutants in cocultures of anaerobic and aerobic bacterial consortia. Antonie Van Laeuwenhoek. 67: 47-77

Gibson, S.A and Sewell, G.W. (1992). Stimulation of reductive dechlorination of tetrachloroethane in anaerobic aquifer microcosms by addition of short chain organic acids or alcohols. Appl. Environ. Microbiol. 58: 1392-1393

Gibson, S.A and Suflita, J.M. (1990). Anaerobic biodegradation of 2,4-5-trichlorophenoxyacetic acid. In samples form a methanogenic aquifer: Stimulated by short chain organic acids and alcohols. Appl. Environ. Microbiol. 56: 1825-1832

Glaser, J.A. (1988). The development of water and soil treatment technology based on the utilization of a white-rot, wood rotting fungus. Presented at the international conference on physico-chemical and biological detoxification of Hazardous waste, Wu, X.C (Ed). Technomic publications, Lancaster, Pa, Vol. II, 633-644

Haggblom, M.M., Rivera, M.D and Young, L.Y. (1991). Anaerobic degradation of chloroaromatic compounds under different reducing conditions. EPA symposium on bioremediation of hazardous wastes, Falls church, va., Microbial degradation of organic compounds, Gibson, David, T. (Ed), Marcel Dekker, New York pP. 439-485

Holliger, C., G. Schraa, A.J.M. Stams and A.J.B. Zehnder. (1992). Enrichment and properties of an anaerobic mixed culture reductively dechlorinating 1,2,3-trichlorobenzene to 1,3-dichlorobenzene. Appl. Environ. Microbiol. 58: 1636-1644

Haggblan, M.M., Rivera, M.D., and Young, L.Y. (1991). Anaerobic degradation of chloroaromatic compounds under different reducing conditions, EPA Symposium on bioremediation of Hazardous wastes, Falls Church, Va.

Hutzinger, O and Veerkamp, W. (1981). In microbial degradation of xenobiotic and recalcitrant compounds, ed. T. Leisinger, A. Cook, R. Hutter and J. Nuesch. Academic press, London Pp. 3-8

Henschler, D. (1985). In bioactivation of foreign compounds. Ed., M.W. Anders. Academic press, New York. Pp. 317

Hinchee, R.E., Miller, R.N and Johnson, P.C. (1995). *In situ* aeration: air sparging, bioventing and related remediation processes, Battella, Columbus, OH

Hamer, G. (1993). Bioremediationn: a response to gross environmental abuse. Trends in Biotechnol. 11: 317-319

Hendriksen, H.V, S. Larsen, and Ahring, B.K. (1992). Influence of supplemental carbon source on anaerobic dechlorination of pentachlorophenol in granular sludge. Appl. Environ. Microbiol. 58: 365-370

John T. Cookson, Jr. (1995). Bioremediation engineering design and application. McGraw-Hill. Inc. New York. Pp. 51-93

John T. Cookson. Jr. (1995). Bioremediation Engineering, Design and Application, McGrow-Hill. Inc. New York. Pp. 135-164

Kaufman, D.D. (1977). Degradation of PCP in soil and by soil microorganisms, USEPA Proceedings of a symposium held in Pensacola, FL Plenum Press, New York. Pp. 27-39

Kohler, H.P. E., Kohler-Staub, D and Focht, D.D. (1988a). Degradation of 2-Hydroxybiphenyl and 2',2'-Dihydroxy biphenyl by *Pseudomonas* sp. strain HBP1, Appl. Environ.Microbiol, 54: 2683-2688

Kohler, H.P. E., Kohler-Staub, D and Focht, D.D. (1988b). Cometabolism of polychlorinated biphenyls: Enhanced Transformation of Aroclor 1254 by Growing Bacterial cells, Appl. Environ. Microbiol. 54: 1940-1945

Lamar, R.T and Dietrich, D.M. (1990). *In-situ* depletion of PCP from contaminated soil by *Phanerochaete* Sp. Appl. Environ. Microbiol. Vol. 56: 3093-3100

Lu, C.J. (1994). Effects of hydrogen peroxide on the in situ biodegradation of organic chemicals in a simulated ground water system, in Hydrocarbon bioremediation (Hinchee, R.E., Allemann, B.C., Hoeppel, R.E., and Miller, R.N., eds), Lewis, Boca Raton, FL. Pp.140-147

Litchfield, C.D. (1991). Practices, potential and pitfalls in the application of biotechnology to environmental problems, in environmental biotechnology for waste treatment. (Sayler, G.S., Fox, R.D and Blackburn, J.W. eds.) Plenum, New York, Pp. 147-157

Liu, S and Suflita, J.M. (1993). Ecology and evolution of microbial populations for bioremediation. Trends in Biotechnol. 11: 344-352

Lawrence P. W. (1991). Dehalogenation reactions catalyzed bacteria. In Biological Degradation of Wastes, edited by A.M. Martin. Elsevier Applied Science. London, Pp. 187-205

Mercer, J.W., Skipp, D.C., Griffin, D and Rose, R.R. (1990). Basics of pump and treat ground water remediation technology, U.S. Environmental Protection Agency, Robert S. Kerr, Environmental Research Laboratory, Ada, OH, EPA. 600/8-90/003

Middledrop, P.J., Briglia, M and Salkinoja-Salkonen, M. (1990). Biodegradation of pentachlorophenol in natural soil by inoculated *Rhodococcus chlorophenolicus*. Microbial. Ecol. 20: 123-139

McCarty, P.L and Semprini, L. (1994). Ground water treatment for chlorinated solvents, in Hand Book of bioremediation, Norris, R.D. et al., eds., Lewis publishers, Boca Raton, FL

Morris, P.J., Mohn, W.W., Quensen, III, J.F., Tiedje, J.M and Boyd, S.A. (1992). Establishment of a polychlorinated biphenyl degrading enrichment culture with predominantly meta dechlorination. Appl. Environ. Microbiol. 53: 3088-3094

Mohn, W.W. and Tiedje, J.M. (1990). Strain DCB-1 conserves energy for growth from reductive dechlorination coupled to formate oxidation. Arch. Microbiol. 153: 267-271

Morgan, P and Watkinson, R.J. (1989). Microbiological methods for the cleanup of soil and ground water contaminated with halogenated organic compounds. FEMS Microbiol. Rev. 63: 277-300

Pardick, D.L., Bovwer, E.J and Stone, A.T. (1992). Hydrogen peroxide use to increase oxidant capacity for in situ bioremediation of contaminated soils and aquifers: a review. J. Contaminant Hydrology. 9: 221-227

Quensen, J.F., Boyd, S.A. and Tiedje, J.M. (1990). Dechlorination of four commercial polychlorinated biphenyl mixtures (Aroclors) by anaerobic microorganisms form sediments. Appl. Environ. Mirobiol. 56: 2360-2369

Quensen, J.F., Boyd, S.A and Tiedje, J.M. (1990). Dechlorination of four commercial polychlorinated biphenyl mixtures (Aroclors) by anaerobic microorganisms from sediments. Appl. Environ. Microbiol. 56: 2360-2369

Rasmussen, G.S.J. Komisar and J.F. Ferguson. (1993). Transformation of tetrachloroethane to ethane in mixed methanogenic cultures. Effect of electron donor, biomass levels, and inhibitors in bioremediation of chlorinated and polycyclic aromatic hydrocarbon compounds. Levies Publishers. Pp. 309-313

Reineke,W and Knackmuss, H.J. (1988). Microbial degradation of haloaromatics. Ann. Rev. Microbiol. 42: 263-287

Reineke, W. (1984). Microbial degradation of halogenated aromatic compounds, Microbial degradation of organic compounds, Gibsen, D.T. (Ed), Marcel Dekker, New York. Pp. 319-360

Slater, H.J and Lovatt, D. (1984). Biodegradation and the significance of microbial communities, Microbial degradation of organic compounds, Gibson, David T. (Ed), Marcel Dekker, New York. Pp 439-485

Suthan S. Sutherson. (1997). Remediation engineering, design concepts. CRC Lewis Publishers, New York. Pp. 136-156

Smith, A.E and Lafond, G.P. (1990). Effects of long-term phenoxyalkanoic acid herbicide field applications on the rate of Microbial Degradation. Enhanced biodegradation of pesticides in the environment, Racke, K.D and Coats, J.R. (Ed) American Chemical Society, Washington, DC Pp. 13-22

Saez, P.B and Rittman, B.E. (1991). Biodegradation kinetics of 4-chlorophenol, an inhibitory cometabolite. Res. J. Water Pollut. Control. Fed. 63: 838-847

St. John, W.D and Sykes, D.J. (1988). Complex industrial waste sites, in environmental biotechnology: reducing risks from environmental chemicals through biotechnology. Vol. 45 (Omenn G.S, ed) Plenum, New York. Pp. 237-252

Singleton, I. (1994). Mirobial metabolism of xenobiotics: fundamental and applied research. J. Chem. Tech. Biotechnol. 59: 9-23

Sherwood, J.L., Petersen, J.N., Skesm, R.S. and Hooker, B.S. (1995). Effect of nitrate availability on chloroform production during CT destruction, in bioremediation of chlorinated solvents (Hinchee, R.E., Leeson, A and Semprini, L. eds) Battelle, Colombus, OH. Pp. 85-89

Tiedje, J.M. and Stevens, T.O. (1987). The ecology of an anaerobic dechlorinating consortium. Environmental Biotechnology, Omenn, G.S. (Ed), Plenum Press, New York. Pp. 3-14

Unterman R, Bedard, D.L, Brennan, M.J., Bopp, L.H., Mondello, F.J., Brookes, R.E., Mobley D.P., McDermott., J.B, Schwartz. C.C and Dietrich, D.K. (1988). Biological approaches for PCB degradation, reducing risks from environmental chemicals through biotechnology. G.S.Omenn, et.al., Plenum press, NY

Wittch, R.M., Wilkes, H., Sinwell, V., Franke, W and Forntages,P. (1992). Metabolism of dibenzo-p-dioxin by Sphingomonas sp. strain RW1. Appl. Environ. Microbiol. 58: 1005-1010

Woods, S.L., J.F. Ferguso and M.M. Benjamin. (1989). Characterrization of chlorophenol and chloromethoxy benzene biodegradation during anaerobic treatment. Environ. Sci. Technol. 23: 62-68

Wu, W., L. Bhatnagar and J.G. Zeikus. (1993). Performance of anaerobic granules for degradation of penta chlorophenol. Appl. Environ. Microbiol. 59: 389-397

Young, L.Y. (1984). Anaerobic degradation of aromatic compounds, in microbial degradation of organic compounds, Gibson, D.T. (Ed), Marcel Dekker, New York. Pp. 487-523

Yare, B.S. (1991). A comparison of soil-phase and slurry-phase bioremediation of PNA-containing soils, in on site bioreclamation processes for xenobiotic and hydrocarbon treatment (Hinchee,/R.E and Olfenbuttel, R.F., eds.). Butter worth-Heinemann, Stoneham, M.A. Pp. 173-187

Zhang, X and Wiegel, J. (1990). Sequential anaerobic degradation of 2-4, dichlorophenol in fresh water sediments. Appl. Environ. Microbiol. 56: 1119-1127

Fish Processing in India – an Overview

B. A. Shamasundar

Introduction

Ensuring sustainable nutritional security is the major challenge confronting all the developing countries. Fisheries are emerging as one of the fastest growing food production system all over the world. The potential for fisheries development in India is highly promising in view of physical and biological resources available in the country coupled with technological developments.

Fish is a highly perishable commodity and supply of acceptable quality is a challenging task to the all concerned in fisheries sector. This is compounded further with the world fish production from capture fisheries is more or less stagnant at around 100 million metric tons (mt) (Nair, 1998). As human population is ever increasing, the stagnation in fish production implies that less and less fish will be available every year. Nevertheless, a large fraction of this valuable commodity is wasted due to discard at sea and deterioration after landing and this post harvest loss is estimated to be 20-25% of total catches landed (FAO, 1995). Better utilization of aquatic resources should therefore aim primarily at reducing this enormous loss by improving preservation of fish and fishery products and upgrading the discarded low value fish to high value fishery products. At an International Convention of 100 maritime nations at Tokyo in 1996 discussed future plans on 'Sustainable contribution of fisheries to food security' and adopted a set of immediate actions for conservation and management of fishery resources. These included minimizing post harvest loses and optimal use of the resources to increase the available supply of fish and fishery products for human consumption. In India, the total fish production is 5.3 million mt. during the year 2002-2003 (CMFRI, 2002). The present annual yield from capture fisheries is about 2.7 million mt and only additional production that can be expected in future is about 0.6 million mt from unfished off shore waters (Somavanshi, 1998). Based on projected requirements of 7.2 million mt. by 2020, the deficit has to be made by some other sources. In the present circumstances the only available source of mooting national, as well as global requirements is the reduction in wastage of fish caught and rapid expansion of aquatic farming.

There are several unique features about fish that inhibit or plague trade and development. The major one is that it is very perishable and requires special attention and facilities if quality is to be maintained. This remains as a major problem in the marketing of fish even today and results in major losses throughout the industry in both developed and developing countries. Another handicap is that fish, unlike agricultural produce, must be transported from place of harvest to distant places. It is recognized that protein from aquatic sources could drastically reduce famine in the world today if distribution system could be set up to transport available fish resources to where they are most needed.

In the present paper different aspects of fish processing for food and industries purposes with special reference to constituents and post mortem changes in fish have been discussed. The potential of application of biotechnology for better utilization of aquatic resources have also been explained.

Principal Constituents

The chemical composition of fish varies greatly from one species and one individual to another depending on age, sea environment and season. The principal constituents of fish and mammals may be divided into same categories and examples of the variation between the constituents of fish are shown in Table 1. As can be seen from Table, a substantial normal variation is observed for the constituents of fish muscle. The minimum and maximum values listed are rather extreme and encountered more rarely.

Table 1: Principal constituents (%) of Indian fish muscle (From Nair & Mathew, 2000).

Constituent	Percentage
Protein	10 – 22
Lipid	1 – 20
Carbohydrate	1 – 2
Minerals	0.5 – 5
Water	65 – 90

The main factors for the high perishability of fish are the structure and composition of the muscle and nature of associated micro flora. As compared to mammalian muscle, the fish flesh is markedly tender (Love, 1975). One reason for this tenderness is the low content of stroma proteins. The structural proteins make up the contractile apparatus for the muscle movement and also contribute to eating quality of fish meat. Among different components of structural proteins, myosin molecule is dominant in imparting most of functional properties (Kinsella, 1982: Shamasundar and Prakash, 1994a).

The lipids from fish species are special in nature and differ from mammalian lipids. The main difference is that fish lipids include up to 40% of long chain fatty acids (14-22 carbon atoms), which are highly unsaturated. Mammalian fat will rarely contain more than two double bonds per fatty acid molecule. During handling and processing, these polyunsaturated fatty acids are highly susceptible for auto oxidation leading to impairment in eating quality.

Table 2: Proximate composition and calorific value of red and white meat of Indian oil sardine* (From Mukundan *et al.*, 1985)

Parameters	Red muscle	White muscle
Moisture	64.6	72.8
Carbohydrate	0.9	0.5
Fat	14.1	4.8
Ash	1.4	1.7
Protein	19.1	20.7
Calorific value	206.9	127.2

* (All values are given as percentage, except calorific value which is K.cal. per 100 g)

The color of fish muscle can be white/light or red which is species specific. Many fishes will have certain amount of dark tissue of a brown or reddish color. The dark muscle is located just under skin along the side of the body. The proportion of dark to light muscle varies with activity of fish. In pelagic fish, such as mackerel, sardine and herring, which swim more or less continuously, the proportion of red meat, varies from 37–48% (Mukundan *et al.*, 1985). In demersal fish, i.e., species, which feed on bottom and only move periodically, the amount of dark muscle is small. There are many differences in the chemical composition of two muscle types, some of the more noteworthy having higher levels of lipids and myoglobin in the dark muscle. Proximate composition and calorific value of red and white meat of Indian oil sardine is given in Table 2. From a technological point of view, the high lipid content of dark muscle is important because problems with rancidity. However, nutritionally both red meat have equal amount of protein.

Spoilage of Fish

Biochemical changes in the muscle

Fish is prone to rapid spoilage due to autolytic changes, bacterial activity and other reactions such as lipid oxidation. Immediately after death, fish muscle is relaxed and elastic. Within a few hours, rigor mortis sets in, when the whole body becomes inflexible and hard. The onset of rigor depends upon the temperature, particularly on the difference of temperature between that of water and storage. When the difference is large, the time from death to onset of rigor is short and vice verse. During rigor mortis, aerobic respiration ceases and the anaerobic oxidation of glucose leads to accumulation of lactic acid which results in a drop in the muscle pH from about 6.8 to 6.5. However, most teleost fish and crustaceans contain little carbohydrate, whereas its content is higher in bivalve molluscan shellfish. The final pH depends upon the species and composition of the organism. During rigor, depletion of adenosine triphosphate (ATP) (to a content of about one micromole per g) results in stiffening of the muscle due to the irreversible association of myosin and actin molecules in the absence of the nucleotide. Resolution of the rigor is a slow process essentially due to the low pH-favored hydrolysis of actomyosin by acid proteases such as cathepsins present in the muscle. Rigor mortis of fish has technological significance since the process influences the quality of fillets. Ideally, fish should be filleted post-rigor. Fillets prepared in rigor will be stiff with poor yields, which can cause gaping, when handled roughly (FAO, 1995, Connell, 1995). If the fillets are removed from the bone pre-rigor, the muscle can contract freely and the fillets will shorten following the onset of rigor. For example, studies on salt absorption behavior of salmon and cod conducted by Sorensen *et al.* (1997) showed that post-rigor fish, when immersed in saturated brine, absorbed 3% salt with a 6.5% increase is weight within a period of one hr. On the other hand, pre-rigor fillets required 3 hr to absorb same amount of salt, but was associated with a 7% loss in weight.

Immediately after death, the ATP present in the muscle is degraded by autolytic enzymes by the sequence shown below:

$$ATP \rightarrow ADP \rightarrow AMP \rightarrow IMP \rightarrow HxR \rightarrow Hx \rightarrow X\ U$$

The first five steps of the reaction sequence proceed at a relatively faster rate. The oxidation of hypoxanthine (Hx) to xanthine (X) and ultimately to uric acid (U) is much slower and is the result of microbial enzyme activity. Uric acid (U) is much slower and is the result of microbial enzyme activity. A strong correlation has been observed between nucleotide catabolism and loss of freshness of fish (Ashie, *et al.* 1996, Saito *et al.* 1959).

Saito *et al.* (1959) proposed that the freshness of fish could be determined in terms of 'K-value', by the content of the autolytic degradation products of ATP, as given below :

$$K - value = \frac{[H \times R] + [Hx]}{[ATP] + [ADP] + [AMP] + [Imp] + [Ino] + [Hx]}$$

where [ATP], [ADP], [AMP],]IMP], [HxR] AND [HX] represent the relative concentrations of adenosine triphosphate, adenosine diphosphate, adenosine monophosphate, inosine monophosphate, inosine and hypoxanthine, respectively, at a specific time during storage of the fish. The content of IMP is known to contribute to the fresh fish flavour, and its loss due to subsequent breakdown causes loss of freshness, associated with a higher K-value.

Since the concentrations of ATP, ADP, and AMP significantly change within the first day of death, a simplified K-value (see below) has also been employed.

$$K-\text{value} = \frac{[H \times R] + [Hx]}{IMP + HxR + Hx}$$

The K-value has been found suitable for several species, although it has limitation for some fish such s cod. Instead of K-value, and modified K-value equal to the concentrations of hypoxanthine/adenine has been found suitable for freshness valuation of certain shellfish (FAO, 1995). The enzymes involved in the oxidation of Hx and X are given below :

Nucleoside phosphorylase : Inosine+Pi → Hypoxanthine+ribose–Pi

Xanthine oxidase : Hypoxanthine + Oxygen → X anthine + hydrogen peroxide

Xanthine oxidase : Xanthine + oxygen → uric acid + hydrogen peroxide

Lipid oxidation is associated with early post-modern changes in fish tissue. The process is initiated by the accumulation of active oxygen species, the activation of haemoproteins, an increase in free iron and the consumption of antioxidants (Hultin, 1994). Lipid oxidation is comparatively more during frozen storage, rather than chilled (0° – 2°C) storage, and can be both non-enzymatic as well as enzymatic. Enzymes such as lipoxygenase, peroxidase and microsomal enzymes from animal tissues can potentially initiate lipid peroxidation producing hydroperoxides (German and Kinsella, 1985, McDonald and Hultin, 1987). The breakdown products of these hydroperoxides include aldehydes, ketones, and alcohols which may cause development of off-flavours. Fish lipids rich in n-3 polyunsaturated fatty acids are very susceptible to oxidation, giving rise to n-3 aldehydes which cause distinctive oxidative off-flavours. These compounds include cis-4-heptenal, trans-2-helptenal, trans-2, cis-4-heptadienal etc. and also 1,5-octadien-3-ol, 1-octen-3-ol and hexanal.

Lipoxygenase-dependent oxidative activity has been detected during chilled storage of fatty fish species, sardine (*Sardina pilchardus*) and herring (*Clupea harengus*). Lipoxygenases were concentrated in the skin tissue of fish, and were active for up to 48 h of chilled storage. The pro-oxidative activity due to haem proteins continued for longer than that due to lipoxygenase. The most abundant degradation products of the hydroperoxides formed from arachidonic and docosahexaenoic fatty acids were 12-and 16-hydroxy acids. Trends of fluorescent formation resulting from interaction between oxidation products and biological amino constitutions with quality loss during chilling (Medina *et al.* 1999). The extent of lipid oxidation can be reduced by glutathione peroxidase which reduces unstable lipid hydroperoxides to non-radical, stable products, which are inactive in the oxidative chain propagating mechanism (Hultin, 1994). Other enzymes include superoxide dismutase and catablase, which remove superoxides from the peroxidation mechanism.

Microbial spoilage

Fish are prone to contamination with a variety of microorganisms depending upon environmental habitats, handling conditions and temperature (Gram and Huss, 1996). The low molecular weight peptides and free amino acids produced by the autolysis of proteins not only lower the commercial acceptability of fish, but also provide an ideal medium for growth of contaminant microorganisms. The bacterial flora of fish caught in clean waters include gram-negative organisms such as *Pseudodomonas, Moraxella, Acinetobacter, Shewanella putrefaciens, Flavobacterium, Vibrio* and *Aeromonas spp.* and gram-positive bacteria namely, *Bacillus, Clostridium, Micrococcus,* and *Lactobacillus* spp. In polluted waters a high number of Enterobacteriaceae may be found. The flora of tropical fish often carry a slightly higher load of gram positive and enteric bacteria. During storage, the bacteria will grow with a doubling time of approximately 20 min and one day at ambient and ice temperatures, respectively, to reach a value of 10^8 per gram within a period 24 h at ambient temperatures and 2 to 3 weeks, when iced. The composition of the microflora of several different species of cold water fish at spoilage are reported to be dominated by gram-negative bacteria, while gram-positive bacteria may dominate in warm water fish species (Gram and Huss, 1996). Nevertheless, bacterial spoilage patterns of both cold and tropical water fish during ice storage are comparable. Delay in ice storage can cause a shift in the composition of the microflora.

Bacteriological growth profiles in fish during storage have been studied in detail by several authors (Gram and Huss, 1996, Venugopal, 1990, Kraft, 1992, Liston, 1980). Initially, aerobic microorganisms grow on the fish surface, and when the surface becomes covered and slime builds up, conditions become more favourable for the growth of anaerobes. Associated with the bacterial proliferation, several volatile flavour-affecting compounds are generated on the fish muscle. Some of the bacteria such as *Alteromonas, Vibrio, Proteus* etc. present on fish are able to carry out respiration by using trimethylamine-oxide (TMA-O), as an electron acceptor, which is abundantly present in marine fish. This result in accumulation of trimethylamine (TMA), which is one of the dominant compounds (about 10 to 25 mg TMA per 100 g muscle) of spoiling fish contributing to the characteristic spoiled fishy odour. TMA, in addition to ammonia and other volatile amines, is a major component of the total volatile basic nitrogen (TVBN) compounds during fish spoilage. TMA formation in many fish species during spoilage is paralleled by bacterial production of hypoxanthine. Many spoilage-causing bacteria such as *Shewanella putrefaciens* and *Vibrio spp.* also produce off-smelling volatile sulphur compounds such as hydrogen sulphide, methylmercaptan and dimethylsulpide, formed from sulphur containing amino acids such as methionine.

Shewanella putrefaciens and *Pseudomonas spp.* have been recognised as the specific spoilage causing bacteria of iced fresh fish regardless of the origin of the fish (Dainty, 1996, Gram and Huss, 1996). Modified atmosphere-stored marine fish from temperate waters are spoiled by the CO_2-resistant *Photobacterium phosphoreum,* whereas, gram-positive bacteria are likely spoilers of CO_2-packed fish from tropical waters. *Photobacterium phosphoreum* has also been identified as the main spoiler in vacuum-packed cod (Dalgaard *et al.*, 1993), while in iced, gutted cod, the specific spoilage organism was S. putrefaciens (CAC, 1996). Fish products with high salt contents may spoil due to growth of halophilic bacteria, anaerobic bacteria and/or yeast, whereas in lightly salted fish, spoilage could be because of lactic acid bacteria and certain Enterbacteriaceae (Gram and Huss, 1996). At the advanced stage of microbial growth, extracellular proteases secreted by the microorganisms cause further degradation of the fish muscle.

Flavour changes

The flavour of fish significantly influences its consumer acceptability. Fresh marine fish are nearly odourless, because they contain a small quantity of volatiles, whereas freshwater fish given off earthy-odour compounds, such as pyrrolidine, depending upon the nature of their habitats (Kwai, 1996, Lindsay, 1991). During storage, the action of exogenous enzymes in post-mortem tissue contributes to flavor changes. Fish chilled for one or two days may have better flavor due to their content of inosine monophosphate (IMP formed through the enzymatic degradation of ATP. The flavor of cooked cod, for example, has the strongest intrinsic characteristics after 2 days in melting ice. The rapid oxidation of large amounts of unsaturated lipids present in fish is a major reason for changes in smell, taste, color, texture and nutritional value. Enzymatic as well as non-enzymatic oxidation of polyunsaturated fatty acids of fish muscle lead to a variety of carbonyls, alcohols etc. that are responsible for flavor changes in fish, as mentioned above. Many fatty species such as salmon, ocean perch or halibut improve much in flavor, taste and texture, during the first 2 to 4 days of ice storage (Martinez *et al.*, 1997).

Associated with the lipid oxidation compounds, microbial formation of several volatile nitrogenous compounds results in significant loss of fresh fish flavour. The chemical compounds produced by the action of microorganisms that affect sensory quality of fish include trimethylamine, hydrogen sulphide, dimethyl sulphide and methyl mercaptan, various amines and ammonia, indole, skatole, putrescine and cadaverine (Reineccius, 1991). The factors influencing the flavour quality of farm-raised fish have been discussed by Johnson (1989) and Kawai (1996). In the case of elasmobranchs like sharks and rays, ammonia may form from the action of endogenous urease on urea in the flesh. Similarly, rapid formation of ammonia in shrimp is correlated with adenosine deaminase and AMP deaminase activity on nucleotides and arginase-catalysed formation of urea and its conversion to ammonia.

Changes in texture

Tenderness is, perhaps, the most important quality parameter of muscle foods including fish. Fish is generally more soft than red meat because the former is characterised by a lesser amount of connective tissue and lower degree of its cross linking (Venugopal and Shahidi, 1996). Therefore, tissue softening in fish is faster than in red meat. Tenderisation or flesh softening is associated with the disappearance of Z-disks in the muscle cell with the release of α-actinin, dissociation of the actomyosin complex, destruction of connection, and general denaturation of connective tissue (Bremner, 1992). Haard (1992 a and b) observed that degradation of fish muscle caused by endogenous proteases is a primary cause of quality losses during cold storage. Muscle proteases including cathepsin D and cathepsin L, calcium-activated proteases (calpains), trypsin, chymotrypsin, alkaline proteses and collagenases have all been involved in the softening of fish tissue during storage (Bremner, 1992). Doke *et al.* (1979) reported the role of hydrolytic enzymes in the autolytic spoilage of fishery products. The major source of hydrolytic activity in marine fish (Bombay duck) was concentrated essentially in the drip, while in the freshwater fish *Tilapia mossambica* it was in the skin. Removal of cathepsin D by removal of the drip or deskinning tilapia suppressed enzyme-mediated spoilage. It has been suggested that the proteins of muscle are largely unaffected during post-mortem storage, and that the softening of muscle is due not to the breakdown of myofibrils but to proteolytic digestion of minor cell components that link the major structural units together (Ando *et al.*, 1991). Collagenases have been found to be involved in the development of mushiness in prawns stored on ice (Lindner *et al.*, 1988).

Discolouration

Another major quality problem in seafood industry is the discolouration of items such as shrimps, lobsters etc. The pink/red color of the skin of most seafoods fade during iced/chilled storage due to oxidation of the carotenoid pigments. The extent of the loss of color is dependent upon the fish, availability of oxygen and the storage temperature (Haard, 1992 b). Carotenoid fading may involve several mechanisms including (i) autooxidation of the conjugated double bonds by reaction with atmospheric oxygen; (ii) coupled oxidation with lipids involving free radicals released from fatty acids combining with carotenoids to form lipid hydroxyperoxides and carotenoid radicals; and (iii) enzyme activity (Simpson, 1985). A lipoxygenase-like enzyme capable of catalysing oxidation of carotenoids like astaxanthin, tunaxanthin and β-crotene has been characterised from some fish species (Tsukuda, 1970). Another enzyme involved in carotenoid oxidation has its activity influenced by hydrogen peroxide and halides, particularly bromides and iodides. The bleaching of β-carotene could be acomplished by one of these three possible mechanisms (Kanner and Kinsella, 1983).

Oxidation of muscle pigments, namely myoglobins is another reason for colour change in fish. Oxidation of bright red myoglobin to brown metmyoglobin can occur through both non-enzymatic and enzymatic routes. An enzyme, metmyoglobin reductase, has been isolated from the skeletal muscle of dolphins and bluefin tuna (Hseigh and Kinsella, 1986). Development of black spot due to melanosis or enzymic browning in shrimp and lobster during storage is caused by the phenolases present in their tissues (Yan *et al.*, 1989).

Food safety problems

Apart from spoilage-causing microorganisms, contamination of fish, shrimp, oysters, mussels, crabmeat etc. by pathogenic organisms, viruses and parasites has become a major issue in recent years because of their potential to cause food-borne diseases. Some of the common pathogens found in fishery products include *Salmonella* spp., *Staphylococcus aureus*, different species of *Clostridium botulinum Bacillus cereus*, *Campylobacter jejuni*, *Escherichia coli* O157:H7, *Vibrio parahaemolyticus*, *Yersinia enterocolitica* and *Listeria monocytogens*. The incidence of many of the above pathogens in processed fishery items including frozen products, has been of great concern with respect to their international trade. Furthermore, a variety of parasites, including flatworms, roundworms, tapeworms, and protozoa, infest the gills, viscera and skin of marine, freshwater, as well as farm-raised fish species which can cause diseases to consumers (Venugopal *et al.*, 1999).

Several insects, such as flesh files, beetles and mites, also infest fish, particularly during sun drying, the most destructive pest being the hide beetle, *Dermestes maculatus* Deg. Virus contamination is another common quality problem because of the habit of filtering large amounts of water by molluscan shellfish. From the sources of the contamination, public health issues associated with seafood scan be grouped as environment -, process, distribution – and consumer-induced (Garret *et al.*, 1997). In addition to the above contaminants, the presence of toxins, such as ciguatera, paralytic shellfish poisoning (PSP), diarrhetic shellfish poisoning (DSP), amnesic shellfish poisoning (ASP), histamine etc. and environmental pollutant, such as mercury, pesticides residues etc. are other major problems faced by the seafood as well s aquaculture industry (Venugopl, *et al.*, 1999; WHO, 1999). Understanding the profound influence of the above factors on the fishery products, the industry has adopted quality standards in order to safeguard the in tests of the consumer.

Evaluation of fish freshness

Since fish freshness deteriorates fast, its evaluation is of utmost importance for the seafood industry for development of products of acceptable quality. The various freshness quality indices of seafoods have been discussed by Alur *et al.* (1997), Olafsdottir *et al.* (1997), Botta (1994) and Gill (1990). Sensory evaluation is the rapid method for quality evaluation and is defined as the scientific discipline used to evoke, measure, analyse and interpret characteristics of food as perceived by the senses of sight, smell, taste, touch and hearing (Olafsdottir *et al.* 1997). Sensory tests can be divided into three groups : discriminative, descriptive, and affective tests (FAO, 1995). Discriminative testing is used to determine if a difference exists between samples (triangle test, ranking test). Descriptive tests are used to determine the nature and intensity of the differences (profiling and quality tests). The affective test (market test) is based on a measure of preference or acceptance by the consumer.

The most common descriptive tests are structured scaling for quality assessment and profiling for a detailed assessment of one or more attributes. In Europe, the most commonly used method for the quality assessment of raw fish in the inspection service and in the industry is the European Union Scheme (EC, 1996). The quality index method (QIM) is a descriptive method, which has been suggested for raw fish based on sensory parameters. In this method, the descriptions of the individual grades are precise, objective, independent and primary rather than a cluster of terms. The scores for all of the characteristics are then added to give an overall sensory score (quality index), which can also be used to predict storage life. The method can be used to estimate the influence of time, temperature, and handling on raw as well as cooked quality of fish species including sardines, anchovies, saithe, place, and cod. Computer programs based on these evaluations have also been developed to predict the overall quality and hence, the value of these fish species at any given point of ice storage (Neilson, 1993). A descriptive, 10-point hedonic scale can also be used to determine the chilled storage life of fresh or cooked fish, where a score of '10' indicates the maximum freshness characteristic of the fish, and '1' indicates a strong putrid odor. A score of '5' is indicative of borderline acceptability (FAO, 1995; Bilinsky, *et al.*, 1983).

Microbiological and biochemical methods have also been used to determine fish quality. Since the activity of microorganisms is the main factor limiting the shelf life of fresh fish, an estimation of the total viable counts (TVC) or total plate count (TPC) in terms of colony forming units (cfu) helps quality assessment. However, the value is seldom a good indicator of the sensory quality or expected shelf life of the product. Newly caught fish have a TVC of 10^2 to 10^6 colony-forming units (cfu) per g. At the point of sensory rejection, the TVC of fish products are typically 10^7 to 10^8 cfu per g. Nevertheless, TVC counts are used in standards, guidelines and specifications that often use much lower TVC as indices of acceptability. Biochemical methods of measurement of odor-bearing compounds can be used for freshness evaluation. Classical chemical methods for the analysis of total volatile based (TVB) and trimethylamine (TMA) have been used for the determination of fish freshness in the industry. TVB values less than 30 mg N per 100 g are indicative of excellent quality, while values in excess of 45 mg N per 100 g indicate an unacceptable product. There is a good correlation between TVB values and sensory analysis

(Woyewoda and Ke, 1980). Trimethylamine oxide (TMAO) found in appreciable quantities in marine fish is broken down to trimethylamine by the bacterial enzyme trimethylamine oxide reductase. The content of TMA has been related to loss of freshness in marine fish.

Recently there is much interest in the development of biosensors and 'electronic noses' for rapid measurement of odour compounds in fish. The potentials for various biosensors for fish quality evaluation have been discussed recently by Venugopal *et al.* (2000).

Changes in fish freshness can also be determined by measuring the electrical properties of fish muscle. Three different instruments are available to measure the changes in electrical properties: The Torrymeter (Distell Industries Ltd., Fauldhouse, West Lothian, UK), the Fishtester VI (Intelectron International Electronics, Hamburg, Germany) and the RT-Freshness Grader (RT Rafagnataekni, Reykjavik, Iceland), which all show good correlation with sensory scores of fish freshness when used within their applicable range of operation. These meters cannot be used with thawed fish or fish that have been stored in chilled seawater, and their use for fillets is limited to a few days.

Time-temperature indicators (TTIs) are devices that can be attached to foods to indicate the time-temperature history of the food, based on measurement of some biological, physical or chemical process that depends on time and temperature. The temperature record can be used to develop appropriate models to predict the shelf life of fish. There is good potential for use of such devices for fresh, processed and chill stored fish (Labuza and Fu, 1995).

The sensory quality of some fish, like yellowfin, skipjack, and bonito are highly affected by the color of the muscles. Brown discoloration due to oxidation of myoglobin, replacing the original pink or deep red color, makes the fish unsuitable or less appealing for consumption (Sikorski and Pan, 1994). There is need to develop instrumental methods for colour measurement of such fish. Spectroscopic methods have gained acceptance to determine whether a fish has been frozen and also to estimate the storage time of fish on ice. Recently, the use of near-infrared spectroscopy has shown promise of being useful to assess the storage time of fresh fish (Olafsdottir *et al.*, 1996). L*a*b* is an international standard for colour measurements, adopted by the Commission Internal d'Eclairage (CIE) in 1976. L* is the luminance or lightness component, which ranges from 0 to 100, and * (from green to red) and, b* (from blue to yellow) are the two chromatic components which range from –120 to +120. The principles of colour measurement have been discussed by Papadakis *et al.* (2000) and Cydesdle (1978).

Fish Preservation Methods

The principal aim of fish preservation is to delay, reduce or inhibit the microbial spoilage. In the case of fatty fish, the preservation may also aim at reducing or inhibiting oxidation and other undesirable changes in the fish lipids, which are highly unsaturated and capable of going rancid at various stages of processing. Among different methods of fish preservation, it is short-term preservation by chilling which has attracted worldwide attention. Perhaps, this is mainly because of preference for fresh fish by consumers. The other methods of fish preservation, which have commercial relevance, are freezing, canning, salting and drying, smoking and mince production. These preservation techniques are basically for human food and processing of fish into non-food items includes production of fishmeal and oil, silage and other industrial products.

Short term preservation by chilling

The disposition of fish caught in India (Fig. 1) reveals that more than 50% of the total catch is consumed in fresh form. As the starting material for all fish products, fresh fish preservation deserves special consideration. Although much quality deterioration can occur prior to landing the fish on shore, what is important that further deterioration can be minimized by chilling. The major processes responsible for fresh fish decomposition are bacterial spoilage, autolysis and oxidative rancidity. These factors can be minimized effectively by chilling, by use of ice. Apart from ice, mechanically refrigerated or ice-chilled seawater has been demonstrated as an excellent medium for and transporting of whole fresh fish. Although there has been little or no application of chilled seawater to the retailing of fresh fish, it is being used at sea and for the road transport of bulk quantities of whole small fish and salmon (Bligh, 1980). Salt uptake must be monitored and through mixing with compressed air and/or recirculation is essential to avoid thermal stratification.

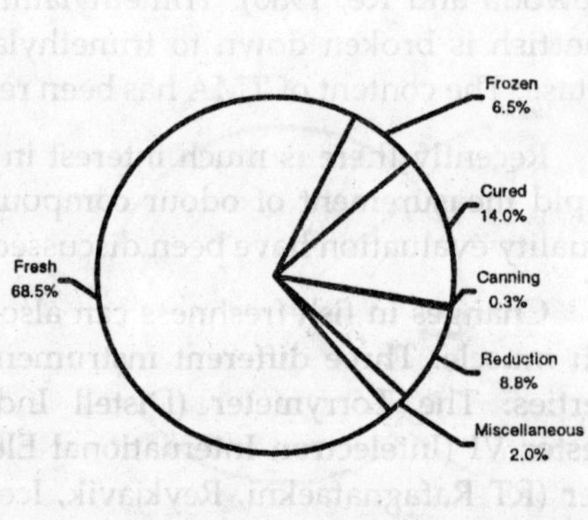

Fig. 1 : Disposition of fish catch in India

In a tropical country like India icing the harvested fish is of paramount importance, if good quality fish were to reach consumer. In order to bring down the temperature of fish, the fish: ice mixing should be proper. The thermal conductivity (i.e., the ability of a material to transfer heat through itself) of the fish mass becomes a major barrier to cooling. With some metabolic activity still taking place within the fish flesh, the mass of fish in the middle might ever heat up. The second problem is that the weight of fish pile crushes the fish on the bottom, leading to a loss of weight (yield) as well as other physical damage. The advantages of ice as chilling medium have distinct advantages, when compared with other cooling methods, including refrigeration by air.

Despite the knowledge on the advantage of fish chilling, ice is not as widely used, as it should be, particularly on board and at landing center. Some of the problems that can be found are as follows: The production of ice is a mechanical process and involves higher cost and energy. To produce ice under tropical conditions, 55 to 85 kWh/ton of the ice (depending on type of ice) are necessary whereas, in cold and temperate countries from 40 to 60 kWh is required for the same purpose. This is certainly large power requirement for many states in India. Also, ice plants require maintenance and hence trained people and spare parts. A cold chain will also require chill rooms, insulated containers, insulated trucks and other auxiliary equipment. All this adds up to the cost of the sale price of fish.

In developed countries, the cost of ice is very cheap and costs only a fraction of the price of fresh fish. In developing countries, ice is very often expensive when compared with fish prices. It is heartening to note that majority of Indian fishermen have realized the importance of chilling with ice and high value fishes are invariably iced to extend the shelf life. But the problem still exists with small pelagics and other by catches. Much effort is being directed to educate the persons involved in handling and short-term preservation by chilling.

Preservation by freezing

Freezing is an energy intensive process. Good air circulation with a blast freezer or good contact with cold surface is essential to ensure quick efficient cooling. The actual phase transition from water to ice requires much more energy than cooling. For example, to get from 5°C to –30°C takes 419 kilojoules per kilogram of water, of which 333.5 kJ/kg is used for the phase transition, that is the latent heat of fusion. It is generally estimated that 40% more energy is necessary to freeze tropical fish because of their higher ambient temperature. The freezing of fish, unlike the freezing of water does not take place at a single unique temperature, but is rather a gradual process taking place between –1°C to –5°C. Only when the freezing process is completed should fish be stowed in the cold store, which is normally not designed for temperature changes but only for the maintenance of temperature.

There are two main kinds of quality defects, which can occur during frozen storage; those affecting the fish surfaces, and those affecting the entirety of the tissues. Lowering storage temperature and avoiding fluctuation. For long-term storage, fish should be kept at temperatures below –30ºC. Surface damage is often associated with evaporation or sublimation causing 'freezer burn', that is drying of exposed surfaces, which thereby becomes spongy. Changes affecting the muscle include texture deterioration and rancidity. Earlier it was difficult to evaluate the quality of frozen fish as the chemistry of protein denaturation, the cause of inferior texture was not understood. Evidence from different researchers suggested a relationship between lipid and protein deterioration (Sikorski *et al.*, 1976; Sikorski and Pan, 1994).The schematic representation of the changes in the native structure of protein in combination with oxidized lipids is presented in Fig. 2 (Kolodziejska and Sikroski, 1979).

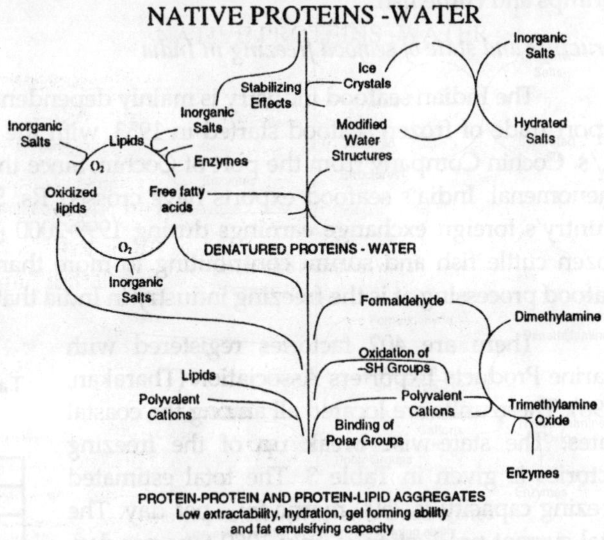

Fig. 2 : Protein changes in frozen stored fish meat (From Kolodziejska and Sikorski, 1979)

Thawing : The technology of the thawing of fish, that is bringing the temperature of the product back above 0°C, is not well developed as that of freezing. Many different thawing techniques are available and routinely used, each with distinctive advantages and disadvantages. It should be noted that the rate of freezing is considered important because of its effect on ice crystal size (along with temperature fluctuations during frozen storage), but there does not seem to be any evidence that the rate of thawing has any effect on fish quality other than the effect of time.

The cheapest way to thaw fish is to put them into a running cold-water bath. Good water circulation is essential. The water can be run into pipes placed along the bottom of the thawing tanks with appropriately scattered holes to insure that many separate water streams are formed. Because of the relatively long period of contact with water, this method is only recommended for whole, dressed or nobbed fish. Air thawing is still being used in some of the plants, but the temperature of the surface of fish must be kept low enough to minimize the growth of microorganism. Dielectric thawing and electric resistance heating are two new techniques that have been developed in laboratory pilot plant. The former uses a high voltage alternating current and the later a low-voltage through the material to be thawed. Although laboratory-based literature has emphasized the advantages of these methods, they have not yet been widely adopted in commercial practice.

Because of frozen storage, certain conformational changes that occur to major protein molecule, the ability to hold the released water during thawing will be altered. Hence, during thawing the exudates that are released termed, as "drip" will vary depending on the condition of species prior to freezing and frozen storage condition. The drip loss has commercial significance in terms of weight and loss of flavoring compounds along with drip (Shamasundar and Prakash, 1994b). Hence all efforts should be directed towards minimize the drip loss, especially in frozen shrimps and cuttle fish.

Structure and state of seafood freezing in India

The Indian seafood industry is mainly dependent on the export trade of frozen products. The export trade of frozen seafood started in 1953, with the first shipment of frozen shrimp to USA by M/s. Cochin Company from the port of Cochin. Since then the growth of frozen seafood industry is phenomenal. India's seafood exports have crossed Rs. 5000 crore mark, contributing nearly 4% of country's foreign exchange earnings during 1999-2000 (Sona, 2000). Frozen shrimp, frozen finfish, frozen cuttle fish and surimi contributing to more than 80% of the earnings. Among the various seafood processing, it is the freezing industry in India that has occupied pre-eminent position.

There are 402 factories registered with Marine Products Exporters Association (Tharakan, 1998). These units are located all among the coastal states. The state-wise break up of the freezing factories is given in Table 3. The total estimated freezing capacity is 7500 metric tons per day. The total current production is only 1050 tons per day, which amounts to 14% as industry's capacity utilization in relation to installed capacity.

Table 3: State-wise break up of the freezing plants in India (From Tharakan, A.J., 1998).

State	Number of plants
Gujarat	56
Maharashtra	41
Goa	07
Karnataka	14
Kerala	124
Tamil Nadu	28
Andhra Pradesh	55
Orissa	21
West Bengal	36

The entire production is for 100% export. There are insignificant sales in domestic market. Most of the factories are processing shrimps, cuttle fish and squids, finfishes and some are freezing fresh water fishes (minor quantity). Hence, the capture fisheries from marine waters still form the bulk of the supply. There are only a handful of organized and clean fishing harbors in the country. Sanitation and hygiene at all these landing centers are extremely poor.

There are no proper organized modern peeling sheds and pre processing centers. These activities are carried out under traditional sanitary conditions. Most factories are old and do not meet the importers requirements. The fact that only about less than 100 units have met the European Union certification are testimony to the outdated plants. The industry feels that there should be proactive participation from the Government to improve the infrastructure need for meeting quality requirements of the importing countries. With stagnation in the marine catches, the only source of supply of raw material for industry is from Aquaculture. Unfortunately, the growth of aquaculture industry has received a setback with a ban by Supreme Court ruling. It is hoped that aquaculture for export will become viable both in terms of environment and economically.

It is heartening to note that several central agencies in consultation with industry, various measures are being adopted to improve the health of seafood freezing industry.

Seafood quality assurance system in India

Export products have to meet the specifications of the importer as well as of the country of import. This is particularly true in the case of fishery products where the developed countries have prescribed certain mandatory requirements to be fulfilled by the exporter. Concern over the safety of food is on the increase because of the probability of the hazardous pesticide residues, harmful chemicals, heavy metals, toxins, antibiotics, pathogenic microorganisms etc. As a result, there has been major shift in the quality control of food from the end product inspection to prevention, elimination or reduction of hazards. This is reflected in mandatory regulations of the European Union (EU) and the United States of America (USA). Quality assurance in respect of products exported from India was introduced systematically with the enactment of Export (Quality Control and Inspection) Act, 1963. Quality assurance in fishery products was introduced in 1965 when frozen and canned shrimp was brought the purview of the Export (Quality Control and Inspection) Act, 1963. Subsequently, frog leg, lobster tail, squid, cuttlefish, canned crab meat, pomfret, beehede-mer, dried shrimp and dried fish also brought under the compulsory quality control and pre-shipment inspection scheme.

The In Process Quality Control (IPQC) system was introduced in early 1978 prescribing the minimum requirements for raw materials, manufacturing process, end product testing, preservation and packaging of the final product. Thus, in addition to the end product specifications, requirements were also introduced for processing, freezing and storage.

By 1995, new quality concepts based on assessment of risk to human health and life, particularly the risks on account of harmful chemicals, antibiotics, pesticide residues, heavy metals, additives and disease causing organisms came into prominence. This quality management termed as Hazard Analysis Critical Control Point (HACCP). The HACCP with stress on safety based on a systematic approach to hazard identification, assessment and control which makes it different from the traditional inspection and quality control procedures. In HACCP system control is transferred from end product testing to on-line checking, that is a change from 'testing to failure' to 'preventing failure'. EC and USFDA have introduced a regulation making HACCP based seafood quality assurance system mandatory for its domestic industry as well as for all countries, which export seafood to EU and USA.

The Government of India, therefore reviewed the existing export inspection system and introduced a new HACCP based Quality Assurance and Monitoring System (QAMS) for fresh, frozen and processed fish and fishery products with effect from August, 21, 1995 in line with the requirements of the major importing countries. The Export of Fresh, Frozen and Processed Fish and Fishery Products (Quality Control, Inspection and Monitoring) Rules, 1995 vest the primary responsibility on the processors to ensure that the fishery products intended for export are handled, processed, stored and transported under proper hygienic conditions so as to meet the stipulated health conditions and that the products conform to the prescribed specifications. It is the responsibility of Export Inspection Council to ensure, by regular monitoring of the processing establishments that processors comply with the requirements.

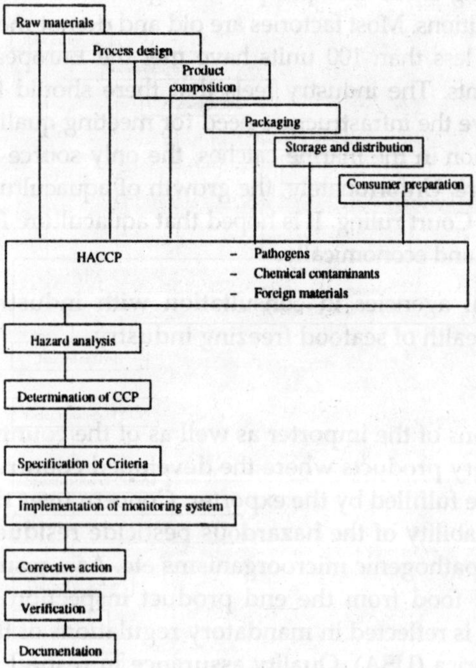

Fig. 3: HACCP concepts and seven principles

Salient features of HACCP: The basic seven principles of HACCP are given in Fig. 3 (Codex, 1991). The system envisages identification of potential hazards in seafood processing at all stages up to the point of consumption. A detailed flow chart of the process covering all aspects from the point of catch through various pre-processing and processing activities will help in identifying the potential hazards and establish the conditions necessary to control the hazards. Based on this suitable remedial measures should be specified.

Constraints in effective implementation of harmonised QAMS

Fish landing centers do not have adequate supply of ice, potable water and electricity. Proper transportation facilities from landing centres to the pre-processing units and from pre-processing units to processing plants are not available.

It is needless to add that Indian seafood industry realizes the urgent need for effective implementation of HACCP based seafood safety management system to ensure quality and safety of fish and fishery products. However, for speedy and proper implementation of this system, seafood industry needs development of common infrastructure such as fishing harbors and landing centers and technical support.

Cured fishery products

Curing includes drying, salting, smoking and combinations of these. It sometimes refers to pickling and fermentation as well, but in this discussion it is restricted to dried, salted and smoked fish. In India, approximately 13–18% of total marine landings are preserved by curing, especially by salting and drying (Ministry of Agriculture, 1996). Over the years preference for salted and dried fish are decreasing. The type of fishes that are subjected to salting and drying are usually small pelagic and other low economic value fishes. Salted and dried shark meat is highly relished and priced high.

The method salt curing may be either wet (bringing) or dry. In wet curing, fishes like sardine and mackerel are used. Here, fishes are put into a wooden or cement vats containing saturated brine. The time of curing varies between 1 to 7 days and it is taken out and sun dried. In some cases the wet cured are packed in bamboo baskets and transported. But it is generally dried and stored. In dry salting, big fishes are split open along the dorsal side and deep cuts are made lengthwise and mixed intimately with dry and placed in cement or wooden tanks. The amount of salt used varies from 25–30% (w/w). Small fishes can be used as it is for salting. The amount of salt used is not exact: usually in excess when fish landings are small and often less when large quantities of fish are available for salt curing. The fish remain inside the tank for at least 16 hours. They are taken out based on demand and conditions suitable for sun drying and dried in open for 2–3 days. After drying the fishes are stored in gunny bags or containers, which will have a shelf life of 3–6 months.

Methods of drying

Drying of salted or wet fishes is traditionally carried out in sun along beaches. The method followed is unhygienic and inefficient leading to low quality salted and dried fishes. Rack drying of fishes (bamboo racks) has distinct advantage over beach drying in terms of drying rate and hygienic point of view. The rack drying has another advantage of saving space and more quantity of fish can be dried per unit area. This will ensure better quality of dried product.

Water activity and growth of microorganisms : Microorganisms require optimum levels of moisture in food products and their levels are expressed in terms of the a_w of the product. At a a_w level of 0.8, growth of bacteria and yeast is normally arrested but certain halophilic bacteria, xerophilic moulds and osmophilic yeast will survive. Cured fish spoiled by mircoorganisms is typically attacked by moulds initially and often at tail region, and onset of mould attack can be considered as end of good quality of storage life of fish (Poulter *et al*, 1982). The pH of the product, temperature of storage, microbial load, salt content and water activity are factors generally considered to influence storage life of cured fish products. For traditionally dried or cured products stored under normal conditions in a tropical climate, a_w can be considered the best index to determine how quickly microorganisms will grow in them. Water activity and proximate composition of several species of salted and dried fish are given in Table 4.

Table 4: Proximate composition and water activity of different varieties of dried fish (From Kalaimani *et al.*, 1988)

Species	Moisture % WWB	Protein % WWB	Fat % WWB	Salt % WWB	Total mineral ash % WWB	Water activity a_w
Saurida tumbil	54.04	18.12	5.87	17.88	20.97	0.78
Sardinella longiceps	51.13	17.19	11.60	21.60	20.08	0.79
Anchoviella heterolobus	24.81	39.77	7.56	11.12	27.86	0.14
Caranx kalla	37.45	27.13	16.05	14.86	19.27	0.75
Loligo sp.	25.72	46.23	12.20	1.43	15.85	0.95
Otolithus ruber	18.93	52.10	8.33	15.65	20.64	0.71
Leiognathus sp.	26.22	44.87	10.14	12.19	18.77	0.74
Leiognathus bindus	14.11	50.30	13.33	15.14	22.26	0.68

Insect infestation of dried fish : In most tropical countries, flies, particularly of the family *Dermestidae*, infest cured fish. Flies are the major carriers of most diseases associated with consumption of dried fish in tropics. They lay eggs when fish are drying. The larvae produced from these eggs invade the flesh causing putrefaction of proteins resulting in extensive damage to the fish meat. It is one of the major problems in curing yards, and recent developments in improved methods of drying like using rack drying will minimize insect infestation, because of efficient drying.

In order to prevent molds and bacterial growth, several chemical treatments have been developed by use of salts of benzoic acid, salts of propionic acid and antioxidants to prevent development of rancid odor (Valsan, 1985). The effect of drying temperature, on nutritional quality of dried fish is influenced by species difference, its composition and experimental conditions of drying. Drying at moderate temperature of 60°C enhances the nutritional quality of threadfin bream (Ragunath *et al.*, 1995).

Smoking

Smoking is a traditional food processing method used for the preservation of fish. Smoking of fishes in India is carried out in only some pockets and potential for export of smoked products appears to be bright. Thermal cracking of wood under reduced oxygen supply results in smoke containing several compounds. Curing of fish by smoking imparts characteristic flavor and color. In modern times smoking process is used chiefly for flavor rather than for preservation of wet fish. The property of smoke depends to a large extent on the type of wood from which it is produced. In tropical countries, oak, teak, mahogany, pine and coconut husk is used to generate smoke. Smoke generated by any type of wood comprises two phases; liquid droplets and a continuous gas phase. Smoke particles having an average diameter of 0.2 mm is best suited for smoke curing (Gopakumar, 1997). The gas phase contains several flavour-producing agents; aliphatic and aromatic numbering over 200.

Some concerns have been raised about smoked products, as it can cause health hazards due to presence of certain carcinogenic compounds such as nitrosoamines and some polynuclear aromatic hydrocarbons, particularly 3, 4, benzophrene, in the hot smoke. High incidence of stomach cancer is reported in countries where large amounts of smoked fish are consumed. This can be prevented to a large extent if liquid smoke is used for curing. Liquid smoke is prepared from wood smoke condensate by fractional distillation and water extraction. The undesirable polynuclear hydrocarbons are insoluble in water and thus absent in liquid smoke. Liquid smoking process, therefore, can be considered safe from the usual health hazards.

Canning

Canning is an excellent way to preserve and transport fish. The product is shelf-stable and needs no refrigeration. However, at best, the quality of the final product is as good as the quality of the fish put into the can. Although, number of different species can be canned, most popular worldwide are salmon, tuna, sardines and mackerels. The fish canning industry in India is very insignificant where the quantity of fish that goes for canning is less than 1% of total catch, in comparison to 10–15% the world average.

The Process of Canning : The heat treatment involved in canning, is designed to provide commercial sterility, that is, a safe product that can be stored for a long period of time. To destroy the bacteria sufficient heat, usually 121°C, is needed for sufficient time. To achieving this temperature, it requires steam under pressure, and for safety reasons, process is carried out in retort.

Based on the known rate of killing of the organisms, which are most difficult to kill, canning equations can be used to determine the heat treatment. Once the lethality is reached, the cans are cooled. This process is used mainly for low-acid foods, that is, products whose pH is higher than 4.6. For high-acid products, less heat is necessary.

Can itself: Traditionally cans were put together from three pieces of metal, the main body of the can and two caps. The seam of the main part of the can was soldered, often with lead. In recent years other materials are being used for solder or a two-piece can is being used. The two-piece can is made by drawing out a piece of steel so that the bottom and the sides are pushed out from a single sheet in a form-and-draw process. Apart from steel and aluminum as material for can making and retort pouches have become very popular. Retort pouches have the advantages of packages that can conform almost perfectly to the shape of the product. The pouch concept is meant to decrease the profile of the product being heated so that the distance from the edge to the center is minimized. This permits the same commercial sterilization effect on the product as with regular cans, but in a noticeably shorter period of time for the heat treatment. In addition to potentially yielding higher-quality canned fish, this type of packaging could also enhance the packing of whole fish, whole steaks, and/or whole fillets.

Potential Problems: One of major serious problems that can be encountered in canned fishery products is 'Botulism'. This is mainly due to improper processing and faulty cans. The botulism is caused by toxin produced by *Clsotridium botulinium*, which can survive in anaerobic condition. The organism can find its entry either through cooling water or can survive under processing condition. The toxin in micro quantity can be fatal for humans. Hence contamination of this deadly bacterium should be avoided at any cost. The other problems associated with canned fishery products are discoloration due to the formation of iron sulfide or tin sulfide, curd formation due to protein precipitation and struvite crystal formation due to salts of magnesium.

Recent developments in use of retort pouches for packing fish curry (Srinivas Gopal *et al.*, 1998) and process for canning of fresh water fishes like rohu and catla (Vijayan *et al.*, 1998) will pave way for reviving the canning industry in India. Also, with emphasis on canning of abundant tuna varieties for export will help in opening of new vistas in export market.

Special Processing Procedures

A number of special processing procedures have been sought and developed to increase the usability of fish as a human food. Though there are number of methods, it is of interest to Indian scenario, that fish mince production (Surimi) and irradiation preservation have a greater relevance.

Surimi Production: Since the beginning of 1980 much interest in surimi has been generated throughout the seafood and food industry by the rapid growth in popularity of surimi-based products, especially crab analogs in the USA and Europe. Surimi is Japanese word, which is nothing but separated fish flesh water washed, mixed with certain additives (cryoprotectants), frozen and stored at below –30°C. The flow chart for the surimi production is given in Fig. 4. The Japanese have a long tradition in the preparation of such products, which was a base material for traditional heat processed product called 'Kamaboko'.

It is necessary to distinguish surimi from minced fish and surimi based products, as these terms are often confused in literature. When fish flesh is separated from bones and skin (usually mechanically) if is called as minced fish, the starting material for surimi production. After minced fish is water washed to remove soluble proteins and fat it becomes 'raw surimi' (Fig. 4). This raw surimi is a wet concentrate of the myofibrillar proteins of fish and possesses enhanced gel forming, water holding, fat binding and other functional properties relative to minced fish. However, myofibrillar proteins in the raw surimi will lose their functional properties rapidly once they are frozen. When the raw surimi is mixed with cryoprotectants such as sugars or sugar alcohols and quick frozen into a block form, 'frozen surimi' is manufactured. The myofibrillar proteins in frozen surimi will retain their functional properties for many months if properly stored.

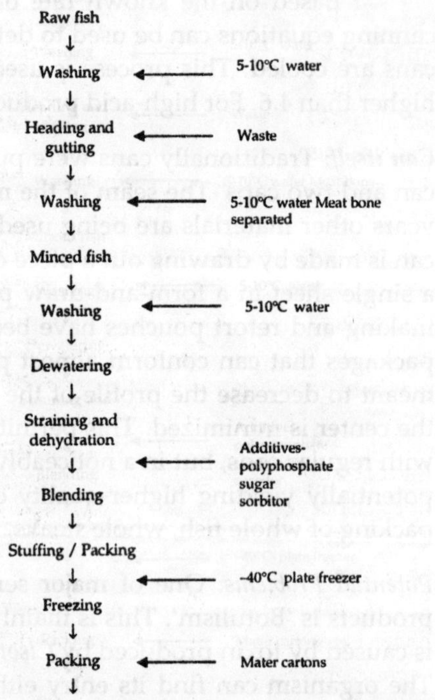

Fig. 4: Flow diagram of Surimi Production

The following characteristics of surimi have stimulated the development.

1. Underutilized and less utilized species can be used successfully as raw materials.
2. Frozen surimi has long shelf life and is a highly functional protein ingredient.
3. A variety of surimi-based products can be manufactured by altering the appearance and eating through application of various processing technologies and ingredients.
4. Current technology permits mass production with consistent quality.

Frozen surimi is produced in compact blocks, which are conveniently and economically transported, stored and handled. These characteristics, in addition to its long shelf life, have made frozen surimi, such that today the term 'surimi' generally refers to 'frozen surimi'.

Surimi in India: Few surimi plants have started production and their growth is very impressive. Nearly 20,000 metric tones of surimi have been exported from Indian shores during 1999-2000 (Sona Philip, 2000). The raw material used is threadfin bream, croaker and certain varieties of deep-sea fishes. All the produce is meant for export and virtually there is no domestic market. The potential for domestic market appears to be bright, as more urban population desires to go in processed fishery products. The utilization of freshwater fishes (other than the major carps) can also be thought of for the production of frozen mince.

Products from Surimi: Production of surimi-based products like Kamaboko, fish sausage, fish ham and seafood analogs are very popular. It is seafood analogs like crabsticks and shellfish analogs that has become popular in USA and perhaps is one of the main reasons for popularity of surimi. This is a process involving extrusion, sheet forming slitting and pasteurization. The process also involves addition of natural or synthetic flavorings to simulate the shellfish products.

Irradiation

Food irradiation is a process for the treatment of food products to enhance their shelf life and to improve microbial safety. The chemistry, technology, and commercial aspects of food irradiation have been well worked out. Electromagnetic radiations, namely, gamma and X-rays having shorter wave length (<300 nm) and higher energy than visible light, can cause ionization by removing electrons from the outershell of atoms and molecules. Generally, ionizing radiations emitted by radioisotpes, Cobalt – 60, and Cesium –137 are used for preservation (Venugopal *et al.*, 1999). From a practical point of view, Cobalt –60 is preferable to Cesium –137 because the later one, having weaker gamma rays, is also water soluble, resulting in environmental hazards. Most of the present day irradiators use Cobalt –60 as source of radiation energy on account of its high penetration and easy availability. Some of public concerns associated with transportation, installation and operation of permanent radiating sources such as Cobalt –60 may lead to increasing use of electrons and X-rays. Both electrons and X-ray being machine generated, can be turned off and on.

Radiation process for fishery products have been standardised both in India and elsewhere. Radurization (synonymous with pasteurization) is the irradiation process for extension of shelf life of fresh fishery products in ice or under refrigeration by reducing the number of spoilage causing bacteria. Two factors are most significant in determining the optimum radiation dose for radurization. These arise from qualitative and quantitative changes in the microbial growth and radiation dose response of tissue constituents that govern the organoleptic attributes of fishery products. Radurization is done in the dose range of 1–3 kgy, which is sufficient to reduce an initial load of spoilage-causing organisms by about 1 to 3 log cycles. The optimum does is selected to give a product with extended shelf life in ice. Shelf life of marine and freshwater fishes irradiated (low does) is given Table 5. The irradiated products had 2–3 times higher shelf life than unirradated samples.

Table 5: Shelf life of marine and fresh water fish by low dose irradiation (From Venugopal *et al.*, 1999)

Fish/ Shell-fish	Radiation dose (kGy)	Storage temperature (°C)	Shelf life (days)
Common carp	5.0	0–2	35 (15) (vacuum packed)
Channel catfish	1–2	0	20 (4)
Shrimp	1.5–2.0	3.0	42 (10)
Freshwater prawn	1.45	0	28
Bombay duck	1–2	0–1	25 (7)
Mackerel, Indian	1.5	0–2	25 (12)
Pomfret, white	1–2	0–1	35 (10)
Ribbon fish	1.0	1.0	28 (8)
Salmon, Indian	1–2	0–1	25 (10)

Note : Numbers in parentheses indicate shelf-life (days) of unirradiated sample.

Recognizing the techno-economic advantages, 40 countries have cleared over 100 different food products for radiation treatments at doses ranging from 1–10 kGy The Indian government has cleared the irradiation process for fish items including fresh, dried (disinfestations and microbial control), frozen products (Venugopal *et al.*, 1999). The major benefit of the application of radiation in fishery products is in the reduction of post harvest losses and the improvement of the hygienic quality of fishery products. Irradiation at appropriate doses and conditions can augment sanitation measures and good manufacturing practices to provide safe and wholesome products.

Value Added Products

The term value added products means 'addition of value to the fish' which otherwise would have fetched lower value. Selling of fish fillets, dressed fishes (be-headed and gutted) and minced meat are all value added products. In other words it is the supply of fish in a more convenient form for further cooking. The most common form of value added products that have become popular for both domestic and export market is Battered and Breaded products. These products are also referred as 'enrobed' products. Fish fillets, shrimps, cuttle fish and squids or semi dressed fishes can be used for enrobing. Here the fish coated with another foodstuff. In general there are two types of coatings, batter and crumb. Batters may be used as an adhesive in order to apply a second or subsequent coating or as a tempura or puffing batter. The crumb is normally a bread-based dry crumb but other coatings such as small potato chips, or puffed grain, such as rice may be used.

Batters: Batters can be broadly classified into two types, adhesive and tempura. For adhesive batter, the ingredients are wheat flour and corn flour and for tempura, wheat flour, corn flour, sodium bicarbonate and acid phosphate. Wheat flour functionality may be considered through the two main components: protein and starch. The viscosity of batters results from the development of gluten protein. Because of the water binding ability of gluten, hard wheat flours with higher protein contents require more water in the mix to produce a batter of given viscosity. The starch swells as gelationisation takes place and the water in the batter is absorbed in this process. Due to insufficient water in the batter, the gelationisation may be limited, affecting the texture of the coating. Addition of water to the batter must be carefully controlled to avoid thinning of viscosity resulting in a reduced pick up. Other ingredients may be added or substituted in order to achieve the desired end product. Some of these ingredients are rice flour, soya flour, starch, seasoning, emulsifiers, salt, sugar, flavoring, gums and oil.

Breadings: The secondary coating is often referred to as a 'breading' though it may not always be derived from bread. The original crumbs were made from ground dry bread and it is still the major type of secondary coating used today. The crumbs or breadings may be defined as thermally processed products. These are normally cereal based though non-cereal products, such as, potato are also used to provide different textures and appearances to the end product. The particle size of the breading is important in terms of appearance, texture and pick-up.

Most of the problems encountered in enrobed fishery products are due to incorrect procedures prior to the freezing stage. Careful control and adherence to specifications for raw materials, fish, batters and breadings may avoid most of these problems. The correct choice of the equipment used in the process lines will, to a great extent, be dictated by the type of finished product desired. However, the raw materials used will also influence the design characteristics of the individual machines incorporated in the process line.

Processing of Fish for Industrial Purposes

Fish meal and oil

The conversion of small pelagic and other fishes into meal oil was flourishing industry world over including Indian in 1970's and early 1980's. At present in India, about 5–7% of total marine catch is being used for fishmeal production. The demand for fishmeal does exists from poultry, pig and fish farmers for feed. But the economics of production of fishmeal cannot sustain the industry; hence there is decline in its production. It is normal practice that when the bulk landings cannot be handled for human food, the fishes will be diverted for fishmeal and oil production. Although there have been continuous pleas to convert more industrial fish into human food, it may not be practical in countries like India, where adequate infrastructure for distribution of all the fishes caught for human food. It is reasonable to expect enormous amount of fish processing waste to generate in the years to come, with more emphasis on marketing of processed products. The generated processing waste can be made use for the production of fishmeal. Such fishmeal can be fortified with press cake of oil seeds (after oil extraction), and marketed as animal feed.

The industrial uses of fish oil have been very well established. Though the share of marine oil in supply of edible oil is only about 2%, their nutritional significance has been recognized from early times. But what has attracted the researchers interests, is the use of fish oil in human nutrition. With the beneficial effects of n-3 polyunsaturated fish oil in reducing the coronary heart deceases, the potential for marketing of good fish oil will be very high.

Chitin and chitosan

Chitin, polymer of N-acetyl glucosamine, is present in the exoskeleton of shrimp, crab, squilla, krill and insects. The major commercial source of chitin is the crustacean waste, mainly shrimp and crab. Shrimp waste contains about 5% chitin and crab waste about 3%. Chitin is isolated from crustacean shell by deproteinisaion using alkali or proteolytic enzymes, followed by demineralization using dilute mineral acids. Chitin is deacetylated using concentrated alkali to produce chitosan. Properties of chitin and chitosan vary considerably depending on the raw materials and the process employed (Madhavan *et al.*, 1986). India has emerged as one of the leading producers of chitin and chitosan.

Properties: Chitin is neutral in nature and chitosan in basic. Chitin is highly hydrophobic and insoluble in water and most organic solvents. Molecular weight of shrimp chitin has been reported as 1.5×10^5 (Mathew *et al*, 1998). Chitosan is also insoluble in water but soluble in dilute acids forming corresponding salts.

Applications: The important applications of chitin and its derivatives have been a matter of considerable research input over the past three decades. The important food related applications are in purifying drinking water, recovery of protein from poultry and fish processing wastes, clarification of fruit juices etc. The non food related applications of chitosan includes removal of mercury from water, improving the functional properties of paper, in making non-woven fabrics and sustained release of liquid fertilizers and pesticides. A thin coating of chitosan on seeds enhances the chitinase activity, which helps to germinate without infection. Chitin increases water retention in soil and controls nematodes.

Commercialization of at least some of the above applications has been able to absorb a sizable quantity of chitin and chitosan. Water treatment, paper and textiles, drugs and pharmaceuticals and cosmetic industry have already started using chitin and chitosan.

Application of Biotechnology in Fish Processing

The *bio* in biotechnology means *life* and refers to microbes and other living cells including animal and plant cells. The *technology* comprises the growth of living cells in fermentor and bioreactors containing nutrients and oxygen (if needed) at the specified conditions, and the processing of biological materials produced by the cells through process integration and optimization at top efficiency for achieving commercialization. Biotechnology has arisen through the integration between various parts of biology and engineering, employing techniques derived from three well-organized disciplines: biochemistry, microbiology and biochemical engineering.

There is a wide array of "biotechnologies" with different techniques and applications. The Convention on Biological Diversity (CBD) defines biotechnology as: *"any technological application that uses biological systems, living organisms, or derivatives thereof, to make or modify products or processes for specific use"*. Interpreted in this broad sense, the definition of biotechnology covers many of the tools and techniques that are commonplace in agriculture and food production. Interpreted in a narrow sense, which considers only the new DNA techniques, molecular biology and reproductive technological applications, the definition covers a range of different technologies such as gene manipulation and gene transfer, DNA typing and cloning of plants and animals. Although biotechnology involves the potential use of all living forms, microorganisms have played a major role in the development of this discipline because of the ease of mass growth, the rapid growth that occurs in media consisting of cheap waste materials and the massive diversity of metabolic types.

Biotechnology is not itself a product or range of products like microelectronics; rather, it is a range of enabling technologies, which will find application in many industrial sectors and fish processing sector is no exception. It is believed that biotechnology will influence the production and preservation of fish to a great extent in years to come. Food processing industry in general has taken advantage of recent developments in biotechnology to improve the processing, production distribution and there is no reason why fish processing industry should not take cue, as most of the unit operations are similar. Some of the areas where biotechnology can have a major role in fish processing have been detailed.

Extension of shelf life of processed fish products

Ready to serve heat-processed fish products like fish sausage, fish ham and other paste products have a shelf life of 1-2 days at ambient temperature, which hinders distribution and marketing. Use of metabolites of lactic acid bacteria, which are collectively referred as bacteriocins have been found to be effective in extending the shelf life of the heat processed products. One such bacteriocin, nisin has been found to be very effective in extending the shelf life of fish paste products. The metabolites from other microorganisms like colicin and pediocin have been found to be effective in meat products. In order to improve the efficiency of these microorganisms in producing required bacteriocins, application of genetic engineering needs to be explored. The biosynthetic genes of nisin production are encoded by a conjugative transposon, and the DNA sequence of these genes has been fully determined. This information will lead to the development of strains that overproduce nisin for commercial purposes. There is a need for similar approach for other bacteriocins for application in extension of shelf life of fish paste products.

Enzymes in fish processing

Enzymes are used in fish processing for the production of functional protein isolates, fermented fish products, texture modification and for quality assessment. Commercial enzymes used by industry may be obtained from plant, animal and microbial sources. Today microorganisms have become a major source of commercial enzymes and the route followed is by fermentation process. Both surface (solid) and submerged (liquid) cultures are used to produce microbial enzymes. Cultures of microorganisms are carried out on the surface of solid media such as wheat bran are very popular, Submerged culture methods using fermentor today dominate in the production of enzymes, since they present less risk of contamination and offer reduced handling costs and yields. However, it should be mentioned that the use of enzymes in fish processing is in its infancy and needs further basic research. The most important future research on enzymes would include 1) use of nonnatural catalytic properties of enzymes 2) elucidation of molecular mechanisms of enzyme inactivation 3) enzyme catalysis in organic solvents and other extreme environments and 4) use of protein engineering to tailor catalytic or other properties of enzymes. The application of recombinant DNA technology and protein engineering will significantly improve the production of commercial enzymes in near future. Protein engineering is a science in which many disciplines such as biology, protein chemistry, structural chemistry and enzymology attempt to define and exploit relationship between enzyme structure and catalytic function. Because current enzymes and microorganisms used in food processing are not ideal catalysts, it would be desirable to redesign enzymes to obtain enhanced stability and catalytic properties to suit particular process. The technique of protein engineering, more specifically termed *in vitro mutagensis or site-directional mutagensis*, involves the alteration of one or a small number of known amino acids in a protein through the use of short oligonucleotide primers coding for the change that permits modification of catalytic or other properties.

Biosensors for biological monitoring of foods

A biosensor is an analytical device in which a biologically derived sensing element is in intimate contact with a physiochemical transducer to given an electrical signals. Biocatalysts for biosensors are enzymes, microbial cells, antibodies, lectins and plant and animal organelles. The immobilization of these materials on these materials on the surface of transducers forms the basis of various biosensors. Two important biosensors that are the result of genetic engineering are DNA probes and monoclonal antibodies. There are considerable prospects for the application of biosensors in fish processing in assessing the quality of fresh fish. The main objective of biosensor research is to develop rapid, specific, reliable, inexpensive and compact devices. There are few such devices on the market and new products are beginning to appear.

Waste management and food processing

Nearly 20-25% of the food produced in the world will go as waste before it reaches consumer. Food wastes are by-products and residues derived from food processes that are relatively low in value. The losses create environmental and economic issues that point to the need for better utilization of food wastes and further reduction of food wastes. The food industry contributes wastes through wasting water, spoiled foods, and nutrients loss in effluents. Fish processing plants contributes substantial wastes in the form of skins, bones, offal, exoskeletons from shrimps and substantial nutrients in wash water from surimi plants. The wastes can be treated by enzymatic, fermentation and microbial process to get value added products like chitoson, pigments, pet foods, flavor enhancers and high protein ingredients. There are many reports of the treatment of treatment of food wastes that produce higher protein SCP products from wastes of citrus peel, agricultural crops, brewing operations, grapes, whey and fish.

Food safety and new biotechnologies

Genetic engineering has reached the stage that permits the relatively straightforward isolation and cloning of the gene for any well-characterized protein. Until recently, only a few recombinant DNA products have gone into the marketplace and more than 100 are under development. Any food ingredient or the process developed using biotechnology as tool; safety should be the prime concern. Consumer perception and ethical and moral issues will need public debate and input as well.

Future Utilization of Fish

Considering the higher demand for fish and fishery products with less availability of raw material, it is imperative that great attention has to be given to appropriate post harvest technology. The fresh fish market is of greater importance in developing countries like India, as infrastructures for distribution of frozen products are not well developed. It is likely distribution of chilled fish will have a major role in catering to fish eating population. Recent advances made in the bulk handling of small fish indicate that the chill preservation at sea may be greatly improved. A number of small and medium size fish, which up to now have been reduced to fishmeal and oil, will be preserved for food in the future. This will require new and improved preservation methods both at sea and ashore.

The domestic market for frozen seafood is at a very low level. It is time that our freezing industry should explore the means to strengthen the domestic market. Though it calls for higher capital investment to establish cold chain system, in the long run it will be beneficial in the event of fluctuations in the International market. With increase in fish production from aquaculture, distribution of these fishes in frozen form will be very much on the cards. Therefore, it is reasonable to except that future frozen seafood trade will be more balanced between domestic and export markets. This will also ensure certain stability in aquaculture.

Nearly 13–15% of the total fish landed goes for salting and drying. There is a strong internal demand for salted and cured products. There is a need to improve the curing process to cater hygienic and wholesome dried product. The exports of salted and dried products from India is on the declining trend, mainly due to quality related problem and there is an urgent need to improve the quality by adopting better available technology to boost the export trade in this sector. Attractive packaging meeting international standards will also help in better trade both for domestic and export markets.

Production of surimi is an efficient process for obtaining fish flesh separate from skin and bones. It is likely to be used more commonly than it is now because it reduces waste through recovery of edible flesh from small fish that do not lend themselves to market as whole fish. The scope for utilizing small varieties of fishes from fresh water aquaculture for surimi production should be though off, as at present there is no demand for these fishes. Strong points in this use for surimi are that such products are consistent with the growing demand for convenience foods such as fish sausage, fish sticks and cakes. Each product should stand on its own and be judged solely on the palatability, nutrition and convenience and price.

Presently the canning industry in India is in poor shape owing to the lack of suitable containers and process. The cost of the canned products is prohibitive even for urban middle class consumers. It is worthwhile to note that recent technology developed by Central Institute of Fishery Technology, Kochi, with locally available reportable pouches, should pave way for popularization of canned fishery products. With exploitation of tuna resources from Indian Ocean region optimally, which is more suited for canning process, there is a higher potential for export market.

Extensive investigations, worldwide, in the last 4 decades have shown the benefits of radiation processing for the preservation and microbial quality of seafood. At the same time food irradiation process is highly capital intensive and hence economic feasibility could be obtained only when there is a fairly large quantity of produce to be processed. To achieve maximum output, the seafood irradiation plant may be located at a central point near landing centers where sufficient amounts of catch are processed. However, unlike other physical processes, irradiation has a low operating cost, especially with respect to the energy required for irradiation. The key issues, plant and worker safety, and future consumer questions with respect to irradiated foods have been addressed. Disinfect ion of dried fish by radiation consumes less energy than fumigation and therefore more economical. More importantly, irradiation can control effectively food borne hazards arising from eating shellfishes. With available technology and extensive research data generated by BARC scientists, irradiation process has a major role to play in reducing post harvest losses considerably in India.

Summary

Fish and seafood are rich sources of high quality proteins, minerals and lipids. The demand for fish is increasing world over, and the supply of acceptable quality of fish is a challenging task to the all concerned. The intrinsic nature of fish makes susceptible for early spoilage and necessitates improved handling methods from the time it is harvested till it reaches to consumers or for further processing. In the present review, an overview of fish processing in India has been presented with special emphasis on the principal constituents, nature of spoilage, methods of preservation, special processing techniques and application of biotechnology in fish processing. Recommendations for better utilization of fisheries resources have been outlined.

References

Alur, M.D., Doke, S.N., Warrier, S.B. and Nair, P.M. (1995) Biochemical methods for determination of spoilage of foods of animal origin. Journal of Food Science 32: 181-189.

Ando, M., Toyoshara, H., Shimizu, Y. and Sakaguchi, M. (1991) Post-mortem tenderisation of rainbow trout caused by gradual disinegration of the extracellular matrix structure. Journal of Science for Food and Agriculture 55: 589-599.

Ashie, I.N.A., Smith, J.P. and Simpson, B.K. (1996) Spoilage and shelf extension of fresh fish and shelfish. Critical Reviews in Food Science and Nutrition 36: 87-121.

Bilinski, E.R., Jonas, E.E. and Peters, M.D. (1983) Factors controlling the deterioration of spiny dogfish (*Squalus acanthias*). Journal of Food Science 48: 808-812.

Bligh, G.E. (1980) Methods of Marketing, distribution and quality assurance, p. 48–55,. In:Advances in fish science and technology, *ed* J.J. Connell; Fishing News Books Ltd., Farnham, England.

Botta, J.R. (1994) Freshness quality of seafoods: a review, p, 140-167,. In: Seafoods:Chemistry, Processing Technology and Quality; *eds* F. Shahidi. And J.R. Jones; Chapman and Hall, Glassgow.

Bremner, H.A. (1992) Fish flesh structure and the role of collagen-its post mortem aspects and implications for fish processing, p,39-52,. In: Developments in Food Science; *eds* H. H. Huss., M. Jackobson, and J. Liston; Elsevier Science, Amesterdam.

CAC, Codex Alimentarius Commission, (1996) Proposed Draft Guidelines for the Sensory Evaluation of Fish and Shellfish, Committee on Fish and fishery products, 22nd Session,

Bergen, May CX FFP96/9, Food and Agricultural Organisation of the United Nations, Rome, Italy.

CMFRI. (2002) Central Marine Fisheries Research Institute, Annual Report (2001-2002), Cochin, India.

Connell, J.J. (1995) Control of fish quality, Fishing News Books, Londoan.

Cydesdale, F.M. (1978) Colourimetry: Methodology and Applications. Critical Reviews in Food Science and Nutrition 10: 243-301.

Dainty, R.H. (1996) Chemical/biochemical detection of spoilage. International Journal of Food Microbiology 33:19-33.

Dalgaard, P., Gram, L. and Huss, H.H. (1993) Spoilage and shelf life of cod fillets packed in vacuum or modified atmospheres. International Journal of Food Microbiology 19: 283-294.

Doke, S.N., Warrier, S.B.K., Ninjoor, V. and Nadkarni, G B. (1979) Role of hydrolytic enzymes in the spoilage of fish. Journal of Food Science and Technology 16: 223-226.

EC, (1996) Council Regulations (EC No. 2406/96), Laying down common marketing standards for certain fishery products. Official Journal of European Community No. L-334/2,23.12.1996.

FAO, (1995) Quality and quality changes in fresh fish. FAO Fisheries Technical paper 348: 1–2.

Garrett, E.S., Jhancke, M.L. and Tennyson, J.M. (1997) Microbiological hazards and emerging food safety issues associated with seafoods. Journal of Food Protection 60: 1409- 1415.

German, J.B. and Kinsella, J.E. (1985) Lipid oxidation in fish tissue. Enzymatic initiation via lipoxygenase. Journal of Agricultural and Food Chemistry 33: 681-686.

Gill. T.A. (1990) Objective analysis of seafood quality. Food Reviews International 6: 681-714.

Gopakumar, K. (1997) Tropical Fishery Products, Oxford and IBH Publishing Co. Pvt. Ltd., New Delhi, 190 pp.

Gram, L. and Huss, H.H. (1996) Microbial spoilage of fish and fish products. International Journal of Food Microbiology 33:121-138.

Haard, N.F. (1992a) Technological aspects of extending prime-quality of seafood, a review. Journal of Aquatic Food Product Technology 1: 9-15.

Haard, N.F. (1992b) Biochemical reactions in fish muscle during frozen storage. P, 1-176,. In: Advances in Seafood Biochemistry: Composition and quality; *eds* G.J. Flick. And R.E. Martin; Technomic Publishing Co, New York.

Hseigh, R.J. and Kinsella, J.E. (1986) Lipoxygenase-catalyzed oxidation of N-6 and N-3 polyunsaturated fatty acids. Relevance to activity in fish tissue. Journal of Food Science 51: 940-945.

Hultin, H.O. (1994) Ixidation of lipids in seafoods, p, 47-74,. In: Seafoods: Chemistry, Processing Technology and Qualiaty. *eds* F. Shahidi. and J.R. Botta; Blackie Acdemic, London.

Jhonson, P.B. (1989) Factors influencing the flavor quality of farm - raised catfish. Food Technology 43: 94-97.

Kalaimani, N., Gopakumar, K. and T.S.U. Nair, (1988) Quality Characteristics of Cured Fish of Commerce, Fishery Technology, 25 : 54–56.

Kanner, J. and Kinsella, J.E. (1983) Lipid deterioration initiated by phagocytic cells in muscle foods: beta carotene destruction by hydrogen peroxide-halide system. Journal of Agricultural and Food Chemistry 31: 370-376.

Kinsella, J.E. (1982) Relation between structural and functional properties of food proteins, p 51–103,. In :Food Protein; *eds* R.F. Fox and J.J. Condon; Applied Science Publishers, New York.

Kolodziejska, I. and Sikorski, Z. E. (1979) Inorganic salts and functional properties of fresh and frozen muscle proteins, proceedings of 25th European Meeting of Meat Research Workers, Budapest.

Konosu, S. and Yamagachi, K. (1982) The flavor components in fish and shellfish, p.367-404,. In: Chemistry and Biochemistry of Marine Food Products; *ed* R. E. Marten; AVI Publishing Co., Westport, Connecticent.

Kostuch, S. and Sikorski, Z. E. (1977) Interaction of formaldehyde with cod proteins during frozen storage. Institute of International Refrigeration Communication. C1 and C2 Karlsruhe.

Kraft, A.A. (1992) Spoilage of eggs and fish, p, 99-111,. In: Psychrotrophjc Bacteria in Foods, Disease and spoilage. CRC Press, Boca Raton, Florida.

Kwai, T. (1996) Fish Flavors. Critical Reviews in Food Science and Nutrition 36: 257-298.

Labuza, T.P. and Fu, B. (1995) Use of time/temperature integrators, predictive microbiology and related technologies for assessing the extent and impact of temperature abuse on meat and poultry products. Journal of Food Safety 15: 201-227.

Lindner, P., Angel, S., Weinberg, Z.G. and Granit, R. (1988) Factors inducing mushiness in stored prawns. Food Chemistry 29: 119-123.

Lindsay, R.C. (1991) Chemical basis of the quality of seafood flavors and aromas. Marine Technological Society Journal 25: 16-22.

Liston, J. (1980) Bacterial spoilage of seafood, p.138–157., In: Advances in Fishery Science and Technology; *ed* J.J. Connell; Fishing News Book Ltd., Farnham, England,.

Love, R.M. (1975) Variability of Atlantic cod (*Gadus morhua*) from the north east atlantic : a review of seasonal and environmental influences on various attributes of fish. Journal of Fisheries Research Board Canada. 32: 2333–2342.

Madhavan, P.,. Nair, K.G.R.,. Thankappan, T.K., Prabhu, P.V. and . Gopakumar, K. (1986) Production of chitin and chitosan, Project report, Central Institute of Fisheries Technology, Cochin.

Martinez, I., Olsen, R.J., Nilesen, H. and Sorensen, N. K. (1997) Seafood: Fulfilling market demands. Outlook in Agriculture 26: 107-114.

Mathew, P.T.,. Nair, K.G.R. and Madhavan, P. (1998) Emerging trends in the application of chitin and chitosan, p 210–216., In: Advances and priorities in Fisheries Technology; *eds* K.K. Balachandran., T.S.G. Iyer., P. Madhavan., Jose Joseph., P.A. Perigreen., M.R. Raghunath.. and M.D.Varghese; Society of Fisheries Technologists (India), Cochin

McDonald, R.E. and Hultin, H.O. (1987) Some characterstics of the enzymic lipid peroxidation system in the microsomal fraction of flounder skeletal muscle. Journal of Food Science 52: 15-19.

Medina, I., Saeed, S. and Howell, N. (1999) Enzymatic oxidative activity in sardine and herring during chilling and correlation with quality. European Food Research Technology 210: 34-38.

Ministry of Agriculture (Department of Agriculture and Cooperation), (1996) Hand book on Fisheries Statistics, Fisheries Division, Government of India, New Delhi, 217 pp.

Mukundan, M.K., Radhakrishnan A.G., James M.A, and . Nair M.R. (1985) Nutritive value of Red and White meat of oil sardine (*Sardinella longiceps*), p 426–428,. In: Harvest and post-harvest Technology of Fish; *eds* Ravindran, K., Unnikrishnan Nair, N., Perigreen, P.A., Madhavan, P., Gopalakrishna Pillai, A.G., Panicker, P.A. and Mary Thomas; Society of Fisheries Technologists (India), Cochin, India.

Nair, K.G. R (1998) Diversification in Fishery Byproducts, p 203–209,. In: Advances and Priorities in Fisheries Technology;*eds* K.K Balachandran., I T.S.G. Iyer., P. Madhavan., J. Joseph., P.A. Perigreen., M.R. Raghunath. and M.D.Varghese; Society of Fisheries Technologists (India), Cochin.

Nair,V.P.G. and Mathew S. (2000) Biochemical composition of fish and shell fish. Central Institute of Fisheries Technology, Technology Advisory Series, Cochin.

Neilson, J. (1993) Quality management of the raw material in the food fish sector: European Economic Community Contract No. UP-2-452.

Olafsdottir, G., Martinsdottir, E., Oehlenschlager, J., Dalgaard, P., Jensen, B., Underland, I., Mackie, I.M., Henehan, G., Neilsen, J. and Nilsen, H. (1997) Methods to evaluate fish freshness in research and industry. Trends in Food Science and Technology 8: 258-265.

Papadakis, S.E., Abdul-Malek, S., Kamdem, R.E. and Yam, K.I. (2000) A versatile and inexpensive technique for measuring color of foods. Food Technology 54: 48-51.

Poulter, R.G., Doe P.E. and . Olley J. (1982) Isohalic sorption isotherms. Use in prediction of storage life of dried salted fish. Food Technology, 17: 201–210.

Ragunath, M.R., Shankar, T. V., Ammu K.and Deadasan K. (1995) Biochemical and nutritional changes in fish proteins during drying. Journal for Science of Food and Agriculture 67: 197–204.

Reineccius, G (1991) Off flavors in foods. Critical Reviews in Food Science and Nutrition 29:381-412.

Saito, T., K. Arai, K. and Matsuyoshi, M.(1959) A new method for estimating the freshness of fish. Bulletin of Japanese Society of Scientific Fisheries 24:749–750.

Shamasundar, B.A. and Prakash V. (1994a) Physico-chemical and functional properties of proteins from prawn (*Metapenaeus dibsoni*). Journal of Agricultural and Food Chemistry 31:459–482.

Shamasundar, B.A. and Prakash V. (1994b) Properties of proteins from drip of frozen prawns (*Metapenaeus dibsoni*). Journal of Food Science and Technology 31: 459-482

Sikorski, Z.E. and Pan B.S. (1994) Preservation of seafood quality, p. 165-193,. In: In:. Seafoods: Chemistry, Processing Technology and Qualiaty. *eds* F. Shahidi. and J.R. Botta; Chapman and Hall, London.

Sikorski, Z.E., Olley J. and Kostuch, S. (1976) Protein changes in frozen fish. Critical Reviews in Food Science and Nutrition 8: 97–129.

Simpson, K.L. (1985) Chemical changes in natural food pigments, p. 409-418,. In: Chemical changes in food processing; *eds* T. Richardson and J.W. Finely; Van Nostrand Reinhold Co., New York.

Somvanshi, V.S. (1998) Marine Fishery Resources and Sustainable Utilization, p, 1-5,. In Advances and Priorities in Fisheries Technology; *eds* K.K Balachandran., T.S.G. Iyer., P. Madhavan., J. Joseph., P.A. Perigreen., M.R. Raghunath. and M.D.Varghese; Society of Fisheries Technologists (India), Cochin.

Sona P. (2000) Seafood Export Exceeds record. Seafood Export Journal, 5: 3–5.

Sorensen, N.K., Helgasson, I.G. and Brataas, R. (1997) Pre-rigor fish: handling and processing, p. 135-140,. In: Making the most of the catch; Proceedings of Symposium, Brisbane, July 1996.

Srinivasa Gopal, T.K., Vijayan P.K.,. Balachandran K. K. and Madhavan P.(1998) Heat penetration in fish curry in retort pouch, p, 236-241,. In Advances and Priorities in Fisheries Technology; *eds* K.K Balachandran., T.S.G. Iyer., P. Madhavan., J. Joseph., P.A. Perigreen., M.R. Raghunath. and M.D.Varghese; Society of Fisheries Technologists (India), Cochin

Tharakan, A.J. (1998) Seafood industry reasons for impasse and solutions, Seafood Export Journal, 29: 5–22.

Tsukuda, N (1970) Studies on the discoloration of red fishes VI, Bulletin of Japanese Society for Scientific Fisheriès 36: 325-328.

Valsan, A.P. (1985) Control of insect infestation in dried Bombay duck, p.554-555,. In: Harvest and post-harvest Technology of Fish; *eds* Ravindran, K., Unnikrishnan Nair, N., Perigreen, P.A., Madhavan, P., Gopalakrishna Pillai, A.G., Panicker, P.A. and Mary Thomas; Society of Fisheries Technologists (India), Cochin, India.

Venugoapl, V., Doke, S.N. and Thomas, P. (1999) Radiation processing to improve the quality of fishery products, Critical Reviews in Food Science and Nutrition 39: 341-400.

Venugoapl, V., Lakshmanan, S.N., Doke, S.N. and Bongriwar (2000) Enzymes in fish processing, biosensors and quality control. A review, Food Biotechnology 14: 21-77.

Venugopal, V. (1990) Extracellular proteases of contaminant bacteria in fish spoilage: A review, Journal of Food Protection 53: 341-350

Venugopal, V. and Shahidi, F. (1995) Value added products from under utilized fish species, Critical Reviews in Food Science and Nutrition 35: 431-453.

Venugopal, V. and Shahidi, F. (1996) Structure and composition of fish muscle, Food Reviews International 12: 175-197.

Venugopal, V., Doke S. N. and Thomas P.(1999) Radiation Processing to improve the quality of Fishery Products. Critical Reviews in Food Science and Nutrition, 39: 391–440.

Vijayan, P.K., Joseph J. and . Balachandran K. K (1998) Partial drying and smoking to improve the quality of canned *Catla catla,* p,241-245,. In Advances and Priorities in Fisheries Technology; *eds* K.K Balachandran., T.S.G. Iyer., P. Madhavan., J. Joseph., P.A. Perigreen., M.R. Raghunath. and M.D.Varghese; Society of Fisheries Technologists (India), Cochin.

WHO. (1999) Food safety issues associated with products from aquaculture. Report of a joint FAO/NACA/WHO study group, WHO Technical Report Series # 883, World Health Organization.

Woyewoda, A.D. and Ke, P.J. (1980) Laboratory quality assessment of Canadian Atlantic squid, Fisheries ad Marine Service Technical Report No. 902, Department of Fisheries and Oceans, Halifax, Nova Scotia.

Yan, X., Taylor, K. D. A. and Hanson, S.W. (1989) Studies on the mechanism of blackspot development in Norway lobster, Food Chemistry 34: 273-278.

Bioremediation

Biotechnology towards Assimilative Capacity and Health of Aquatic Ecosystem

T.K.Ghosh, A.V. Kulkarni and T. Chakrabarti

Preamble

Economic growth is not a panacea for human development. The contents of growth—the composition of inputs (including ecological resources) and outputs (including waste products)—determined by, among other things, the economic institutions within which human activities are conducted, is critical for human development. These institutions need to be designed so that they provide the right incentives for protecting the resilience of ecological systems. Such measures will not only promote greater efficiency in the allocation of ecological resources at all income levels, but would also assure a sustainable scale of economic activity within the ecological life-support systems. Protecting the capacity of ecological systems to sustain welfare is of as much importance to poor countries as it is to the rich.

Environment can be divided into several physical and living compartments. The main compartments of an aquatic ecosystem are water, sediment and biota, although all can be subdivided. Pollution is defined as either the presence of products of human activity, which have harmful or objectionable effects, or as the introduction of these products in the environment. Pollutants enter in a compartment, through which these can be transported and dispersed. These may also be transformed into other compounds or transferred to another compartment. The transport of pollutants through environmental compartments occurs as a consequence of three processes, such as advection, dispersion and diffusion. Pollutant enters in one compartment but may reach equilibrium in another depending on the assimilative potential or capacity of the medium (water, sediment or biota). The equilibrium distribution of a chemical pollutant between compartments is controlled by partitioning laws, which reflect the properties of pollutant and the compartment (e.g. Henry's law). Such laws can predict the assimilative capacity of various media towards different pollutants. Assimilation of pollutant in biota may degrade or transform the pollutant in more dangerous molecules (anaerobic degradation of tetrachloroethane to chloroethane or vinyl chloride). Many organic pollutants undergo reversible transformations in the environment, and are not completely destroyed. Furthermore, the sediment biogeochemical processes are responsible for the amount of nutrients present in water column. Excess nutrients pose a threat to the aquatic ecosystem by facilitating the overgrowth of algae (eutrophication).

Assimilative Capacity is defined as the ability of an area to maintain a "healthy" environment and "accommodate" wastes [GESAMP (IMO/FAO/UNESCO-IOC/WMO/WHO/IAEA/UN/UNEP Joint Group of Experts on the Scientific Aspects of Marine Environmental Protection), 1986].

Aquatic systems that support microorganisms vary considerably, encompassing habitats as diverse as small ponds, lakes, reservoirs, estuaries etc. Although aquatic microbial populations differ by system, they have some common characteristics. An aquatic system can be roughly divided into a sediment fraction, which is suspended or settled in a bottom layer, liquid fraction, and the normal flora and fauna of the respective system. Vertical zonation of the liquid fraction is found in most standing freshwater bodies deeper than two meters with well-distinguished layers differing in temperature, oxygen content, and nutrient distribution. Fig. 1 shows the microhabitat distribution in a freshwater aquatic ecosystem. The assimilation of organic and inorganic pollutants in such a system primarily depends on the activities of microorganisms present at various zones.

Rapid urbanization and industrialization is responsible for generation of considerable amounts of waste effluents including nutrients, pesticides, organic and inorganic pollutants. These effluents can have undesirable impacts on the local environment, depending on the amounts released, the time-scale over which the release takes place, and the assimilation capacity and flushing ability of the local recipient water body (Axler *et al.*, 1996; Kelly *et al.*, 1996). The environmental impact of dissolved constituents (ammonia, phosphorus, nitrogen, dissolved organic carbon, chemicals and pesticides released from industrial units) depends on the rate at which those products are diluted before being assimilated by the pelagic ecosystem (Black, 2001). In restricted exchange environments, there is a risk of high levels of nutrients accumulating in one area, causing hypernutrification and potentially creating undesirable effects (Midlen and Redding, 1998 and MacGarvin, 2000).

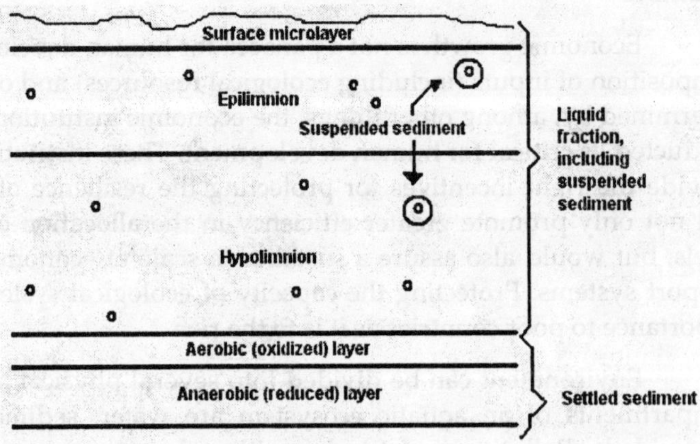

Fig. 1 : Microbial Microhabitats in a Generalized Freshwater System

Pollutants

Types of pollutant

Environmental pollutants can be biological, physical or chemical. Physical pollutants include energy forms, such as, heat, noise and radioactivity, and particulate matters such as, smoke and dust. For example, the waste heat from power plants is often disposed of by discharging warmed water into an open water body. The rise in temperature will alter the local ecosystem and so is a form of pollution. Radionuclides cause harm to living organisms by the energy released in their disintegration deep within the organism. It may be noted that some radioactive substances, such as, plutonium, are also chemically harmful. Most naturally occurring substances can be pollutants, if their concentrations in the environment increase to the level at which they cause health problems or environmental (ecosystem) damage. For example, chloride is one of the most common ions dissolved in water. The high concentration seen in seawater is in harmony with marine ecosystems, as bathers expect it if they swim in the sea, and so chloride is not viewed as a marine pollutant. However, a much weaker solution leached out of a road salt depot into a stream would cause pollution if, as is likely, it damaged the fresh water ecosystem. Similarly, carbon dioxide is an abundant substance causing no harm at natural concentrations. However, the apparently slight rise in concentration in the atmosphere due to fossil fuel consumption seems likely to harm and change the global ecosystem through the greenhouse effect.

However, chemical pollutants are usually thought of as substances that are intrinsically harmful, including heavy metals and a wide range of man-made organic chemicals like pesticides. Organic chemicals are a particular problem because of the large numbers of compounds – over 10 million have been synthesized and perhaps 10,000 are in regular use. Many have been created specifically for their biological effects, while others have such useful properties, including for example resistance to degradation, that users consider their polluting effects acceptable. Toxicological data are not available for many compounds, although modern testing and licensing arrangements are fairly strict, and the information shortages are more for older compounds.

Organic pollutants: The mechanisms by which microorganisms affect changes in pollutant speciation and mobility are fundamental components of biogeochemical cycles for elements, including carbon, nitrogen, and phosphorus, with additional implications for primary and plant productivity and human health (Gadd, 2002). The biogeochemical processes associated with major biogeochemical cycles play an important part in deciding the fate of organic matter in an aquatic ecosystem. The normal microbial flora of sediment and water column of a system decides the rates at which these pollutants will be remineralized. High organic pollution load to an aquatic system is a major problem. The rate of utilization of this organic matter by aerobic or anaerobic microorganisms decides the overall health of system, as the nutrient release rates are directly responsible for the overall productivity of the system.

The result of normal sediment biogeochemical processes is the nutrient release to the water column above. These processes are governed by various factors, which include physico-chemical and biological factors. The nutrients released from sediment are required primarily by primary producers (phytoplankters), which are the source of energy for consumers (zooplankton, fish etc.). The enhanced nutrient release is considered as pollution as it results in algal blooms and release of toxic compounds by various algae there by deteriorating the water quality.

Inorganic pollutants: The majority of inorganic pollutants are compounds or derivatives of nitrogen, phosphorus, sulfur, and carbon. These pollutants include but are not restricted to:

- Nitrogen, nitrates (NHx), oxides (NOx), ammonia (NH_3) nitrogen fixation
- Phosphorus, phosphates (Px)
- Sulfates (SOx), sulfur dioxide
- Carbon monoxide, carbon dioxide

They originate from point sources like industry and municipalities, and also from diffuse sources like agriculture. Much of the inorganic pollution load from industry and municipalities reaches the water body via wastewater drainage into the rivers and reservoirs. The river load accounts for 86 percent of the nitrogen and 73 percent of the phosphorus, for example. These pollutants may serve as nutrients in the water body resulting in eutrophication. Concentrations of metals are comparatively low in the open areas where human activities have not yet had any alarming effects. In inland waters, however, accumulation is evident. Maximum levels of trace metals are found in the sediment offshore from the points of discharge of municipal and industrial wastewater runoff. Metals can also reach the water body by air from industrial smoke stack discharges and traffic exhausts.

Sediment pollution: Though a natural component of all water bodies, sediment itself is often considered a pollutant. However, sediment pollution, as described here refers to the disposition of a variety of contaminants that become attached to mineral and organic sediments. Contaminants of primary concern include pesticides and other organic chemicals, radionuclides, heavy metals and petroleum hydrocarbons (Olsen, 1984). In some instances, fertilizers and nutrients released from sediments can also create problems. Sediment pollution commonly occurs when contaminated sediments are supplied directly to water bodies. However, pollution can also occur when contaminants are applied to soils, which are subsequently eroded and delivered to water bodies as sediment, or when contaminants are introduced directly to water that contains sediments. Contaminants become attached to sediments simply by coating the sediments or by various sorption forces that depend on the nature of the sediment and the contaminant as well as the chemistry of the water. Most contaminant sorption and desorption occurs on the smaller clay size sediments less than about 0.004 mm in size. Sediment pollution may also occur from natural contaminants such as, heavy metals in sediments derived from mine ores and may be found in sediments of considerably larger sizes.

Contaminated sediments, particularly the smaller sediment sizes, are often transported long distances from their sources, thus greatly expanding the potential for pollution damage. Sequential deposition of sediments occurs from large to small particle sizes. In some instances, flocculation of fine mineral and organic sediments can influence sediment deposition rates. Deposition of larger contaminated sediments tends to occur near to the source and thus may be found in more active waters, such as rivers. Deposition of smaller sediments requires relatively still waters and so is most likely to occur in lakes, reservoirs, estuaries, bays and harbors. In many locations, sediments are considered to be the main source of toxic contaminants that affect lakes, rivers and coastal waters. The sediments can adversely affect bottom-dwelling animals and plants and in some cases tend to bioaccumulate into the food chain of higher animals and humans as well. In addition, some pollutants are released directly to the water from the sediments as the result of biological activity and changes in the chemical equilibria of the sediment. Numerous examples of pollution problems caused by contaminated sediments have been documented throughout the United States and elsewhere in the world (Baker, 1980; Hadley and Ongley, 1989; DePinto *et al.*, 1993).

Left undisturbed, hazards from contaminated sediments tend to decrease over time, as the various pollutants decay and contaminated sediments are buried under new sediments. Pollutant decay rates vary considerably from a few years for some petroleum hydrocarbons to millennia for some radionuclides. Pollutants can become concentrated in interstitial waters surrounding sediments but tend to remain in place as long as the sediments remain undisturbed. Unfortunately, deposits of contaminated sediments are commonly disturbed by natural events such as, floods and large storms and by dredging, leading to resuspension and further spread of pollutants. Dredging of rivers, lakes and harbors is a common practice throughout the world in order to maintain navigation. In the United States alone, an average of about 285 million m^3 of sediment is dredged from water bodies each year and transported to disposal sites on land or water where contaminated sediments may create new problems (Engler, 1980). Such anthropogenic activities are also responsible for changing the assimilative capacity of sediments.

Pollution sources and events

Virtually all-human activities generate potential pollutants, because they involve the creation or concentration of substances and energy, in a context in which accidents can occur, and the generation of wastes, which must then be disposed of. Many of these activities are discussed elsewhere. Research on industrial sites has shown that virtually all are polluted to some degree while even relatively benign activities can, when concentrated together by urbanization, create sufficient loads to become polluting. Pollutants may be released accidentally or deliberately into the environment in a single slug, such as an oil spill, or continuously in an effluent, a leak, or by leaching of wastes. Whether such releases constitute pollution will depend on the critical load for the receiving ecosystem, or on the acceptable standards for such discharges.

Response of microbes to pollutants

The most important role of microorganisms in an aquatic environment is to accommodate the waste (organic matter) in the system and simultaneously its efficient recycling. Microorganisms can decompose variety of pollutants including excreta of fish, reaming food materials and other organic materials to CO_2, nitrate and phosphate. These inorganic salts provide the nutrition for the growth of micro algae, which by the mechanism of photosynthesis provide dissolved oxygen for oxidation and decomposition of the organic materials and for the respiration of the microbes. This kind of cycle improves the nutrient cycle creating a balance between bacteria and micro algae and maintains a good water quality.

Response of plankton to pollutants

When an organic polluting load is discharged into a water body it is gradually eliminated by the activities of bacterioplankters (bacteria, phyto and zoo-plankton) in a way very similar to the processes in the sewage treatment works. This process requires sufficient concentrations of oxygen, and involves the breakdown of complex organic molecule into simple inorganic molecules. Algae play a very important role in supplying oxygen to the other aquatic organisms, which require it for the degradation of organic pollutants. Organic waste entering the pond is broken down by bacterial decomposition to form simple nutrients, which are taken up by the algae and used to synthesize new algal cell material in presence of light. CO_2 produced during the oxidation of organic matter enhances the photosynthesis of algae while the oxygen produced by algae during photosynthesis promotes the aerobic degradation process of bacteria and other microorganisms (nutrient release). The high nutrient release from sediments due to the functioning of biogeochemical processes to the water column above may lead to eutrophication.

Acceptable levels of pollutants

Decisions as to whether soil, air or water are polluted are often made by reference to standards for acceptable concentrations, such as the advisory World Health Organization limits for drinking waters (Anon., 1984) or the legally binding limits in Europe (CEC, 1980). Organic or inorganic input to the sediment, whether from natural or artificial sources, lead to a series of changes in chemical and physical parameters, which in turn have both direct and indirect effects on the faunal communities in the area. The extent of the environmental impact on the sea or lake bottom is a function of local assimilative capacity and amount of organic waste introduced from industrial activities (GESAMP, 1986). The former is largely dependent on local environmental conditions (bottom depth, topography, and current patterns) while the latter is strongly related to management practices (effluent treatment, effluent release rates etc).

Trying to determine the assimilative capacity for industrial effluents is extremely difficult unless an integrated approach is taken for all of the activities that go on in a body of water. In addition, environmental impacts will depend on the sensitivity of the receiving system. The concept of assimilative capacity should also include the goal being aimed at and a clear description of what is being protected by not exceeding the limit and capacity of the receiving ecosystem. Estimating the assimilative capacity of receiving ecosystems is difficult due to their dynamic nature. Therefore, there is a need to include uncertainty in the decision-making process (the "precautionary principle").

Tracing assimilation of pollutant (organic compounds by using $\delta^{13}C$)

Various techniques are employed according to the need to evaluate the assimilation of pollutant in different environmental compartments. Stable isotope ratios have been used to examine carbon, nitrogen, and sulfur cycling and to determine the major sources of organic matter that support the food web in a variety of ecosystems (Fry, 1986 and Peterson *et al.*, 1985). Further development of this technology has provided the ability to identify cycling of carbon at a molecular level (Hullar *et al.*, 1996; Meier-Augenstein, 1995 and Coffin *et al.*, 1989) allowing specific identification of microbial roles in the biogeochemical cycling of carbon and nitrogen. In most ecosystems, a complex mixture of physical, chemical and biological factors control the carbon sources that are available to support bacterial production. Stable carbon isotope analysis has provided a more thorough understanding of microbial processing in complex ecosystems, as well as assisted in the development and interpretation of bioremediation strategies for contaminated ecosystems (Aggarwal *et al.*, 1991 and Coffin *et al.*, 1997).

Role of Microorganisms on different types of pollutants

Organic pollution

Nutrient pollution: Nutrients are essential for survival of all organisms, but substantial increase of nutrient levels pose a threat to the ecosystem health, by facilitating the overgrowth of algae (algal blooms). The main practical approaches for the mitigation of the environmental impacts of nutrient enrichment are: bacterial dissimilation into gases and plant assimilation into biomass. Bacterial biofilters are dissimilative. Through a series of oxidation and reduction processes, they break the pollutants down into harmless gaseous N_2 and CO_2. Biofiltration by plants, such as algae, is assimilative, and therefore adds to the assimilative capacity of the environment for nutrients (Krom, 1986). With solar energy and the excess nutrients (particularly C, N and P), plants photosynthesize new biomass. The operation recreates in the culture system a mini-ecosystem, wherein, if properly balanced; plant autotrophy counters fish (or shrimp) and microbial heterotrophy, not only with respect to nutrients but also with respect to oxygen, pH and CO_2 (Hirata *et al.*, 1994 and Rai *et al.*, 2000). Plant biofilters can thus, in one step, greatly reduce the overall environmental impact and stabilize the environment. Algae are most suitable for biofiltration because they probably have the highest productivity of all plants and can be economically cultured or managed (Gao and McKinley, 1994). Algae use sunlight to build their biomass while assimilating dissolved inorganic nutrients, removed from the water. If properly cultured, the organisms of both extractive groups can turn pollutant nutrients into commercial crops and loaded effluents into clean water. Understanding of these processes is of utmost importance for better management of an aquatic ecosystem. The recent advances in biotechnological methods have opened new avenues to mitigate the environmental challenges. Thus the use of biotechnological methods can be employed for increasing assimilative capacity of the region. The major biogeochemical cycles include following biogeochemical processes

- **Carbon utilization rate: -** The sediment –water interface is characterised by steep biological and physico-chemical gradients that foster changes in the nature and quantity of the deposited organic pollutants (Berner, 1980). Biogeochemical reactions, taking place at the benthic boundary layer, determine the rate of OM remineralization, which leads to nutrient and CO_2 regeneration. The fraction of organic carbon (OC) that escapes oxidation processes is possibly buried and preserved within the sediment column, and thus lost from the ocean–atmosphere system over geological time scales (Hedges and Keil, 1995). The organic monomers originated from OM degradation represent the main source of carbon and energy to benthic heterotrophic organisms, which can mineralize them into dissolved inorganic carbon (IC) via respiration processes, or incorporate them into cellular constituents. This assimilation of organic monomers prevents the reuse of organic compounds thereby influencing the nutrient (inorganic carbon) release from sediments. This phenomenon is of great concern to the overall productivity and health of the aquatic ecosystem. Monomers can also undergo abiotic transformations by random polymerization and consequent formation of humic substances (Rashid, 1985), which are highly refractory geopolymers, precursors of kerogens, and are responsible for a great part of the OC preservation (Tissot and Welte, 1984).

- **Nitrification rate: -** The utilizable form of nitrogen for higher organisms is nitrate (NO_3). Biological nitrification has been widely used for nitrifying ammonia to nitrite and nitrate. The prerequisite for maximum nitrification rate is the low organic carbon environment. The organic pollution of an environment provides a favorable environment for the growth of heterotrophic microorganisms. Under such circumstances, heterotrophic microorganisms can rapidly out-compete nitrifiers due to their growth rates and biomass yields (Harremos, 1982). Thus the ability of heterotrophic microorganisms to assimilate the organic monomers greatly lowers the nitrification rate influencing the nitrate release.

- **Denitrification rate: -** Denitrification plays a major role in the nitrogen (N) cycle of aquatic systems. In this processes, facultative anaerobic bacteria transform nitrate or nitrite into nitrogen gas, which escapes into the atmosphere (Knowles, 1982). The nitrogen and phosphorous have the lowest supply: demand ratio of the nutrients found in freshwaters, the removal of nitrogen by denitrifiers could limit primary production (Seitzinger, 1988). Furthermore, the ability of denitrifying bacteria to reduce high levels of ambient N is becoming increasingly important as N loads to freshwater increase due to human activities. Denitrification primarily occurs in the sediment (Seitzinger, 1988). A number of environmental factors have been shown to influence denitrification in freshwater sediment. Bacterial activity is stimulated by an increase in water temperature (Messer and Brezonik,

1983; van Luijn *et al.*, 1999) and is disproportionately high in warm zones. Both nitrate and organic matter supply can be considered as limiting factors for denitrifiers (Seitzinger, 1988).

- **Phosphatases:** - Anthropogenic alterations of carbon (C) and nitrogen (N) cycles may alter key processes controlling phosphorus (P) cycling in an ecosystems (Cole and Heil, 1981). Chronic deposition of N compounds decreases soil pH and base saturation, increases mineral weathering rates and affects a wide range of chemical, biochemical and microbial processes (Cook, 1983 and Psenner and Catalan, 1994). While weathering rate and soil physico-chemical processes mainly affect P transformations and accumulation in deeper mineral layers, shifts in biochemical and microbial processes are key factors affecting P transformation in the litter and humus layers, in which organic P is a dominant form. Organic P enters the cycle of P transformations during microbial and biochemical transformations of organic matter (Stewart and Tiessen, 1987). The last step in microbial mineralization of organic P is the hydrolysis of esters and release of orthophosphates by extracellular phosphatases. Microorganisms can exclusively take up orthophosphates, as is widely accepted in aquatic microbial ecology (Currie and Kalff, 1984). Microorganisms assimilate a portion of the released orthophosphates and the remainder can be fixed by chemical reactions with free ions (Fe^{3+}, Al^{3+}, Ca^{2+}) or Fe and Al oxides or complexed with dissolved organic compounds in soil solution (Brady and Weil, 2002). While microbial P transformation is directly related to microbial mineralization of organic compounds, biochemical hydrolysis of organic P by extracellular phosphatases might be independent of organic matter mineralization (McGill and Cole, 1981, Stewart and Tiessen, 1987 and Olander and Vitousek, 2000).

Inorganic and metal pollutants

The ability of microorganisms to affect pollutant speciation stems from their ability to effect and/or mediate mobilization or assimilation processes that influence the balance of metal or pollutant species between soluble and insoluble phases. Mobilization of metals can be achieved by, example, protonation, chelation, and chemical transformation, while immobilization can occur by precipitation or crystallization of insoluble organic or inorganic compounds or by sorption, uptake and intracellular sequestration. Redox reactions can mobilize or immobilize pollutants depending on the pollutant species involved, as well as, being an integral component of biogeochemical cycles for metals, these processes may be exploited in environmental biotechnology (Lloyd and Lovley, 2001 and McLean *et al.*, 2002). Almost all metal–microbe interactions have been examined as a means for removal, recovery or detoxification of inorganic and organic metal or radionuclide pollutants (Stephen and Macnaughton, 1999 and Eccles, 1999). In such a context, solubilization may enable removal from solid matrices, such as, soils, sediments, dumps and industrial wastes. Alternatively, immobilization processes may enable metals to be transformed *in situ* into insoluble and chemically inert forms and are also applicable to removing metals from aqueous solution.

Mobilization: Microorganisms can mobilize metals through autotrophic and heterotrophic leaching, chelation by microbial metabolites and siderophores, and methylation, which can result in volatilization. Such processes can lead to dissolution of insoluble metal compounds and minerals, including oxides, phosphates, sulfides and more complex mineral ores, and desorption of metal species from exchange sites on, for example, clay minerals or organic matter in the soil.

These processes are therefore become important in the assimilation of various pollutants at various compartments, thereby making the known assimilative potential highly unpredictable. Mobilization generally takes place through following processes.

- **Heterotrophic (chemoorganotrophic) leaching: -** Microorganisms can acidify their environment by proton efflux via plasma membrane H^+-ATPases, maintenance of charge balance or as a result of respiratory carbon dioxide accumulation. Acidification can lead to metal release via a number of obvious routes, for example, competition between protons and the metal in a metal–anion complex or in a sorbed form, resulting in the release of free metal cations. Heterotrophic metabolism can also lead to leaching as a result of the efflux of organic acids and siderophores. Organic acids can supply both protons and metal-complexing anions (Burgstaller and Schinner, 1993;Gadd, 1999; Gadd and Sayer, 2000). Citrate and oxalate anions can form stable complexes with a large number of metals. Many metal citrates are highly mobile and are not readily degraded (Francis *et al.*, 1992). Oxalic acid can also act as a leaching agent for those metals that form soluble oxalate complexes, including Al and Fe (Strasser *et al.*, 1994). Effective leaching of a variety of wastes and low-grade minerals has been demonstrated, for example, soils and muds, filter dust/oxides, lateritic ores, copper converter slag, fly ash and electronic waste materials (Sukla *et al.*, 1992; Burgstaller and Schinner, 1993; Tzeferis *et al.*, 1994; Bosshard *et al.*, 1996; Brandl *et al.*, 1999 and Gadd, 1999). A strain of *Penicillium simplicissimum* has been used to leach Zn from insoluble ZnO contained in industrial filter dust. The production of citric acid (>100 mM) was induced by the presence of the filter dust (Schinner and Burgstaller, 1989; Franz *et al.*, 1991 and Franz *et al.*, 1993). Organic acid production is also an important agent of mineral deterioration, playing a role in both biogenic chemical weathering and soil formation (Gadd, 1999). Heterotrophic solubilization can also have consequences for other remedial treatments for contaminated soils. Pyromorphite [$Pb_5(PO_4)_3Cl$] is a stable lead mineral and can form in urban and industrially contaminated soils. The formation of pyromorphite has been suggested as a remediation technique for lead-contaminated land, if necessary, by means of phosphate addition. However, pyromorphite can be solubilized by phosphate-solubilizing fungi, for example, *Aspergillus niger*, and during fungal transformation of pyromorphite, biogenic production of lead oxalate dihydrate can occur (Sayer *et al.*, 1999).

- **Autotrophic leaching:** Most autotrophic leaching is carried out by chemolithotrophic, acidophilic bacteria which fix carbon dioxide and obtain energy from the oxidation of ferrous iron or reduced sulfur compounds, which cause the solubilization of metals because of the resulting production of Fe (III) and H_2SO_4 (Rawlings, 1997 and Schippers and Sand, 1999). The microorganisms involved include sulfur-oxidizing bacteria, for example, *Thiobacillus thiooxidans*, iron- and sulfur-oxidizing bacteria, e.g., *Thiobacillus ferrooxidans* and iron-oxidizing bacteria, e.g., *Leptospirillum ferrooxidans* (Ewart and Hughes, 1991 and Bosecker, 1997). As a result of sulfur- and iron-oxidation, metal sulfides are solubilized concomitant with the pH of their immediate environment being decreased, therefore resulting in solubilization of other metal compounds. Such leaching of metal sulfides by *Thiobacillus* species and other acidophilic bacteria is well established for industrial scale biomining (Ewart and Hughes, 1991; Bosecker, 1997; Rawlings, 1997 and Rawlings and Silver, 1995). In a bioremediation context, autotrophic production of sulfuric acid has also been used to solubilize metals from sewage sludge and soils (Sreekrishnan and Tyagi, 1994). In a two-stage process, sulfur-oxidizing bacteria were used to acidify soil and solubilize toxic metals before metal removal from the metal-contaminated liquid leachate using sulfate-reducing bacteria (White *et al.*, 1997). Studies have also been conducted on the bioremediation of red mud, the main waste product of Al extraction from bauxite. Here, autotrophic leaching by indigenous *Thiobacillus* spp. was more efficient than heterotrophic leaching by a range of fungal strains, and decreased the toxicity of the waste product (Vachon *et al.*, 1994). *T. ferrooxidans* has also been used to treat air pollution control residues (APCR: fly ash and used lime) which contain high amounts of toxic metals. Although growth of the *Thiobacilli* was poor, removal of up to 95% of Cd (from APCR of ~270 mg Cd kg−1) was achieved with ~69% Pb removal (from APCR of ~5 g Pb kg−1; Mercier *et al.*, 1999).

- **Siderophores:** - Siderophores are highly specific Fe (III) ligands (formation constants often >1030). These low molecular weight coordination molecules are excreted to aid iron assimilation (Gadd, 2001). Such assimilation may be improved by attachment to solid Fe minerals, for example, Fe oxides to facilitate contact with the Fe substrate. Although primarily produced as a means of obtaining iron, siderophores are also able to bind other metals such as, magnesium, manganese, chromium (III), gallium (III) and radionuclides like plutonium (IV) (Birch and Bachofen, 1990). One method for the treatment of metal-contaminated sandy soil relies on siderophore-mediated metal solubilization by *Alcaligenes eutrophus*. Solubilized metals were adsorbed to the biomass and/or precipitated with biomass, separated from soil slurry by flocculation. This resulted in a complete decrease in the bioavailability of Cd, Zn and Pb (Diels *et al.*, 1999).

- **Biomethylation:** - A range of bacteria and fungi under aerobic and anaerobic conditions can mediate methylation of Hg, As, Se, Sn, Te and Pb. Methyl groups are enzymatically transferred to the metal, and a given species may transform a number of different metal (loid) s. Methylated metal compounds, formed by these processes, differ in their solubility, volatility and toxicity. Volatile methylated species, for example, $(CH_3)_2Se$ and $(CH_3)_2Se_2$, are often lost from the soil (Karlson and Frankenberger, 1993). Many environmental and soil factors, for example, organic amendments and frequent tillage, can be optimized to increase diffusive transport through soil and enhance Se volatilization (Stork *et al.*, 1999 and Zhang and Frankenberger, 1999). Microbial methylation of selenium, resulting in volatilization, has been used successfully for *in situ* bioremediation of selenium-containing land and water at Kesterson reservoir, California, reducing the selenium concentrations to acceptable levels (Thompson-Eagle and Frankenberger, 1992). Several bacterial and fungal species can methylate arsenic compounds such as, arsenate [As (V), AsO_4^{3-}], arsenite [As (III), AsO_2^-] and methylarsonic acid ($CH_3H_2AsO_3$) to volatile dimethyl- [$(CH_3)_2HAs$] or trimethylarsine [$(CH_3)_3As$] (Tamaki and Frankenberger, 1992).

- **Redox transformations:** - Microorganisms can mobilize metals, metalloids and organometallic compounds by reduction and oxidation processes (Gharieb *et al.*, 1999 and Lovley, 2000). For example, metal solubilities increase on reduction of Fe (III) to Fe (II) and Mn (IV) to Mn (II) (Lovley and Coates, 1997; Lovley, 2000 and McLean *et al.*, 2002). Most iron reduction is carried out by specialized anaerobic bacteria that use iron (III) as a terminal electron acceptor. Dissimilatory metal-reducing bacteria can use a variety of metal (loid)s with an appropriate redox couple, including Fe (III), Mn (IV), Se (IV), Cr (VI) and U (VI) (Oremland *et al.*, 1991 and Stolz and Oremland, 1999). While Fe and Mn increase their solubility upon reduction, the solubility of other metals such as U (VI) to U (IV) and Cr (VI) to Cr (III) decreases, resulting in immobilization (Phillips *et al.*, 1995 and Smith and Gadd, 2000). Reduction of Hg (II) to Hg (0) by bacteria and fungi results in diffusion of elemental Hg out of cells (Silver, 1996; Silver, 1998 and Hobman *et al.*, 2000). Mercuric reductase, from a recombinant *Escherichia coli* strain, has been immobilized on a chemically modified diatomaceous earth support with immobilization enhancing stability and reusability: maximal activity was 1.2 nmol Hg mg^{-1} protein s^{-1} at an initial [Hg^{2+}] of 50 μmol dm^{-3} (Chang *et al.*, 1999). Fe (III) and Mn (IV) oxides strongly absorb metals and this may hinder metal extraction from contaminated soils. Microbial reduction of Fe (III) and Mn (IV) may be one way for releasing such metals and this process may be enhanced with the addition of important soil polymers like humic materials, or related compounds. Such compounds act as electron shuttles, for example, U (VI) and Cr (VI), especially if located in tight pore spaces away from microbial activity (Lovley and Coates, 1997). Bacterial Fe (III) reduction resulted in release of Mn and Co from goethite where 5% of the iron was substituted by these metals (Bousserrhine *et al.*, 1999).

Immobilization or Assimilation: A number of processes lead to immobilization or assimilation of metals or pollutants. Although assimilation reduces the external free metal or pollutant species, it may also promote solubilization in some circumstances by shifting the equilibrium to release more metal into solution.

- **Biosorption and intracellular accumulation: -** Biosorption can be defined as the microbial uptake of organic and inorganic metal species, both soluble and insoluble, by physicochemical mechanisms, such as adsorption. In living cells, metabolic activity may also influence this process because of changes in pH, E_h, organic and inorganic nutrients, and metabolites. Biosorption can also provide nucleation sites for the formation of stable minerals (Beveridge and Doyle, 1989; Southam, 2000 and McLean *et al.*, 2002) as well as sorption to cellular surfaces, some cationic species can be accumulated within cells via membrane transport systems of varying affinity and specificity. Once inside cells, metal species may be bound, precipitated, localised within intracellular structures or organelles, or translocated to specific structures depending on the element concerned and the organism (White *et al.*, 1997). Peptidoglycan carboxyl groups are the main binding sites for cations in Gram-positive bacterial cell walls with phosphate groups contributing significantly in Gram-negative species (McLean *et al.*, 2002). Chitin is an important structural component of fungal cell walls and this is an effective biosorbent for radionuclides, as are chitosan and other chitin derivatives (Tobin *et al.*, 1994). In *Rhizopus arrhizus*, U biosorption involves coordination to the amine N of chitin, adsorption in the cell wall chitin structure and further precipitation of hydroxylated derivatives (Tsezos and Volesky, 1982). Fungal phenolic polymers and melanins possess many potential metal-binding sites with oxygen-containing groups including carboxyl, phenolic and alcoholic hydroxyl, carbonyl and methoxyl groups being particularly important. Both freely suspended and immobilized biomass from bacterial, cyanobacterial, algal and fungal species have received attention with immobilized systems appearing to possess several advantages including higher mechanical strength, easier biomass/liquid separation and, for living cells, increased metal tolerance (Macaskie,

1991 and Tobin *et al.*, 1994). Immobilized living biomass has mainly taken the form of bacterial biofilms on inert supports and is used in a variety of bioreactor configurations including rotating biological contactors, fixed bed reactors, trickle filters, fluidized beds and air-lift bioreactors (Gadd, 2001; Macaskie and Dean, 1989 and Schiewer and Volesky, 2000). Metal-binding peptides, proteins, polysaccharides and other biomolecules: - Microorganisms produce a range of specific and nonspecific metal-binding compounds. Nonspecific metal-binding compounds range from simple organic acids and alcohols to macromolecules, such as polysaccharides, humic and fulvic acids (Sayer and Gadd, 2001). Extracellular polymeric substances (EPS), a mixture of polysaccharides, mucopolysaccarides and proteins (Zinkevich *et al.*, 1996), are produced by bacteria, algae and fungi and also bind potentially toxic metals (Beech and Cheung, 1995). Extracellular polysaccharides can also adsorb or entrap particulate mattets such as, precipitated metal sulfides and oxides (Flemming, 1995 and Vieira and Melo, 1995). Specific, low molecular weight (6000–10,000 Da) metal-binding proteins, termed metallothioneins, are produced by animals, plants and microorganisms in response to the presence of toxic metals (Howe *et al.*, 1997). Other metal-binding proteins, phytochelatins and related peptides, contain glutamic acid and cysteine at the amino-terminal position, and have been identified in plants, algae and several microorganisms (Rauser, 1995). Eukaryotic metallothioneins and other metal-binding peptides have been expressed in *E. coli* as fusions to membrane or membrane-associated proteins such as, LamB, an outer membrane protein. Such *in vivo* expression of metallothioneins provides a means of designing biomass with specific metal-binding properties (Chen *et al.*, 1999).

- **Metal precipitation by metal- and sulfate-reducing bacteria: -** Where reduction of a metal to a lower redox state occurs, mobility and toxicity may be reduced offering bioremediation applications (Lovley, 2001 and Finneran *et al.*, 2002). Such processes may also accompany other indirect reductive metal precipitation mechanisms, for example, in sulfate-reducing bacterial systems, where reduction of Cr (VI) can be a result of indirect reduction by Fe2+ and the produced sulfide. Aerobic or anaerobic reduction of Cr (VI) to Cr (III) is widespread in microorganisms (Smith and Gadd, 2000 and McLean and Beveridge, 2001) and both *ex situ* reactor systems and *in situ* treatment approaches have been documented (Lovley and Coates, 1997 and Wang, 2000). U (VI) can be reduced to U (IV) by certain Fe (III)-dissimilatory microorganisms and this reduction in solubility can be the basis of U removal from contaminated waters and leachates (Lovley and Coates, 1997; Lovley, 2001 and Finneran *et al.*, 2002). Sulfur and sulfate-reducing bacteria are geochemically important in reductive precipitation of toxic metals, for example, U (VI) and Cr (VI), a process mediated by multiheme cytochrome *c* proteins (Aubert *et al.*, 1998). *Desulfovibrio desulfuricans* can couple the oxidation of a range of electron donors to reduction of Tc (VII), which is precipitated as an insoluble oxide at cell peripheries. Resting cells, immobilized in a flow-through membrane bioreactor accumulated substantial quantities of Tc when supplied with formate as electron donor, suggesting potential of this organism for treatment of Tc-contaminated wastewater. *D. desulfuricans* can also reduce Pd (II) to cell-bound Pd (0) with hydrogen-dependent reduction being O_2-insensitive, thus providing a means of Pd recovery under oxic conditions (Lloyd and Macaskie, 1998). Some sulfate-reducing bacteria, like *Desulfotomaculum reducens* share physiological properties of both sulfate- and metal-reducing groups of bacteria, and can grow with Cr (VI), Mn (IV), Fe (III) and U (IV) as sole electron acceptors (Tebo and Obraztsova, 1998). Se (VI) reduction to elemental insoluble Se (0) has been employed to remediate contaminated waters and soils. *In situ* removal of SeO_4^{2-} due to reduction to Se^o by sediment bacteria in agricultural drainage regions of Nevada has also been described (Oremland *et al.*, 1990 and Oremland *et al.*, 1991). Flooding of exposed sediments at Kesterson reservoir with water (to create anoxic conditions) resulted in reduction and immobilization of large quantities of selenium that was present in the sediment (Long *et al.*, 1990). Some bacteria can use such reduction to support growth, making this a natural process for *in situ* applications. Although reduction of oxyanions of As and Se can occur by different mechanisms, the most environmentally significant process is dissimilatory reduction. Oxyanions of arsenic and selenium can be used in microbial anaerobic respiration as terminal electron acceptors providing enough energy for growth and metabolism. Their reduction can be coupled to a variety of organic substrates, e.g., lactate, acetate and aromatics, with the bacteria found in a range of habitats and not confined to any specific genus. These organisms, and perhaps even the enzymes themselves, may have applications for bioremediation of selenium- and arsenic-contaminated environments (Stolz and Oremland, 1999 and Oremland and Stolz, 2000). The incidental ability of a variety of microorganisms from all major groups to reduce Se (VI) and Te (VI) by additional often uncharacterised mechanisms offers additional scope for bioreactor-based approaches (Gharieb *et al.*, 1999). Sulfate-reducing bacteria oxidize organic compounds or hydrogen coupled with the reduction of sulfate, producing sulfide (White and Gadd, 1997). The solubility products of most heavy metal sulfides are very low, in the range of 4.65×10^{-14} (Mn) to 6.44×10^{-53} (Hg) so that even a moderate output of sulfide can remove metals to levels permitted in the environment with metal removal being directly related to sulfide production. Sulfate-reducing bacteria can also create extremely reducing conditions, which can chemically reduce metals such as uranium (VI). In addition, sulfate reduction partially eliminates acidity from the system as a result of the shift in equilibrium when sulfate (dissociated) is converted to sulfide (largely protonated). This can result in the further precipitation of metals as hydroxides as well as increasing the efficiency of sulfide precipitation. Sulfate reduction can provide both *in situ* (Uhrie *et al.*, 1996) and *ex situ* metal removal from contaminated wastes (Barnes *et al.*, 1991; Barnes *et al.*, 1994; Hammack and Edenborn, 1992; Lyew *et al.*, 1994 and Christensen *et al.*, 1996) and contribute to removal of metals and acidity in artificial and natural wetlands (Hedin and Nairn, 1991 and Perry, 1995). A process integrating bacterial sulfate reduction with bioleaching by sulfur-oxidizing bacteria has been developed to remove contaminated toxic metals from soils. In this process, sulfur- and iron-oxidizing bacteria were employed to liberate metals from soils by the breakdown of sulfide minerals and production of sulfuric acid (White *et al.*, 1998). Metals are liberated in the form of an acid sulfate solution, which enables both a large proportion of the acidity and almost the entirety of the metals to be removed by bacterial sulfate reduction (White and Gadd, 1997 and White *et al.*, 1998). Sulfate-reducing bacterial biofilm reactors may offer a means of process intensification and entrap or precipitate metals, for example, Cu and Cd at the biofilm surface (White and Gadd, 2000). Mixed sulfate-reducing bacterial cultures were more effective than pure cultures with metal removal being enhanced by the production of exopolymers (White and Gadd, 2000).

- **Phosphatase-mediated metal assimilation: -** In this process, metal or radionuclide assimilation by bacterial biomass is mediated by a phosphatase, which liberates inorganic phosphate from a supplied organic phosphate donor molecule, for example, glycerol 2-phosphate. Metal/radionuclide cations are then precipitated as phosphates on the biomass (Yong and Macaskie, 1995). Most work has been carried out with *Citrobacter* sp., and a range of bioreactors, including those using immobilized biofilms (Tolley *et al.*, 1995).

Mathematical Modeling

A model may be simply defined as an abstraction and, therefore, necessarily a simplification of reality. Models are as necessary "to theoretical problems as experiments are to empirical problems" (Caswell, 1988). Models can be qualitative (expressed in words or diagrams) or quantitative (expressed in mathematical equations or computer language). The latter are called formal models. Although it may take many forms other than the mathematical, much of ecological theory today is expressed in largely mathematical terms because mathematics provides ecologists the most precise language to describe complex phenomena and is also an ideal tool for prediction (Roughgarden *et al.*, 1989). Indeed, mathematical formulations usually force us to make the assumptions and conclusions clear and unambiguous.

There are two major types of quantitative ecological models: analytical models and simulation models, each of which uses quite different mathematical methods. The analytical modeling approach is characterized by finding exact solutions to a set of mathematical equations with relatively complex mathematics. The roles of mathematical models in ecology include generating and testing hypothesis, predicting the dynamics of complex systems, synthesizing knowledge and ideas, and serving as a powerful tool for assessment and optimization (Fahrig, 1988 and Roughgarden *et al.*, 1989). Generating hypothesis is one of the most important uses of ecological models. Lindeman's (1942) trophic-dynamic model has had profound influence on the development of ecosystem ecology and systems ecology. Models become inevitable when one deals with both spatial and temporal dynamics of systems simultaneously and when the scale (spatial and/or temporal) of the phenomenon is very large. The rapid development of computer simulation modeling has greatly enhanced the understanding of nature and effects of human interventions and holds the promise to unify ecological theory (Meadow *et al.*, 1972).

Fundamental concepts for construction of a mathematical model are

- Dynamics of single component or factor of a system (which includes all biotic and abiotic components)
- The entire system considering multiple or all components with their interrelationship

The first concept deals with the study of individual component of the system, while the second deals with the interactions of processes occurring in the system in totality. Both the concepts have some merits and demerits as delineated in Table 1.

Table 1 : Comparison between two modeling approaches

Factors	Single component model	Multiple component model
Resources required	Less	More
Time required	Less	More
Data collection	Less	More
Error in Data collection	Less	More
Accuracy in Prediction of impact of various factors on the system	High	Low
System impact evaluation	Low	High

Considering the general properties of different modeling approaches, and available literature, an attempt to evaluate phosphorus flow (single element, single component) through different compartments/ trophic levels of aquatic ecosystem could be an ideal approach for construction of a mathematical model. This would be a simplified model to evaluate the flow of phosphorus through the various components in a freshwater lake/ river, comprising fish, phytoplankton, sediment, etc. In order to understand the complex interactions of all of the components, and to experiment with the effects of changing the various parameters of the ecosystem, phosphorus can be measured in six components:

- Dissolved inorganic phosphorus in the water
- Phosphorus in the sediment of the lake
- Phosphorus in fish
- Phosphorus in lake detritus
- Phosphorus in phytoplankton
- Phosphorus in zooplankton

Phosphorus enters the water as dissolved inorganic phosphorus. From the water, phosphorus is absorbed by the phytoplankton at an uptake rate of *x* day^{-1}.

From the phytoplankton, phosphorus can take two paths:

- Phytoplankton die, and form detritus. Death of phytoplankton (actually the amount of release of phosphorus due to their deaths) depends upon the death rate of phytoplankton.
- Phytoplankton are also food for zooplankton, and the phosphorus in the phytoplankton gets transferred to zooplankton.

Zooplankton can transfer their phosphorus in three ways:

- Zooplankton are eaten by fish at a predation rate of *y* day^{-1}
- Zooplankton also deposit phosphorus back into the water in the form of dissolved inorganic phosphorus in the form of excrement, according to
- Zooplankton also die, becoming detritus

Fish also contribute phosphorus back to the water through excrement and death.

Thus, phosphorus in the detritus can change in following six ways:

i. Regular contribution through death of fauna

ii. Regular contribution through death of flora

iii. Certain fauna, being detritus feeders, decrease the amount of phosphorus in the detritus by feeding on it

iv. Detritus also settles, becoming part of the sediment

v. Some of the phosphorus returns to the water through dissolution of phosphorus from the detritus

vi. Some of the phosphorus is taken up from the sediment, by microorganisms, which utilize the detritus as energy source (mineralization), there by releasing the bound P in water column

Sediment also returns phosphorus to the water (in addition to the changes in the sediment in its interaction with the detritus):

Construction of mathematical model for an aquatic ecosystem needs to consider both the systems, that is, lotic and lentic. The nutrient dynamics in both the systems vary considerably due to various factors such as, water level, flow, allochthonous input etc. Two typical models to assess the assimilative capacity in lentic and lotic systems are delineated hereunder.

Lentic system (stagnant water body):

Phosphorus (P) is a globally important pollutant of freshwater lakes and reservoirs. The quantitative information on both the external pollutant inputs (allochthonous) and the processes affecting pollutant dynamics within the ecosystem (autochthonous) are necessary for setting an appropriate guideline for total maximum load, which can be accommodated by the system. Setting a maximum load input for P generally requires identification of a water body concentration standard above which the uses of the resource are considered to be impaired. Empirical relationships or models then help to quantify the external loading rate, necessary to achieve the standard. It is necessitated to consider the extent to which the ecosystem can process and assimilate P into long-term (sediment) storage. Quantifying this lake "assimilative capacity" is difficult; as it is determined by a complex function of physical, chemical and biological processes that vary over time and space. Assimilative capacity is particularly difficult to quantify in lakes where there are many different types of sediment-water interactions. Nevertheless, a mass balance approach of relationships between P assimilation and maximum load input can be employed to predict the overall assimilative capacity of the ecosystem (Havens and Schelske, 2001).

For a lake or reservoir, the total mass input of P can be related to the mass output and net storage or assimilation as per Eqn. 1. Under steady state conditions, the maximum load can be determined by expressing output and storage terms as concentration times volume (Eqn. 2), and then substituting into the equation the in-lake P concentration goal (Eqn. 3 and 4). Here the outflow P concentration is assumed to be that of lake water.

$$M_{in}+M_{atmos}=M_{out}+M_{sed} \quad (1)$$

$$M_{in}+M_{atmos}=P_{lake}\times Q_{out}+P_{lake}\times A\times K_{net} \quad (2)$$

$$M_{in}+M_{atmos}=P_{target}\times Q_{out}+P_{target}\times A\times K_{net} \quad (3)$$

$$ML=P_{target}\times Q_{out}+P_{target}\times A\times K_{net} \quad (4)$$

In these equations, M_{in} is the P mass input (metric tons year^{-1}) from tributaries, M_{atmos} is the input from the atmosphere, M_{out} is the output to tributaries, and M_{sed} is the net loss of P to sediment storage. The term P_{lake} is the lake water P concentration (mg m^{-3}), Q_{out} is the outflow volume (km^3 year^{-1}), A is lake area (km^2), ML is the maximum load and K_{net} is the net P sedimentation rate (m year^{-1}). All of the model terms can be directly measured except K_{net}, which is calculated by solving Eqn. (2) using data from the lake's yearly budget. The M_{sed} term is rarely measured and is generally calculated by difference using Eqn. (1).

Based on the lake or reservoir P budget, a decrease or increase in K_{net} trend, seasonally or annually, an approximate assessment of P assimilative capacity can be calculated. In a virgin aquatic body, the decrease in K_{net} value suggests that the water body (lake or reservoir) has lost some of its capacity to assimilate P, which may be attributed to saturation of P binding sites on lake sediment particles and/ or other processes such as decline in microbial flora, which is responsible for effective recycling of P in the ecosystem. However, in a system where sediment is saturated with P, there will be increase of *Knet* due to constant supply of P through allochthonous sources. Gradually accumulation of P will increase in the detritus, causing altered abiotic characteristics of the surroundings, leading to incomplete microbial activity/ mineralization and there by ultimately reducing the assimilative capacity of the water body.

Understanding how K_{net} has changed over time and how it is anticipated to change under reduced P loads is critical to setting the ML. However, the linkage between K_{net} and P loads may be complicated where biological processes produce feedback loops and non-linear responses.

Biological processes that affect lake phosphorus assimilation: - A major loss process for P from the water column of lakes is the settling of plankton (Stabel, 1985). When P is deficient, as is the natural state of many freshwater lakes, very little soluble P remains in the water column and most occurs within the phytoplankton cells, or when there is heavy grazing pressure, in the tissues of zooplankton (Gulati *et al.*, 1992). Under these conditions, the P is contained in particles, which can readily transport it into sediment storage when they settle due to gravity. In contrast, lakes with many years of high P loading can develop a P surplus, where the supply of soluble P exceeds demands of phytoplankton. Under these circumstances some other factors, such as nitrogen (N) may become "secondarily limiting" (Schelske, 1984) and a large portion of the total P occurs in the dissolved form. The net loss of P from the water column is reduced, lowering a lake's K_{net}. Another process that can affect the net loss of P from a shallow lake's water column is storage of polyphosphates in the algal cells. The taxonomic composition of phytoplankton may also influence the extent to which P is lost from the water column by sedimentation, and hence the ability of a lake to assimilate external P loads. In future, this model could be modified to include: (1) the effects of submerged plant macrophyte; (2) a better consideration of the ecosystem's spatial variability and (3) other biological processes, which include interactions between various organisms.

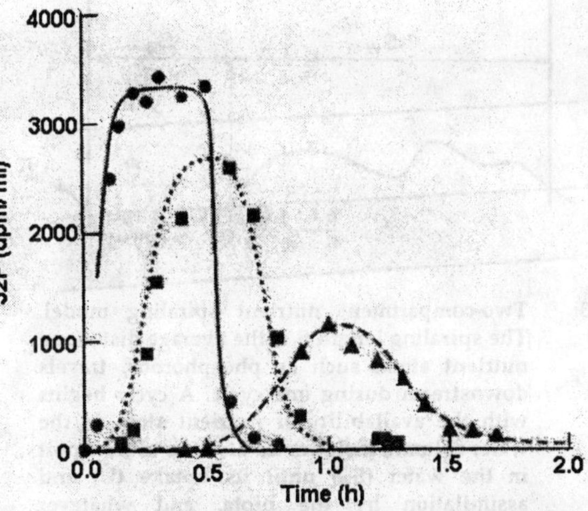

Fig. 2: **Concentration of ^{32}P in stream water at 15 (●), 47 (■) and 120 m (▲) locations downstream of a 30 min. release of radioactive phosphorous in a small woodland stream (From Newbold *et al.*, 1983).**

Lotic system (running water):

Nutrient dynamics in large rivers are less complex compared with small rivers because large rivers generally have low slopes, are deeper than the roughest bed feature and have relatively uniform and perhaps regulated flows. In the streams, solute or nutrient disperses from its point of release due to the unidirectional force of current (advection) and diffusion and turbulent mixing throughout the stream.

The nutrient spiraling concept: - In a stream ecosystem, nutrient cycling describes the passage of an atom or element from a phase where it exists as dissolved available nutrient, through its incorporation or assimilation into living tissue and passage through perhaps several links in the food chain, to its eventual release by excretion and decomposition and re-entry into the pool of dissolved available nutrients. In running waters, an atom or element occurring in the water column, as dissolved available nutrient, is transported some distance as a solute, then becomes incorporated into the biota, and eventually is returned to the water column in dissolved form or gets assimilated in any of the compartment of the system. Since the cycle involves downstream transport, it is described as the spiral (Webster and Patten, 1979).

The major requirement for the application of spiraling concept is a method to quantify the distance traveled by an atom in completing a cycle. One such method is the use of ^{32}P, where the known amount of ^{32}P is released from the point and is measured at the downstream locations. This will provide the data about the assimilation of P in various compartments of the system. Fig. 2 shows the disintegrations per minute (dpm) values in water at various distances. The higher dpm values at 15m indicate the assimilation of ^{32}P in the area near discharge.

Because the uptake of dissolved available nutrient from the water column is relatively easier to quantify than its subsequent release, most studies measure assimilation rate and distance. The theoretical framework for nutrient spiraling provides a useful measure of spiraling distance and thus allows investigation of the factors that influence this quantity.

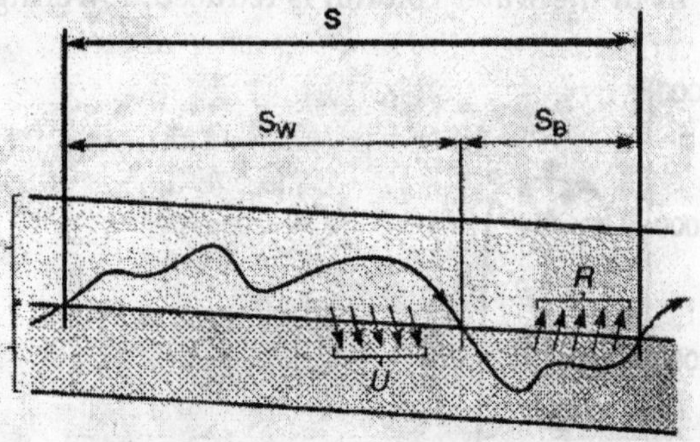

Fig.3: Two-compartment nutrient spiraling model. The spiraling length S is the average distance a nutrient atom, such as phosphorous, travels downstream during one cycle. A cycle begins with the availability of nutrient atom in the water column, includes its distance of transport in the water (S_W) until its uptake (U) and assimilation by the biota, and whatever additional distance the atom travels downstream within the biota (S_B) until that atom is eventually re-mineralized and released (after Newbold, 1992)

The nutrient spiraling concept model in its simplest form, considers, spiraling length (S, in meters) to the sum of transport in two compartments: water column, where the element occurs as inorganic solute; and the biota, where the element occurs in particulate form (Fig.3). Thus,

$$S = S_W + S_B$$

S_W is a measure of distance traveled by an atom as inorganic solute, from when it becomes available in the water column until its uptake and assimilation into biota. S_B is a measure of distance traveled by the atom within the biota until its eventual release back into the water column, thus completing the spiral passage downstream. The biota, generally associated with streambed, are attached microorganisms, periphyton and benthic invertebrates. Typically an atom will travel the greatest distance in the water column, and so S_W is expected to be greater than S_B. Low values (short distances) for S_W reflect greater demand for the nutrient in question indicating higher assimilative capacity of biota and more frequent cycling of each atom over a given length of stream.

Management of Aquatic Ecosystem

The above models with suitable site-specific modifications, if required, and validation would be immensely helpful for understanding the flow of elements or compounds and will aid in determining the assimilative capacity of various components of the system. Considering the various aspects of modeling and overall knowledge of the system, the components and various processes occurring in the system, and assimilation of pollutants (metals, nutrients etc.) in different compartments, can be evaluated. This information of assimilative capacity can then be used along with other management practices to mitigate the environmental problems.

A lot of effort is needed to put in for better management and conservation of the aquatic ecosystems. Various aspects, in relation to this, are considerations of developing an effluent management plan, which will aid in reducing the pollution load and will be helpful in keeping the ecosystem below its assimilative capacity threshold limits. Other considerations for developing an effluent management plan are:

- Consideration to improve production efficiency and reduction of generated waste per unit production
- Offering economic incentives for better management procedures (BMPs)
- Identification of beneficial bacteria (probiotics) for application in the ecosystem, so as to enhance *in situ* mineralization, thereby improving assimilative capacity of the water body
- Monitoring of systems to assess effectiveness, verify compliance, or convince concerned parties towards effectiveness
- Education and information exchange with the industries and the general public

Probiotics

The newest attempt to improve water quality is the application of probiotics. Probiotics refers to beneficial bacteria, which, when administered, help in better management of the assimilative capacity of an aquatic system. The organic matter load to water body generally disturbs the dynamic equilibrium in one-way or other. The assimilative capacity of a closeed ecosystem has some threshold limits, beyond which the accumulation of toxic compounds and pathogenic organisms persists in the system. Recent advances in biotechnology have come up with some techniques where cultures of beneficial bacterial are used to improve the ecosystem health. The central concept to manage any ecosystem is to manage its assimilative capacity, that is, to prevent the system components from saturation. The introduction of probiotic bacteria to the system would increase the potential of system to prevent accumulation of toxic compounds and pathogenic organisms. The probiotic bacteria compete with the pathogenic organisms for the substrate (organic matter) and there by eliminate them and return the inorganic nutrients for use of primary producers.

Probiotics and their role: - Few examples of probiotics with their role in various ecological processes (Rao, 2003) are as follows:

Bacillus sp. → Mineralization and breakage of proteins
Nitrosomonas sp. → Oxidation of ammonia
Nitrobacter sp. → Oxidation of nitrites
Aerobacter sp. → Reduction of organic matter
Cellulomonas sp. → Breakage of plant material

The administration of probiotics aids in less accumulation of organic matter at bottom of the water body, which helps in increasing the assimilative capacity of the system, and preventing the system to go beyond the threshold levels. Such biotechnological (bioremediation) approach, that involves manipulation of microorganisms in water bodies to enhance mineralization of organic matter and get rid of undesirable pollutants, appears to be most environment friendly and sustainable tool for conservation of ecosystem health.

Future Research Needs

The existing information, available to evaluate functioning of surface water bodies and estimation of assimilative capacities of different ecosystems with reference to wastewater discharges, are inadequate, and there is an ample scope, as delineated hereunder, to generate data in this regard.

- Improvement of methods of assessing assimilative capacity of wastewater receiving water bodies
- Investigation towards role of microorganisms on various kinds of effluents
- Identification and application of suitable probiotics for different wastewaters, so as to reduce the risk of pathogens and enhancing organic matter mineralization rates
- Development of models to evaluate and prioritize impacts from all activities and industries
- Establishment of database of various ecosystems with respect to their assimilative capacities
- Application of suitable methods to utilize resources from effluents by using probiotics, rather than simply discharging the waste in the surface waters

Summary

Rapid urbanization and industrialization is responsible for generation of considerable amounts of waste effluents including nutrients, pesticides, and organic and inorganic pollutants. These effluents can have undesirable impacts on the local environment, depending on the amounts released, the time-scale over which the release takes place, and the assimilation capacity and flushing ability of the local recipient water body.

The most important role of microorganisms in an aquatic environment is to accommodate the waste (organic matter) in the system and simultaneously its efficient recycling. Microorganisms can decompose variety of pollutants including excreta of fish, excess food materials and other organic materials to CO_2, nitrate and phosphate.

Considering the general properties of different modeling approaches, and available literature, an attempt to evaluate phosphorus flow (single element, single component) through different compartments/ trophic levels of aquatic ecosystem could be an ideal approach for construction of a mathematical model. Two typical models to assess the assimilative capacity in lentic and lotic systems have been explained in detail. The models with suitable site-specific modifications, if required, and validation would be immensely helpful for understanding the flow of elements or compounds and will aid in determining the assimilative capacity of various components of the system. Considering the various aspects of modeling and overall knowledge of the system, the components and various processes occurring in the system, and assimilation of pollutants (metals, nutrients etc.) in different compartments, can be evaluated.

References

Aggarwal, P.K. and Hinchee, R.E. (1991). Monitoring in situ biodegradation of hydrocarbons by using stable isotopes. Environ. Sci. Technol. 25, 1178–1180.

Aubert, C., Lojou, E., Bianco, P., Rousset, M., Durand, M.-C., Bruschi, M. and Dolla, A., (1998). The *Deşulfuromonas acetoxidans* triheme cytochrome c_7 produced in *Desulfovibrio desulfuricans* retains its metal reductase activity. *Appl. Environ. Microbiol.* **64**, pp. 1308–1312.

Axler, R. Larsen, C. Tikkanen, C. McDonald, M. Yokom, S. and Aas, P. (1996) Water quality issues associated with aquaculture: a case study in mine pit lakes. *Water Environ. Res.* **68**, pp. 995–1011.

Baker, R. A. (ed.), (1980). Contaminants and Sediments, Volume 1: Fate and Transport, Case Studies, Modeling, Toxicity. Ann Arbor, Mich.: Ann Arbor Science Publishers, 558 pp.

Barnes, L.J., Janssen, F.J., Sherren, J., Versteegh, J.H., Koch, R.O. and Scheeren, P.J.H., (1991). A new process for the microbial removal of sulphate and heavy metals from contaminated waters extracted by a geohydrological control system. *Trans. Inst. Chem. Eng.* **69**, pp. 184–186.

Barnes, L.J., Scheeren, P.J.M. and Buisman, C.J.N., (1994). Microbial removal of heavy metals and sulphate from contaminated groundwaters. In: Means, J.L. and Hinchee, R.E., Editors, 1994. *Emerging Technol. for Bioremediation of Metals*, Lewis Publishers, Boca Raton, pp. 38–49.

Beech, I.B. and Cheung, C.W.S., (1995). Interactions of exopolymers produced by sulphate-reducing bacteria with metal ions. *Int. Biodeterior. Biodegrad.* **35**, pp. 59–72.

Berner, R.A. (1980) Early Diagenesis: a Theoretical Approach. , Princeton University Press, Princeton p. 241.

Beveridge, T.J. and Doyle, R.J. (1989). Metal Ions and Bacteria. , Wiley, New York.

Birch, L. and Bachofen, R., (1990). Complexing agents from microorganisms. *Experientia* **46**, pp. 827–834.

Black, K.D. Editor, *Environmental Impacts of Aquaculture*, Sheffield Academic Press, Sheffield, UK (2001) 214 pp.

Bosecker, K., (1997). Bioleaching: metal solubilization by microorganisms. *FEMS Microbiol. Rev.* **20**, pp. 591–604.

Bosshard, P.P., Bachofen, R. and Brandl, H., (1996). Metal leaching of fly-ash from municipal waste incineration by *Aspergillus niger*. *Environ. Sci. Technol.* **30**, pp. 3066–3070.

Bousserrhine, N., Gasser, U.G., Jeanroy, E. and Berthelin, J. (1999). Bacterial and chemical reductive dissolution of Mn-, Co-, Cr-, and Al-substituted geothites. *Geomicrobiol. J.* **16**, pp. 245–258.

Brandl, H., Bosshard, R. and Wegmann, M., (1999). Computer-munching microbes: metal leaching from electronic scrap by bacteria and fungi. *Process Metall.* **9B**, pp. 569–576.

Burgstaller, W. and Schinner, F. (1993). Leaching of metals with fungi. *J. Biotechnol.* **27**, pp. 91–116.

Caswell, H. (1988). Theory and models in ecology: a different perspective. Ecological and supercomputer models. *Ecological Modelling* **43** pp. 111-124.

Chang, J.S., Hwang, Y.P., Fong, Y.M. and Lin, P.J. (1999). Detoxification of mercury by immobilized mercuric reductase. *J. Chem. Technol. Biotechnol.* **74**, pp. 965–973.

Chen, W., Brühlmann, F., Richins, R.D. and Mulchandani, A. (1999). Engineering of improved microbes and enzymes for bioremediation. *Curr. Opin. Biotechnol.* **10**, pp. 137–141.

Christensen, B., Laake, M. and Lien, T., (1996). Treatment of acid-mine water by sulfate-reducing bacteria—results from a bench-scale experiment. *Water Res.* **30**, pp. 1617–1624.

Coffin, R.B., Cifuentes, L.A. and Pritchard, P.H. 1997. Effect of remedial nitrogen applications on algae and heterotrophic organisms on oil contaminated beaches in Prince William Sound, AK. *Mar. Environ. Res.* **1**, 27–39.

Coffin, R.B., Fry, B., Peterson, B.J. and Wright, R.T. (1989). Carbon isotopic compositions of estuarine bacteria. *Limnol. Oceanogr.* **34**, 1305–1310.

DePinto, J. V., Lick, W., and Paul, W. F. (1993). Transport and Transformation of Contaminants Near the Sediment—Water Interface. Boca Raton, Fla.: CRC Press, 339 pp.

Diels, L., DeSmet, M., Hooyberghs, L. and Corbisier, P., (1999). Heavy metals bioremediation of soil. *Mol. Biotechnol.* **12**, pp. 149–158.

Eccles, H., (1999). Treatment of metal-contaminated wastes: why select a biological process?. *Trends Biotechnol.* **17**, pp. 462–465.

Engler, R. M. (1980). Prediction of pollution potential through geochemical and biological procedures: development of regulation guidelines and criteria for the discharge of dredged and fill material. In Baker, R. A. (ed.), Contaminants and Sediments, Volume 1: Fate and Transport, Case Studies, Modeling, Toxicity. Ann Arbor, Mich.: Ann Arbor Science Publishers, pp. 143–69.

Ewart, D.K. and Hughes, M.N. (1991). The extraction of metals from ores using bacteria. *Adv. Inorg. Chem.* **36**, pp.103–135.

Fahrig, L. (1988). Nature of ecological theories. *Ecological Modelling* **43** pp.129-132.

Finneran, K.T., Housewright, M.E. and Lovley, D.R., (2002). Multiple influences of nitrate on uranium solubility during bioremediation of uranium-contaminated subsurface sediments. *Environ. Microbiol.* **4**, pp. 510–516.

Flemming, H.-K., (1995). Sorption sites in biofilms. *Water Sci. Technol.* **32**, pp. 27–33.

Fogel, M.L., Sprague, E.K., Gize, A.P. and Frey, R.W. (1989). Digenesis of organic matter in Georgia salt marshes. *Estuarine Coastal Shelf Sci.* **28**, 211–230.

Francis, A.J., Dodge, C.J. and Gillow, J.B., (1992). Biodegradation of metal citrate complexes and implications for toxic metal mobility. *Nature* **356**, pp. 140–142.

Franz, A., Burgstaller, W. and Schinner, F. (1991). Leaching with *Penicillium simplicissimum*: influence of metals and buffers on proton extrusion and citric acid production. *Appl. Environ. Microbiol.* **57**, pp. 769–774.

Franz, A., Burgstaller, W., Muller, B. and Schinner, F. (1993). Influence of medium components and metabolic inhibitors on citric acid production by *Penicillium simplicissimum*. *J. Gen. Microbiol.* **139**, pp. 2101–2107.

Fry, B. (1986). Sources of carbon and sulfur nutrition for consumers in three meromictic lakes of New York State. Limnol. Oceanogr. 31, 79–88.

Gadd, G.M. and Sayer, J.A. (2000). Fungal transformations of metals and metalloids. In: Lovley, D.R., Editor, , 2000. *Environmental Microbe–Metal Interactions*, Am. Soc. Microbiol, Washington, pp. 237–256.

Gadd, G.M. (1999). Fungal production of citric and oxalic acid: importance in metal speciation, physiology and biogeochemical processes. *Adv. Microb. Physiol.* **41**, pp. 47–92.

Gadd, G.M. (2001). Accumulation and transformation of metals by microorganisms. In: Rehm, H.-J., Reed, G., Puhler, A. and Stadler, P., Editors, 2001. *Biotechnology, A Multi-volume Comprehensive Treatise Volume 10: Special Processes*, Wiley-VCH Verlag, Weinheim, Germany, pp. 225–264.

Gadd, G.M., (2002). Interactions between microorganisms and metals/radionuclides: the basis of bioremediation. In: Keith-Roach, M.J. and Livens, F.R., Editors, 2002. *Interactions of Microorganisms with Radionuclides*, Elsevier, Amsterdam, pp. 179–203.

Gao, K. and McKinley, K.R. (1994) Use of macroalgae for marine biomass production and CO_2 remediation—a review. *J. Appl. Phycol.* **6**, pp. 45–60.

GESAMP (IMO/FAO/UNESCO-IOC/WMO/WHO/IAEA/UN/UNEP Joint Group of Experts on the Scientific Aspects of Marine Environmental Protection), (1986). Environmental capacity: an approach to marine pollution prevention. Rome, Italy. Rep. Stud. GESAMP No. 30, 49 pp.

Gharieb, M.M, Kierans, M. and Gadd, G.M. (1999). Transformation and tolerance of tellurite by filamentous fungi: accumulation, reduction and volatilization. *Mycol. Res.* **103**, pp. 299–305.

Hadley, R. F., and Ongley, E. D. (eds), (1989). Sediment and the Environment. Proc. Symp. Baltimore, Maryland, 18–19 May 1989. Publication no. 184, Washington, DC: International Association of Hydrological Sciences, 218 pp.

Hall, C. A. S. and J. Day (editors). (1977). Ecosystem Modeling in Theory and Practice.

Hammack, R.W. and Edenborn, H.M., (1992). The removal of nickel from mine waters using bacterial sulphate-reduction. *Appl. Microbiol. Biotechnol.* **37**, pp. 674–678.

Harremoes, P., (1982). Criteria for nitrification in fixed film reactors. *Water Sci. Tech.* **14**, pp. 167–185.

Hedges, J.I. and Keil, R.G. (1995). Sedimentary organic matter preservation: an assessment and speculative synthesis. *Mar. Chem.* **49**, pp. 81–115.

Hedin, R.S. and Nairn, R.W., (1991). Contaminant removal capabilities of wetlands constructed to treat coal mine drainage. In: Moshiri, G.A., Editor, , 1991. *Proceedings of the International Symposium on Constructed Wetlands for Water-Quality Improvement*, Lewis Publishers, Chelsea MI, pp. 187–195.

Hirata, H. Yamasaki, S. Maenosono, H. Nakazono, T. Yamauchi, T. and Matsuda, M. (1994). Relative budgets of pO_2 and pCO_2 in cage polycultured red sea bream, *Pagrus major* and sterile *Ulva* sp. *Suisanzoshoku* **42**, pp. 377–381.

Hobman, J.L., Wilson, J.R. and Brown, N.L. (2000). Microbial mercury reduction. In: Lovley, D.R., Editor, , 2000. *Environmental Microbe–Metal Interactions*, Am. Soc. Microbiol, Washington, pp. 177–197.

Howe, R., Evans, R.L. and Ketteridge, S.W. (1997). Copper-binding proteins in ectomycorrhizal fungi. *New Phytol.* **135**, pp. 123–131.

Hullar, M.A.J., Fry, B., Peterson, B.J. and Wright, R.T. (1996). Microbial utilization of estuarine dissolved organic carbon: a stable isotope tracer approach tested by mass balance. *Appl. Environ. Microbiol.* **62**, 2489–2493.

Karlson, U. and Frankenberger, W.T., (1993). Biological alkylation of selenium and tellurium. In: Sigel, H. and Sigel, A., Editors, 1993. *Metal Ions in Biological Systems*, Marcel Dekker, New York, pp. 185–227.

Kelly, L.A. Stellwagen, J. and Bergheim, A. (1996). Waste loadings from a fresh-water Atlantic Salmon farm in Scotland. *Water Res. Bull.* **32**, pp. 1017–1025.

Knowles, R. (1982). Denitrification. *Microbiol. Rev.* **46**, pp. 43-70.

Krom, M.D. (1986). An evaluation of the concept of assimilative capacity as applied to marine waters. *Ambio* **15**, pp. 208–214.

Lindeman, R. L. (1942). The trophic-dynamic aspect of ecology. *Ecology* **23**, pp. 399-418.

Lloyd, J.R. and Lovley, D.R. (2001). Microbial detoxification of metals and radionuclides. *Curr. Opin. Biotechnol.* **12**, pp. 248–253.

Lloyd, J.R. and Macaskie, L.E., (1998). Enzymatic recovery of elemental palladium using sulfate-reducing bacteria. *Appl. Environ. Microbiol.* **64**, pp. 4607–4609.

Lovley, D.R. and Coates, J.D. (1997). Bioremediation of metal contamination. *Curr. Opin. Biotechnol.* **8**, pp. 285–289.

Lovley, D.R., 2000. Fe(III) and Mn(IV) reduction. In: Lovley, D.R. Editor, (2000). *Environmental Microbe–Metal Interactions*, Am. Soc. Microbiol, Washington, pp. 3–30.

Lovley, D.R. (2001). Anaerobes to the rescue. *Science* **293**, pp. 1444–1446.

Lyew, D., Knowles, R. and Sheppard, J. (1994). The biological treatment of acid-mine drainage under continuous-flow conditions in a reactor. *Process Saf. Environ. Prot.* **72**, pp. 42–47.

Macaskie, L.E. (1991). The application of biotechnology to the treatment of wastes produced by the nuclear fuel cycle—biodegradation and bioaccumulation as a means of treating radionuclide-containing streams. *Crit. Rev. Biotechnol.* **11**, pp. 41–112.

Macaskie, L.E., Jeong, B.C. and Tolley, M.R. (1994). Enzymically-accelerated biomineralization of heavy metals: application to the removal of americium and plutonium from aqueous flows. *FEMS Microbiol. Rev.* **14**, pp. 351–368.

MacGarvin, M. (2000). Scotland's Secret? Aquaculture, Nutrient Pollution, Eutrophication and Toxic Blooms. , World Wildlife Fund, Perth, Scotland 21 pp.

McIsaac, G. F., Hirschl, M. C., and Mitchell, J. K. Nitrogen and phosphorus in eroded sediment from corn and soybean tillage systems. In Hadley, R. F., and Ongley, E. D. (eds), Sediment and the Environment. Proc. Symp. Baltimore, Maryland, 18–19 May 1989. Publication no. 184, Washington, DC: International Association of Hydrological Sciences, pp. 3–10.

McLean, J. and Beveridge, T.J. (2001). Chromate reduction by a pseudomonad isolated from a site contaminated with chromated copper arsenate. *Appl. Environ. Microbiol.* **67**, pp. 1076–1084.

McLean, J.S., Lee, J.U-. and Beveridge, T.J. (2002). Interactions of bacteria and environmental metals, fine-grained mineral development, and bioremediation strategies. In: Huang, P.M., Bollag, J.-M. and Senesi, N., Editors, 2002. *Interactions Between Soil Particles and Microorganisms*, Wiley, New York, pp. 227–261.

Meadows, D. L. Randers, J. and Behrens, W. W. The Limits to Growth. Universe Books, (1972). New York.

Meier-Augenstein, W. (1995). On line recording of $^{13}C/^{12}C$ ratios and mass spectra in one gas chromatographic analysis. *High Resol. Chromatogr*. **18**, 28–32.

Mercier, G., Chartier, M., Couillard, D. and Blais, J.F. (1999). Decontamination of fly ash and used lime from municipal waste. *Environ. Manage.* **24**, pp. 517–528.

Messer, J. and Brezonik, P. L. 1983. Comparison of denitrification rate estimatin techniques in a large, shallow lake. *Water Res.* **17**, 631- 640.

Midlen, A. and Redding, T. Environmental Management for Aquaculture. , Chapman & Hall, London, UK (1998) 223 pp.Olsen, L. A., 1984. Effects of Contaminated Sediment on Fish and Wildlife: Review and Annotated Bibliography. FWS/OBS-82/66. Washington, DC: US Fish and Wildlife Service, 103 pp.

Oremland, R. and Stolz, J. (2000). Dissimilatory reduction of selenate and arsenate in nature. In: Lovley, D.R., Editor, 2000. *Environmental Microbe–Metal Interactions*, Am. Soc. Microbiol, Washington, pp. 199–224.

Oremland, R.S., Steinberg, N.A., Maest, A.S., Miller, L.G. and Hollibaugh, J.T. (1990). Measurement of in situ rates of selenate removal by dissimilatory bacterial reduction in sediments. *Environ. Sci. Technol.* **24**, pp. 1157–1164.

Oremland, R.S., Steinberg, N.A., Presser, T.S. and Miller, L.G., (1991). In situ bacterial selenate reduction in the agricultural drainage systems of Western Nevada. *Appl. Environ. Microbiol.* **57**, pp. 615–617.

Perry, K.A. (1995). Sulfate-reducing bacteria and immobilization of metals. *Mar. Georesour. Geotechnol.* **13**, pp. 33–39.

Peterson, B.J., Howarth, R.W. and Garritt, R.H. (1985). Multiple stable isotopes used to trace the flow of organic matter in estuarine food webs. *Science* **227**, 1361–1363.

Phillips, E.J.P., Landa, E.R. and Lovley, D.R. (1995). Remediation of uranium contaminated soils with bicarbonate extraction and microbial U(VI) reduction. *J. Indust. Microbiol.* **14**, pp. 203–207.

Rai, L.C., Kumar, H.D., Mohn, F. H. and Soeder, C.J. (2000). Services of algae to the environment. *J. Microbiol. Biotechnol.* **10**, pp. 119–136.

Rashid, M.A. (1985). Geochemistry of marine humic compounds. , Springer-Verlag, New York.

Rauser, W.E., (1995). Phytochelatins and related peptides. *Plant Physiol.* **109**, pp. 1141–1149

Rawlings, D.E. and Silver, S. (1995). Mining with microbes. *Biotechnology* **13**, pp. 773–1338.

Rawlings, D.E. (1997). Mesophilic, autotrophic bioleaching bacteria: description, physiology and role. In: Rawlings, D.E., Editor, , 1997. *Biomining: Theory, Microbes and Industrial Processes*, Springer-Verlag, Berlin, pp. 229–245.

Roughgarden, J., R. M. May and S. A. Levin (editors). (1989). *Perspectives in Ecological Modelling* **43** pp. 33-44.

Sayer, J.A. and Gadd, G.M. (2001). Binding of cobalt and zinc by organic acids and culture filtrates of *Aspergillus niger* grown in the absence or presence of insoluble cobalt or zinc phosphate. *Mycol. Res.* **105**, pp. 1261–1267.

Sayer, J.A., Cotter-Howells, J.D., Watson, C., Hillier, S. and Gadd, G.M. (1999). Lead mineral transformation by fungi. *Curr. Biol.* **9**, pp. 691–694.

Schinner, F. and Burgstaller, W. (1989). Extraction of zinc from an industrial waste by a *Penicillium* sp. *Appl. Environ. Microbiol.* **55**, pp. 1153–1156.

Schippers, A. and Sand, W. (1999). Bacterial leaching of metal sulfides proceeds by two indirect mechanisms via thiosulfate or via polysulfides and sulphur. *Appl. Environ. Microbiol.* **65**, pp. 319–321.

Schiewer, S. and Volesky, B. (2000). Biosorption processes for heavy metal removal. In: Lovley, D.R., Editor, *Environmental Microbe–Metal Interactions*, Am. Soc. Microbiol, Washington, pp. 329–362.

Seitzinger, S. P. (1988). Denitrification in freshwater and coastal marine ecosystems: ecological and geochemical significance. *Limnol. Oceanog.* 33, 702-724.

Silver, S. (1996). Bacterial resistances to toxic metal ions—a review. *Gene* **179**, pp. 9–19.

Silver, S. (1998). Genes for all metals—a bacterial view of the periodic table. *J. Ind. Microbiol. Biotech.* **20**, pp. 1–12.

Smith, W.L. and Gadd, G.M. (2000). Reduction and precipitation of chromate by mixed culture sulphate-reducing bacterial biofilms. *J. Appl. Microbiol.* **88**, pp. 983–991.

Southam, G. (2000). Bacterial surface-mediated mineral formation. In: Lovley, D.R., Editor, *Environmental Microbe–Metal Interactions*, Am. Soc. Microbiol, Washington, pp. 257–276.

Sreekrishnan, T.R. and Tyagi, R.D. (1994). Heavy metal leaching from sewage sludges: a techno-economic evaluation of the process options. *Environ. Technol.* **15**, pp. 531–543.

Stephen, J.R. and Macnaughton, S.J., 1999. Developments in terrestrial bacterial remediation of metals. *Curr. Opin. Biotechnol.* **10**, pp. 230–233.

Stolz, J.F. and Oremland, R.S. (1999). Bacterial respiration of arsenic and selenium. *FEMS Microbiol. Rev.* **23**, pp. 615–627.

Stork, A., Jury, W.A. and Frankenberger, W.T. (1999). Accelerated volatilization rates of selenium from different soils. *Biol. Trace Element Res.* **69**, pp. 217–234.

Strasser, H., Burgstaller, W. and Schinner, F. (1994). High yield production of oxalic acid for metal leaching purposes by *Aspergillus niger*. *FEMS Microbiol. Lett.* **119**, pp. 365–370.

Sukla, L.B., Kar, R.N. and Panchanadikar, V. (1992). Leaching of copper converter slag with *Aspergillus niger* culture filtrate. *BioMetals* **5**, pp. 169–172.

Tamaki, S. and Frankenberger, W.T. (1992). Environmental biochemistry of arsenic. *Rev. Environ. Contam. Toxicol.* **124**, pp. 79–110.

Tebo, B.M. and Obraztsova, A.Y. (1998). Sulfate-reducing bacterium grows with Cr(VI), U(VI), Mn(IV), and Fe(III) as electron acceptors. *FEMS Microbiol. Lett.* **162**, pp. 193–198.

Thompson-Eagle, E.T. and Frankenberger, W.T., (1992). Bioremediation of soils contaminated with selenium. In: Lal, R. and Stewart, B.A., Editors, 1992. *Advances in Soil Science*, Springer-Verlag, New York, pp. 261–309.

Tissot, B.P. and Welte, D.H. (1984). Petroleum formation and occurrence. second ed. , Springer-Verlag, Heidelberg.

Tobin, J., White, C. and Gadd, G.M. (1994). Metal accumulation by fungi: applications in environmental biotechnology. *J. Indust. Microbiol.* **13**, pp. 126–130.

Tolley, M.R., Strachan, L.F. and Macaskie, L.E. (1995). Lanthanum accumulation from acidic solutions using a *Citrobacter* sp. immobilised in a flow-through bioreactor. *J. Indust. Microbiol.* **14**, pp. 271–280.

Tsezos, M. and Volesky, B. (1982). The mechanism of uranium biosorption by *Rhizopus arrhizus*. *Biotechnol. Bioeng.* **24**, pp. 385–401.

Tzeferis, P.G., Agatzini, S. and Nerantzis, E.T. (1994). Mineral leaching of non sulphide nickel ores using heterotrophic micro-organisms. *Lett. Appl. Microbiol.* **18**, pp. 209–213.

Uhrie, J.L., Drever, J.I., Colberg, P.J.S. and Nesbitt, C.C. (1996). In situ immobilisation of heavy metals associated with uranium leach mines by bacterial sulphate reduction. *Hydrometallurgy* **43**, pp. 231–239.

Vachon, P.R., Tyagi, D., Auclair, J.C. and Wilkinson, K.J. (1994). Chemical and biological leaching of aluminium from red mud. *Environ. Sci. Technol.* **28**, pp. 26–30.

Van Luijn, f., Boers, P. C. M., Lijklema, L. and Sewwrts, J. P. R. A. (1999). Nitrogen fluxes and processes in sandy and muddy sediments from a shallow eutrophic lake. *Water Res.* **33**(1), 33-42.

Vieira, M.J. and Melo, L.F. (1995). Effect of clay particles on the behaviour of biofilms formed by *Pseudomomonas fluorescens*. *Water Sci. Technol.* **32**, pp. 45–52.

Wang, Y. T. (2000). Microbial reduction of chromate. In: Lovley, D.R., Editor, (2000). *Environmental Microbe–Metal Interactions*, Am. Soc. Microbiol, Washington, pp. 225–235.

Webster, J. R. and Patten, B. c. (1979) Effects of watershed perturbation on stream potassium and calsium dynamics. *Ecol. Mongr.*, **49**, pp. 51-72.

White, C. and Gadd, G.M. (1997). An internal sedimentation bioreactor for laboratory-scale removal of toxic metals from soil leachates using biogenic sulphide precipitation. *J. Indust. Microbiol.* **18**, pp. 414–421.

White, C. and Gadd, G.M. (2000). Copper accumulation by sulphate-reducing bacterial biofilms and effects on growth. *FEMS Microbiol. Lett.* **183**, pp. 313–318.

White, C., Sharman, A.K. and Gadd, G.M. (1998). An integrated microbial process for the bioremediation of soil contaminated with toxic metals. *Nature Biotechnol.* **16**, pp. 572–575.

Yong, P. and Macaskie, L.E. (1995). Enhancement of uranium bioaccumulation by a *Citrobacter* sp. via enzymically mediated growth of polycrystalline $NH_4UO_2PO_4$. *J. Chem. Technol. Biotechnol.* **63**, pp. 101–108.

Zhang, Y.Q. and Frankenberger, W.T. (1999). Effects of soil moisture, depth, and organic amendments on selenium volatilization. *J. Environ. Qual.* **28**, pp. 1321–1326.

Zinkevich, V., Bogdarina, I., Kang, H., Hill, M.A.W., Tapper, R. and Beech, I.B. (1996). Characterization of exopolymers produced by different isolates of marine sulphate-reducing bacteria. *Int. Biodeterior. Biodegrad.* **37**, pp. 163–172.

27

Exploitation of Microbial Diversity for Bioremediation Using Genetic Tools

Naveen Kumar, Rekha Sharma, Bharani Kumar, S.P.S.Ahlawat & Dinesh Pandey

Introduction

Bioremediation has become an accepted and important technology for restoration of contaminated environment. More over the application of bioremediation to contaminated sites is an important step in preventing further pollution of outlying ecosystem in close proximity to the polluted site. The goal of bioremediation is to stimulate microbes with nutrients and other chemicals that will enable them to destroy the contaminants. The use of microbial metabolic potential for eliminating environmental pollutants provides a safe and economic alternative to their disposal in waste dumpsites and to commonly used physicochemical strategies. Microorganisms capable of mineralizing a variety of toxic compounds under laboratory conditions have been isolated. The accumulation in the environment of highly toxic and persistent compounds however, emphasizes the fact that the natural metabolic diversity of the autochthonous microbes is insufficient to protect the biosphere from anthropogenic pollution. Various genetic approaches have been developed and used to optimize the enzymes, metabolic pathways and organisms relevant for biodegradation. New information on the metabolic routes and bottlenecks of degradation is still accumulating, enlarging the available toolbox. With molecular methods allowing the characterization of microbial community structure and activities, the performance of microorganisms under *in situ* conditions and in concert with the indigenous microflora will become predictable.

Pollutants

Rising environmental contamination day-by-day is causing the ecological disturbances and risks of health problems. The pollutants may include numerous organic and inorganic substances such as municipal sewage, petroleum products, gaseous emissions, hazardous wastes from industries, chemicals, fertilizers, heavy metals, radioactive materials causing land and water unusable (Keith and Telliard, 1976). The entry of these toxic substances into the food chain finally claims the survival of the living beings.

Major pollutants includes sewage sludge, petroleum products (chain and aromatic hydrocarbons like Polyaromatic hydrocarbons; PAHs), industrial wastes from paper & pulp industry like chlorolignins, chlorophenolic residues, lignocelluloses residues, highly toxic dioxin and furans, textile & tannery industry contaminants like colored carcinogenic azo dyes, aromatic halogen contaminants. The soil and water pollutants also includes insecticides (Asana, Guthion, Lannate, Lorsban and Penncap-M), herbicides, pesticides, fertilizers, explosive materials such as TNT-Trinitro toluene, RDX-Hexahydro-1,3,5-trinitro-1,3,5-Triazine, 2,4-dinitrotoluene) and many other toxic metals and metalloids.

The textile and tannery industry effluents are with carcinogenic non-degradable azo dyes and are also affecting the photosynthesis of the aquatic plants. Pulp and paper industry effluent pollution reported the cases of poisoning of the aquatic life and people through aquatic food. Fertilizers, pesticides used in the crop production, heavy metals from industrial wastes, and few organic, inorganic compounds are hazardous and causative of health problems (Srivastava and Dumka 2001).

Metal contamination in the soil is another prevailing intensifying problem in the environment. The various metal pollutants and their effects are mentioned in Table 1.

Table 1 : Metal pollutants and related health problems

Metal	Source	Effects
Arsenic	Occurs naturally	Chronic poisoning leads to loss of appetite & weight, diarrhoea, alternating with constipation, gastro intestinal disturbances, peripheral neuritis, conjunctivitis and sometimes skin cancer.
Cadmium	Mining, metallurgy, chemical industry and electroplating	Lead to chronic poisoning and affects the proximal tubules of the kidney, causing formation of kidney stones.
Lead	Lead smelters storage battery	Lead poisoning can lead to severe mental retardation or death.
Mercury	Industrial wastes	Methyl mercury compounds are much more toxic than other forms of mercury, causes neurological problems and damages renal glomeruli and tubules.
Cyanides	Waste from heat treatment of metals, dismantling of electro plating shops, etc.	Rapid death may follow due to exposure to cyanide as a result of inhibition of cellular respiration.

Remediation

The transformation of toxic chemicals or substances to less toxic or unharmful substances is the process of remediation. Environmental remediation technologies use physical, chemical or biological processes to eliminate or reduce the effect of contaminants. The physical methods involved in treatment of the pollutants are land filling, thermal desorption, incineration, soil vapour extraction etc. Chemical methods include soil flushing, solidification, solvent extraction, dehalogenation, chemical reduction and oxidation. Biological methods comprises of bioremediation using the microorganisms, phytoremediation, bioreactors and land farming. The various physicochemical methods presently under use for the degradation of the contaminants are effective but are expensive, time consuming and are also have few drawbacks. The land filling and incineration methods used in treatment of soil pollutants are highly effective but are expensive. The incineration causes air pollution and land farming spreads soil contamination by moving the contaminated soil from one location to another. Employing the usage of microbes in cleaning up of the environment is faster, efficient, safer, less expensive and is devoid of above drawbacks.

Bioremediation

The use of biological systems in treatment of the environment is known as bioremediation. Biodegradation leaves less toxic, stable chemical forms of contaminants along with the byproducts like water, Carbon-di-oxide, Methane and Hydrogen (Head, 1998). The process is beneficial to environment as it minimizes the environmental damage and removes toxic chemicals. It is widely accepted and popular technology and is environmental friendly for restoration of the contaminated environments.

The process of bioremediation include *bioaugmentation* in which microbes and nutrients are added to the contaminated sites and *biostimulation* is the process in which nutrients and enzymes are added to stimulate intrinsic microbes. Many microorganisms use the pollutants as substrate for their growth and biodegrade the pollutants completely or transform them into intermediate compounds that can be less toxic than the parent compound.

Microbial degradation of organic pollutants can be due to direct metabolism or an indirect effect of microbes on the chemical or physical environment, resulting in a secondary transformation reaction. The efficiency of the microbes in the bioremediation depends on different enzymatic pathways followed by them to degrade the hydrocarbons and the genes encoding or regulating the enzymes in degradation forms the main criteria for the efficiency of the microbes participating in the biodegradation (Samanta *et al.*, 2002).

The key processes involved in the microbial transformation of organic pollutants are biodegradation or *mineralization, cometabolism, polymerization* or *conjugation, accumulation* and *secondary effects of microbial activity*. Primarily bacteria and fungi mediate the degradation of organic materials in natural environments (Table 2). Degradation of organic pollutants has been investigated mostly in bacteria. It has also been proved that xenobiotic degradation by the lignin degrading system of the white rot fungi has several advantages compared to bacterial processes.

Recent studies reveal the promising performance of the Algae mainly cyanobacteria in bioremediation of metals and few fungi like rhizopus in degradation of chromium. The use of algae and plants in the remediation (Phytoremediation) are with successful outputs in degradation of the metals and other contaminants.

Table 2 : Common organisms involved in Bioremediation

Contaminants	Microorganisms involved in bioremediation
Aromatic compounds	*Arthobacter, Achrmobacter, Aspergillus, Bacillus, Fusarium, Phanerochate Pseudomonas, Penicillium,*
Azo Dyes	*Geotrichum candidum sp., Klebsiella, rhizopus sp., white* rot fungi like *Phanerochaete sp.*
Cadmium	*Staphylococcus, Bacillus, Pseudomonas, Citrobacter, Klebsiella, Rhodococcus*
Copper	*Escherichia, Pseudomonas*
Chromium	*Alcaligenes, Pseudomonas*
Cyanide	*Fusarium sp*
Petroleum	*Bacillus, Cunninghamella, Pseudomonas, Proteus, Penicillium*
Pentachlorophenol	*Sphingomona sp.*
Sulphur	*Thiobacillus*
Uranium	*Rhizopus sp., Penicillium chrysogenum, Saccharomyces*

Microbial performance

Many synthetic and organic compounds are highly toxic to the higher animals and plants, which can be digested by microbes. Nevertheless the widespread distribution of microbes, thousands of hectares of land and large stretch of water still remain contaminated. The reason may be the absence or low number of microbes or genes responsible for degradation or may be due to unfavourable chemical conditions for microbial growth, activity and inadequate nutrition. The microbial growth, performance of the microbes in degradation of the pollutants such as sewage, metal contaminated soils differs basing on different parameters like contaminant concentration, presence of the other impurities, pH, temperature etc.

New molecular methods of probing the microbes involved in soil decontamination are being used to trace and examine microorganisms involved in bioremediation. Improved understanding of the relevant microbial populations could result in more efficient and reproducible systems for disposal. The microorganisms, inspite of presence of toxic anti microbial agents, are able to use the components of metal working fluids as nutrients and play main role in biodigestion of them. Bioreactors seeded with microorganisms from sewage are used in disposal procedures, which are having disadvantage of loss of microbial activity and diversity. It is better to select optimum source of inocula to seed bioreactors only after studying various surviving and activity parameters of the microbe. This approach allows collecting systematically bacteria resistant to various chemical components present in contaminated sites or reactors and digests the contaminants.

The presence of the microbial population in the contaminated site doesn't pave the way for the clean up of the site as few constraints like chemical and environmental conditions may also affect the degradation process. The parameters like pH, temperature, catalytic components in the soil, nutrient concentration also play major role in the process of biodegradation. Once the parameters significant for the growth of the microbial population are present then it is concerned with the presence of microbes responsible for the contaminant degradation.

Genetically modified organisms (GMOs)

The genetically modified organisms created the genetic approach; rational design of the genes provides the effective strategy for the bioremediation and better applications. The application of the genetically engineered microorganisms for bioremediation is in the constructing potential microorganisms by introducing the selected foreign genes responsible for coding enzymes involved in bioremediation (Dietmar and Reineke, 2000)

In the degradation of the Polychlorinated Biphenyls (PCB), the enzyme biphenyl dioxygenase (Bph Dox) encoded by bphA1 of the halorespiring bacteria plays the key role. *E.coli* with hybrid oxygenase (Tod C1- BphA2- BphA3-BphA4) are more efficient in degradation of Trichloroethylene (TCE) than those bacteria with original Bph Dox and Tol Dox. The hybrid pseudomonas strain in which bphA1 gene is replaced by tod c1 produced Bph-Tod dioxygenase enzyme is also effective in degradation of Tetrachloroethane (Furukawa, 2000).

Few bacteria even though are capable of biodegrading the contaminant, however the solvents present in the site are toxic to the bacteria by damaging the cell membrane. Successful experiments were conducted in developing recombinant strains *Denicoccus radiodurans* capable of oxidizing toluene and chlorobenzene in highly irradiating environments (Lange et al.1998). Prof. Wong (1996) reported that *Pseudomonas* sp. K-1 isolated from dye-contaminated sludge could absorb large quantities of dyes very quickly but could not degrade the toxic substances in these dyes. Recombinant strains of Pseudomonas are trailed for the efficient degradation of the toxic substances of the dye.

In this approach of use of genetic engineered microbes, the stability, persistence, efficacy, mobility of the organism, effects within the soil community, threat as vectors of foreign genes are to be considered.

Fate of genetic engineered microbes

The application of genetically engineered microorganisms for bioremediation involves the construction of indigenous species with previously unavailable genes that are responsible for efficient degradation of the pollutants, acquiring new expression potential (Ronchel et al. 1995).

Many organic compounds including numerous pollutants are broken down by microbes. In the removal of toxic polychlorinated biphenyls (PCBs) from soils and sediments, *in situ* degradation is an effective approach. The aim is to transfer the PCB degradation pathway from PCB-degrading bacteria, which survive poorly in soil, into bacteria found associated with plant roots and which could survive in soils. By dissecting the regulatory circuits involved in the control of a bacterial pathway for degrading chemical pollutants and introducing the genes responsible for this pathway into the host chromosome of indigenous bacteria isolated from various ecological niches it is demonstrated that the recombinant bacteria survive and behave like the wild types in the degradation of the pollutants. (Reineke 1998)

Transfer of natural plasmids, isolated from soil and water, and of recombinant plasmid vectors between species of *Escherichia coli, Pseudomonas* and *Alcaligenes* have been studied by using heavy metal resistance genes and lactose-utilization genes as markers (De Rore et al. 1994). Transfer between a wide variety of bacteria, including heterotrophs, sulphate reducing bacteria and cyanobacteria, is possible. Retrotransfer allows movement of small recombinant plasmids directly into natural bacterial communities. Plasmids from aquatic bacteria are able to capture recombinant DNA by retrotransfer. Plant growth beneficial pseudomonads survive well in agricultural soils, especially when a dry inoculum is used. Pseudomonad mutants with impaired survival also show the potential for DNA containment.

Among the possible ways of gene dissemination, plasmid-mediated bacterial conjugation is highly relevant, as environmental conjugative plasmids are transferred into recipient strains when introduced at the sites. As few Broad Host Range (BHR) plasmids display the phenotype of gene capture (retrotransfer), to the advantage of their hosts and also carry degradative genes that allow their hosts to degrade man-made chlorinated chemicals, special attention has to be given to these plasmids and their potential for gene dissemination and gene rearrangements.

BHR plasmid-mediated gene transfer plays a role in the capture and the subsequent dissemination of recombinant DNA cloned in current vectors carried by standard GMOs. On plate matings or in sterile microcosms, the frequency of such events is measurable but generally low, and they virtually escape any detection under real environmental conditions that accompany the release of GMOs. New BHR plasmids could easily be isolated from all the tested microbial communities. These plasmids have a measurable potential to disseminate genes across taxonomic barriers. These findings are of importance in the context of natural horizontal gene transfer as an evolutionary force and as a part of the "genetic landscape" in which long-term biosafety issues must be considered.

Monitoring of the microbial population

Monitoring of the microbial inoculation and their growth is to be carried out, as they are potentially powerful agents for manipulation of processes and components in the soil. The microbial count and the availability of the substrate in the soil play main role in the stimulation of biodegradation of xenobiotic compounds (bioaugmentation).

The successful application of bioremediation technology for degradation of environmental pollutants may require that genetically modified microorganisms be introduced into the environment. Because of their perceived novelty, it is necessary to evaluate such GMOs for any potential risks associated with their release into the environment. It is also necessary to assess their survival and ability to degrade pollutants.

Traditional culture based methods of bacterial detection are of limited use for environmental applications. Molecular methods, which can overcome the limitations in the detection and quantification of the microorganisms, are essential. To accurately predict the process of natural attenuation and to maximize the potential for the bioremediation it is necessary to understand the microbial processes at work. Ideally, it would be useful to know the composition of the bacterial population at a site and to be able to correlate particular microbes with the reduction in the target chemical and with the amendments of the site. This would lead to improved strategies for *in situ* bioremediation where treatments of the site could be designed to stimulate the activity of those bacteria most effective in degradation.

Molecular methods like PCR amplification of genes responsible for enzymes involved in degradation pathways and hybridization studies can be used to monitor the bacteria inoculated or present in the site. An improved molecular approach, Direct PCR (DPCR) facilitates the detection of specific bacteria and quantification of range of bacteria present at the site, which even may not be detected by culture-based tests (Fode-vaughan *et al.* 2001). Indication of the activity of the genes involved in the degradation can be studied by mRNA isolation and characterization. The fingerprinting data of the microbes at the site, hybridization methods provides better way for the monitoring of the microbes responsible for the degradation of the pollutants (Wilson et al., 1999)

PCR, the sensitive method allows the monitoring and quantification of the microorganisms and even the introduced catabolic genes. (Tchelet et al., 1999) Extraction, characterization and measurement of expression of mRNA provide the indication of the activity of the genes, their switch off on position involved in the bioremediation. These molecular techniques provide a more complete picture of the microbial processes at work and could serve as the basis for improved remediation strategies and monitoring recommendations.

Future perspectives

The bioremediation systems in operation today rely on microorganisms native to the contaminated sites, encouraging them to work by supplying them with optimum levels of nutrients and other chemicals essential for their metabolism. Researches are currently investigating these to augment the contaminated sites with nonnative microbes including GMOs specially suited to degrading the contaminants of the concern at particular sites. Genetic engineering techniques offer the possibility to equip organisms, known to survive and be active when introduced into certain environments, with the desired catabolic potential. These includes

1. Analysis of genes and regulatory elements for critical molecules in bioremediation
2. Structure and function of bioremedial molecules
3. Genetic selection and engineering of improved bioremedial molecules and organisms.
4. Pathway engineering
5. Cell free systems for bioremediation

Summary

New genes, enzymes and metabolic routes involved in bacterial xenobiotic degradation have been discovered, and new methods have been developed, which allow the discovery of the broad flexibility of microorganisms. Besides application of the natural diversity, the artificial evolution of enzymes and pathways will lead, without doubt, to improved biocatalysts and high-throughput methods of screening for the desired phenotypes are becoming available. Strategies to design superior biocatalysts are taking into account more and more the necessity of such organisms to perform in a reliable fashion under environmental conditions. Thus, studies to understand the interaction between xenobiotics and organisms and on the fate, survival and activities of microorganisms in the environment have to intersect with the biochemical and genetic engineering studies. Such a cross feeding will provide the ground for successful interventions into environmental processes and, thereby lead to optimized strategies for bioremediation.

The use of genetic engineered organisms in the environment has been limited by public perception and the concerns of regulatory agencies about the potential risks of releasing genetically altered organisms into the environment. In addition, engineered organism's inability to compete with natural soil populations is an issue that must be addressed. Accordingly, the technology associated with the regulation and containment of such organisms and direct use of biomolecules extracted or removed from cell-free systems are critical areas for research.

References

De Rore H., Top E., Houwen F., Mergeay M., Verstraete W., (1994) "Evolution of heavy metal resistant transconjugants in a soil environment with a concomitant selective pressure". *FEMS Microbiol.* Ecol., 15: 71.

Dietmar H Piepar and Walter Reineke. (2000). Engineered bacteria for bioremediation. Current opinion in Biotechnology, 11: 262-270.

Fode-vaughan, K.A., Wimpee, C.F., Remsen, C.C. and M.L.P. Collins, (2001). Detection of bacteria in environmental samples by Direct PCR without DNA extraction. Biotechniques 31: 598-607.

Furukawa, K. (2000) Biochemical and genetic basis of microbial degradation of polychlorinated biphenyls (PCBs). J. Gen. Appl. Microbiol. 46: 283-296.

Head, I.M. (1998). Bioremediation: towards a credible technology. Microbiology.144: 599-608.

Keith, LH., and Telliard, W.A. (1976). Priority pollutants. I.A. Prospective view. Environm.Sci.technol. 13: 416-423.

Lange, C.C., Wackett, L.P., Minton, K.W., Daly, M,J., (1998) Engineering a recombinant Deinococcus radiodurans for organopollutant degradation in radioactive mixed waste environments. Nature Biotechnol.16: 929-933.

Reineke W. (1998). Development of hybrid strains for the mineralization of chloroaromatics by patchwork assembly. Annu Rev Microbiol. 1998, 52:287-331.

Ronchel M.C., Ramos C., Jensen L.B., Molin S., Ramos J.L.(1995) "Construction and behavior of biologically contained bacteria for environmental applications in bioremediation". *Appl. Environ. Microbiol*, 61: 2990-2994.

Samanta, S.K.,Singh, V. and Jain, R.K. (2002). Polycyclic aromatic hydrocarbons: environmental pollution and bioremediation. Trends Biotechnol 20: 243-248.

Srivastava, A.K., & Dumka, V.K. (2001) Environmental pollution due to heavy metals and their control strategies. P.232-257,.In:Current topics in environmental sciences; *ed* G.Tripathi & G.C.Pandey; ABD publishers, Jaipur,India.

Tchelet R, Meckenstock R, Steinle P, van der Meer J.R. (1999) Population dynamics of an introduced bacterium degrading chlorinated benzenes in a soil column and in sewage sludge. Biodegradation.10: 113-125.

Wilson M.S, Bakermans C, Madsen E.L. (1999) *In situ* real-time catabolic gene expression: extraction and characterization of naphthalene dioxygenase mRNA transcripts from groundwater. Appl Environ Microbiol. 65: 80-87.

Wong, P.k. & Yuen, P.Y. (1996). Decolourization and biodegradation of methyl red by *Klebsiella pneumonia* RS-13. Water research 30: 1736-1743.

Remediation of Aquaculture Environment : A Biotechnological Approach

G. K. Sehgal and H.S. Sehgal

Introduction

In this century, basic and applied research in genomics, biotechnology and genetic engineering is sparking a new scientific revolution. The endeavor to decode and manipulate the genetic make up of living matter can radically transform agricultural and food production sectors. Already, the development and diffusion of disease-, insect-, and drought-, resistant crops, and more productive farm animals have been witnessed.

An important aspect of biotechnology that has received wider attention in recent years is its multifaceted role in meeting national goals for innovation, environmental protection and sustainable development. The United Nations Environment Programme (UNEP) (founded in 1972 in Stockholm) has laid a special attention on biotechnology as it could be applied for environmental purposes. The UNEP, in collaboration with United Nations Education Science and Cultural Organization (UNESCO) and the International Cell Research Organization (ICRO), developed a project "Development of an Integrated Programme in the Use and Preservation of Microbial Strains for Deployment in Environmental Management" (July 1975- October 1985; total budget US$ 6.3 million), which called for the establishment of six regional centres in Brazil, Egypt, Guatemala, Kenya, Senegal and Thailand. Since 1985, this network has been expanded to include 23 centres that are located in developing and developed countries. Recently, the UNEP has associated with the UN Industrial Development Organization (UNIDO), International Centre for Genetic Engineering and Biotechnology (ICGEB), World Health Organization (WHO) and Food and Agriculture Organization (FAO) for support to sound application of biotechnology for environmental management in developing countries.

Modern aquaculture, though traditional in origin, is a relatively new science, and the industry based on aquaculture is not very old. The world aquaculture production has increased from about 5 million metric tonnes (mmt) in 1985 to 26.38 mmt in 1996 (Ayyappan and Jana, 1999), coming from a variety of farms ranging from small-scale owner-operated fish ponds to large-scale cooperative and corporate farms. Today aquaculture is one of the fastest growing sectors of food production, particularly in Asia which accounts for about 90% of the total global production.The increase in production is being achieved by the expansion of areas of land and water under culture and use of more intensive and modern farming technologies, which involve higher usage of inputs such as water, feeds, fertilizers and chemicals. Higher inputs give rise to sedimentation adversely affecting water quality and fish. These pose health risks to culture organisms (Toor *et al.*, 1983; Sehgal and Sehgal, 2002) and farm workers from water-borne diseases, in addition to increased waste discharges degrading the environment. Thus, until a couple of decades ago, when technologies remained less intensive, aquaculture was generally considered as environment-sound activity. A major change in perception has occurred in recent years with the adoption of more intensive techniques of fish/prawn production and the extension of farming activities to coastal and foreshore areas of the sea.

Therefore, to ensure sustainable aquaculture yields, it is in the long-term interest of aquaculture to follow and apply the developments in environmental biotechnology. Application of biologically- engineered solutions to environmental problems has become far more readily acceptable and widely understood. However, there remains some uncertainty amongst practitioners regarding how and where the microscopic functional level fits into the macroscopic practical applications. It is precisely this gap, which the following pages set out to fill.

Environmental Impact of Aquaculture

The nature and extent of environmental impact of aquaculture largely depend on the location and type of aquaculture farms as well as the production technologies adopted. Although, the type of wastes produced in aquaculture farms are basically similar, there are differences in the quality and quantity of the components depending upon the species cultured and the culture practices.

The main types of wastes in hatcheries or production farms can be: i) residual food and fecal matter, ii) metabolic by-products and iii) residues of biocides and biostats. In land-based aquaculture farms, wastes can be built up as a result of the sedimentation, disintegration and under-utilization of fertilizers and manures used for the production of natural food organisms. In tidal water fish ponds, the inflows may bring in additional organic matter in the form of suspended solids and dissolved nutrients. In intensively fed farms, the composition and physical nature of the wastes reflect the composition of the diet and the digestibility of its components. All feed components and by-products of metabolism, which contain organic carbon and organic nitrogen (carbohydrate, lipid, and protein), ammonia, urea, bicarbonate, phosphate, vitamins, therapeutants, and pigments could become waste products, which sink to the sediment and settle at the bottom. Micro-organisms, in the presence of oxygen, decompose this organic matter. Increase in organic enrichment directly leads to an increased oxygen consumption rate, and when the oxygen demand is more than that which is available, the sediment becomes anoxic. The un-decomposed or partially decomposed organic matter creates anaerobic conditions and produces toxic gases like ammonia, nitrite, hydrogen sulphide and methane which may be released into the water column and deteriorate healthy environment for the culture of food organisms.

Decomposition of wasted feed, faeces, metabolic wastes, manures and fertilizers in aquaculture farms leads to hypernutrification, which can result in algal blooms in the farm itself or in the vicinity of the farms (especially in case of cage farms in open waters and pen culture in eutrophic lakes). Oxygen consumption by dense algal communities during night may result in low levels of dissolved oxygen. Malhotra *et al.* (1979) and Sehgal *et al.* (1982) have reported mortalities of wild and farmed fish due to oxygen depletion.

Enrichment of bottom sediments in land-based aquaculture farms, as well as in cage and pen culture systems in open waters, provide favorable conditions for multiplication of bacterial populations and enhancement of specific bacterial communities. This can lead to disease outbreaks. In case one pond becomes infected with pathogens, there is a possibility of infection being carried from one pond to another through inflows from a common source. Korzeniewski and Korseniewska (1982) found higher bacterial counts in the vicinity of rainbow trout cages in lake Letowo in Poland, evidently as a result of environmental changes brought about by waste discharges from cages.

In recent years, substantial attention has been devoted to the impact of discharges of wastewater from aquaculture farms. In South Asian aquaculture, where 'non-drainable' ponds are used, discharge of water from ponds is done only occasionally (once in three or four years), using pumps on to the land areas. The excess detritus that accumulates at the pond bottom are removed and ponds are prepared fresh for the new aqua crop. In most of the intensive aquaculture farms, the wastewater discharges are frequent and mostly go into some water body. In case of cage and pen culture, farms emit wastes into the environment on a more or less continuous basis.

Environmental applications of biotechnology hold a very good promise for the future. The load of wastes either in the culture waters or in the discharges from aquaculture farms can be treated to some extent using biotechnologies, so that the adverse effects on the production organisms as well as on the environment can be considerably reduced.

Bioremediation Technology

Bioremediation uses living processes of microorganisms or plants or animals to clean up the environment. It has a great scope for improving aquaculture environment and is one of the most cost-effective technologies for dealing with environmental remediation problems. The International Workshop on Aquaculture and Environment held in Cochin in July 2001 recommended that bioremediation technologies need to be evolved for improving efficiency of aquaculture production systems and these should be popularized among farmers, development agencies and policy makers. Following areas of bioremediation are important from the viewpoint of improving aquaculture environment.

Biodegradation : In nature, there are specific beneficial non-pathogenic microorganisms, mainly bacteria and fungi, which can continuously degrade a variety of accumulated organic matter through the process of mineralization in a systematic way and release nutrients into the water.

This process has a great significance in context of aquaculture as constant availability of nutrients due to mineralization uf organic manures helps stabilizing the natural food (phytoplankton and zooplankton) production throughout the culture period (Sandhu *et al.*, 1985; Sehgal *et al.*, 1991, 1992; Singh *et al.*, 1992). This ensures better survival and growth of cultured species. The constant biodegradation process also inhibits sludge formation at the pond bottom.

The important strains of bacteria and fungi having capability to degrade and metabolize various organic compounds are *Bacillus, Lactobacillus, Pseudomonas, Nitrosomonas, Nitrobactor, Saccharomyces, Alcaligenes.* These microorganisms when used together act more efficiently because the process of organic degradation started by one group of microorganisms is continued by the others (Fig. 1).

Microbes in water acting on organic matter require dissolved oxygen (DO). Therefore, they are also called aerobic microbes. The organic matter is food for the bacteria, which multiply rapidly in the presence of organic matter and oxygen. They release end products of their metabolism, the CO_2 and water. The key factor to organic load degradation by microbes is DO, The rate at which organic matter is consumed (degraded) depends on the size and variety of

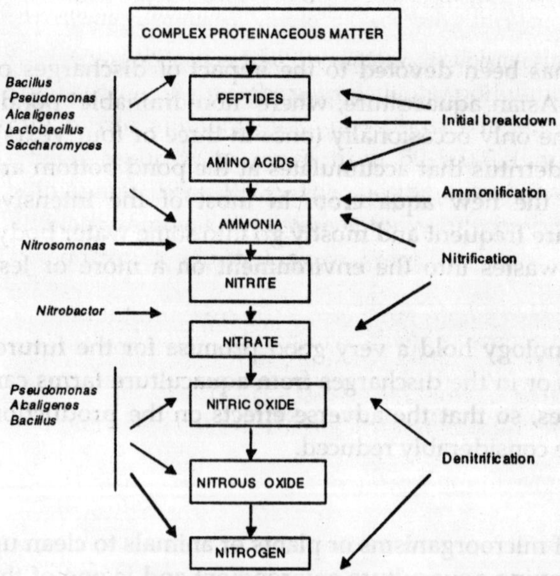

Fig. 1 : Mineralization process of organic matter at the soil-water Interface of the pond bottom.

microbial population. As metabolism of microbes proceeds, the number of microbes increases and the organic matter is degraded at a faster rate. In an aquatic environment, algae use CO_2 produced by aerobic bacteria and release oxygen. This oxygen is taken up by aerobic bacteria to release CO_2. These aerobic bacteria and algal populations in the aquatic environment assist each other to survive (Fig. 2).

Bacteria are the most important biological converters of organic wastes and they generally thrive in a pH range from 6.5 to 8.5 and will not survive at a pH of less than 4.3 or above 10.5. Since aerobic bacteria are both pathogenic and non-pathogenic, it is preferable to introduce and promote growth of a large number of non-pathogenic aerobic bacteria to keep the culture waters safe for the production of food organisms and also to keep the environment safe from effluent discharges of aquaculture farms.

One of the commercial bio-products available in India is ENVIRON- AC™, manufactured by WOCKHARDT BIOSTADT AGRISCIENCES, Mumbai, in collaboration with TBA France, which consists of a consortia of selective, non-pathogenic microorganisms bio-fixed on a highly porous and natural material called cocolith. This product has been successfully demonstrated in Indian aquaculture to continuously degrade a variety of accumulated organic matter at the pond bottom.

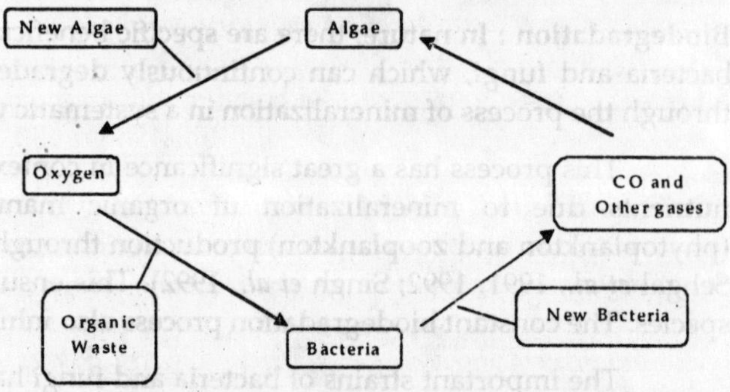

Fig. 2 : Biological cycle of organic waste removal

BACTA-PUR POND is another bio-product containing beneficial bacteria manufactured by Canadian Pond.ca Products Ltd.,. which helps to clean water and removes organic sediments in large ponds and small lakes. BACTA-PUR N-3000 is another bio-product manufactured by the Canadian Pond.ca Products Ltd, which helps nitrification/denitrification activities in aquaculture.

Recently, biofilms (microbial communities attached to a substratum) have been reported to gain importance for removal of dissolved particulate contaminants in natural streams, ponds and in wastewater treatment plants. The biofilm of *Nitrosomonas* and *Nitrobacter* developed on the substratum provided in biofilters are an important group of bacteria involved in metabolizing the nitrogenous wastes and toxic substances into simpler nutrients and nontoxic components (Pandey *et al.*, 2003).

Scientists are now working on ways to enhance the natural ability of degradation of microorganisms. Sangodkar (2000), while unfolding his research findings in the First National Conference on Fisheries Biotechnology at CIFE, Mumbai, said that biodegradation ability of some naturally occurring bacterial strains could be enhanced in the laboratory by maintaining these organisms in a chemostat for several generations.

Biosorption : The potential of cyanobacteria or blue-green algae for remediation of surface waters from organic compounds has been recognized very recently. Cyanobacteria are gram-negative bacteria in their cell structure and chemistry but are considered algae because of their microscopic morphology, pigmentation and oxygen evolving photosynthesis. Chlorinated organic compounds are among primary environmental concerns due to their persistence and toxicity. These compounds are released into the environment due to agricultural practices or industrial processes. Surface waters receive a significant amount of the discharge with chlorinated organic compounds and need to be remediated to prevent environmental and health consequences. Cyanobacteria are common inhabitants of the surface waters and, in some cases, are the dominant component of surface water micro flora. Although, the potential of cyanobacteria to degrade chlorinated organic compounds has been realized, the role and potential application of these microorganisms in remediation of surface waters remain largely unstudied (Kuritz, 1999).

Lindane (hexachlorocyclohexane or HCH) is a chlorinated pesticide, of which India is a leading producer and is one of the main consumers (Farm Chemical Handbook, 1993). The pressure and solubility of HCH are higher than for many other organochloride pesticides, The applied residues volatilize more rapidly and redeposit by rain and snow (Majewski and Lapel, 1995). Lindane is one of the organic pesticides detectable in fog, and rainfall in the United States, Canada and Europe (Goolsby *et al.*, 1997), and in water bodies in Asian countries (Abdullah *et al.*, 1997). Reports also indicate a considerable Lindane contamination in China, where this pesticide is commonly used on crops (Xu *et al.*, 1994). Anaerobic and micro aerobic bacterial catalysis have been reported to be responsible in Lindane degradation in soil, but not in surface water. The surface water, therefore, remains the main medium for Lindane accumulation and transport. Persistence of Lindane in surface waters leads to its piscicidal effect or bioaccumulation in fish (Abdullah *et al.*, *op cit.*).

Recent research showed that two strains of common freshwater cyanobacteria *Anabaena sp.* and *Nostoc ellipsosporium* could degrade Lindane, although this ability depended on the presence of nitrate and light in the medium. These cyanobacteria have now been genetically engineered to produce recombinant strains of *Anabaena* and *Nostoc*, which are able to degrade Lindane both in the presence and absence of nitrate (Kuritz and Wolk, 1995). Other pesticides, which have been reported to be degraded by cyanobacteria are Aldrin, DDT, Diazinon,

Parathion, Melathion, etc. Prominent forms of cyanobacteria involved are *Anabaena, Anacystis, Scenedesmus, Chlorella* and *Euglena*. (Dash and Mishra, 1999).

Cyanobacteria like *Oscillatoria, Halobacterium, Aphanocapsa,* have been reported to reduce the percentage of a number of nutrients/pollutants from the waste water like calcium, chloride, nitrate, nitrite, ammonia, total phosphorus, magnesium, pH, TDS, BOD and COD. The reduction has been reported to range from 50-100% (Uma and Subramaniam 1990; Manoharan and Subramaniam, 1992, 1993). The specific use of cyanobacteria, both free and immobilized forms, in efficient removal of different forms of combined nitrogen and phosphorus has also been reported (Proulx and De la nove, 1988).

Cyanobacteria have also been found to be highly efficient in removing bioaccumulation of soluble and particulate forms of metals, especially from dilute external medium. Genes, whose products are over expressed during heavy metal stress and enable the cyanobacteria to combat the adverse condition, seem to be of biotechnological significance in the task of bioremediation. Survival of organisms in metalliferous environment may be by compartmentalization of metals in vacuoles, cell wall, polyphosphate bodies (Jenson *et al.*, 1982), metal tolerant enzymes/antioxidant enzymes (Bowler *et al.*, 1992; Canini *et al.*, 1992; Chadd *et al.*, 1995), production of metal binding protein (Mallick *et al.*, 1994), and by exclusion mechanism (Foster, 1977). A lot of work has been done in heavy metal uptake and reviewed (Pinas *et al.*, 1995, Angadi *et al.*, 1996).

Bioaccumulation of heavy metals by microorganisms and microbial products is of two types: non-metabolic and metabolic, which operate either in isolation or together. The non-metabolic (biosorption) is rapid and reversible accounting for about 90% of total heavy metal taken up within 1 hour. It is more promising for commercial development as the organism can be used for multiple cycles for sorption and desorption. One of the commercial products available is ALGA SORB™, Biorecovery Inc. New Mexico (*Chlorella + Spirulina*), which is based on biosorption phenomenon. In this phenomenon, metal bind to polyionic cell wall or with ions in cell wall through ion-exchange and the process is not affected by physical conditions such as pH, and ionic strength and can take place in a living or dead cell.

The metabolic (cellular uptake) typically accounts for 10% of the total heavy metal uptake in 1 hour and results in increased permeability, transport, internal compartmentalization, extra cellular precipitation by excreted metabolites. This process is slow, depends on cellular metabolism and is affected by metabolic inhibitors, uncouplers and temperature. Metal binding proteins known as Metallotheoneins (MTs), which are gene products or secondary metabolites (Robinson, 1989), produce metal specific biosensors and they form a complex with metals, thus playing a role in metal detoxification. Metal binding proteins have been reported in cyanobacteria (*Synechococcus*) but the MT gene has not been characterized in detail (Mac Lean *et al.*, 1972). They have also been reported in *Anabaena doliolum* (Mallick *et al.*, 1994) and in *Synechocystis* (Franers *et al.*, 1996). Molecular cloning of metallotheonein genes needs immediate attention for creating biotechnologically and economically more efficient metal detoxifying organisms.

Phytoremediation : Abandoned fallow lands, industrial dumps and other contaminated sites may get a new lease of life: thanks to an innovative form of bioremediation that uses plants to remove toxic substances from soil, sediments and ground and surface waters known as Phytoremediation. This Sun-driven technology is proving to be less invasive, less costly and less energy intensive than existing physical, chemical and thermal remediation methods.

Phytoremediation works on the principle that plants naturally take up trace elements and other substances from soil and water and either store, accumulate, metabolize or volatilize them to the atmosphere. Plant roots also foster the growth of microorganisms that biodegrade organic contaminants. Many plants produce and release substances that help degrade organic pollutants in the root zone. Unlike traditional remediation methods, which often destroy organic material and leave soil inert, planting terrestrial and aquatic species can target specific contaminants and in the process, restore the site to some level of ecosystem health and productivity.

Canada is taking lead in generating a comprehensive database on the use of this technology to stabilize, remove and recover heavy metals such as lead, mercury, iron, manganese, zinc, and copper from contaminated ecosystems. This march, involving two years of intensive study by scientists in Canada's Environmental Technology Advancement Directorate, culminated in the release of PHYTOREM- an interactive electronic database of more than 700 plants including lichens, algae, fungi and bryophytes that have demonstrated an ability to tolerate, accumulate or hyper accumulate a range of 19 different metals. Species that show considerable potential to date include sunflowers, ragweed, cabbage, Indian mustard, geranium and jack pine.

Plants are effective at removing metals because like humans, they require certain trace elements to survive. Some species, known as hyper accumulators, can accumulate and tolerate very high concentrations of metals- some up to five per cent of their dry weight. Another advantage of Phytoremediation is that once a crop has accumulated it's optimum metal content it may be possible to harvest the crop and recover the valuable metal components from the contaminated biomass.

Adler (1998) developed an integrated phytoremediation system for production of rainbow trout and lettuce, and for effluent treatment. The research focused on using plants to remove P from aquacultural effluents, using thin-film technology in a unique production system, called conveyor production strategy (CPS), in which plants grow in water that flows continuously as a thin-film over their roots. The objective of the research was to develop a production system that would allow plants to remove more than 95% of the P in the rainbow trout farm effluent while producing a marketable product. This piece of research has shown that because plants remove nutrients continuously, the effluent storage facilities are also not necessary to treat effluents that are generated 24 hours a day. This system generates income while nutrients are removed to a very low level.

Aquatic macrophytes such as water hyacinth and duckweeds have been grown in effluent- receiving waters for the extraction of nutrients, but the multiplication and dispersal of such plants, which are difficult to control even with repeated use of herbicides, make their use risky (Pillay, 1992). Halophytes, the salt-marsh plants also found in inland saline deserts some of which produce edible oil seeds (similar to those of soybean and canola) have been successfully used as biofilters to remove nutrients from saline aquaculture effluent (Felix, 2003).

Seaweed culture seems to be an efficient means of removing high concentrations of nutrients in the environment. The polyculture of *Laminaria* and other seaweeds, along with scallops and abalone in coastal areas, has been reported to be very effective in controlling hypernurtification in the coastal areas of north-east China. It has been so efficient that additional fertilizers have to be sprayed along the coast to encourage rapid growth of the seaweeds and of the planktonic algae that form the food of the shellfish.

Zooremediation : There are a few reports of ability of some animals to clean excess nutrients from aquaculture waters and wastewater. Mussel culture has been considered to be a means of counteracting eutrophication, as large-scale mussel culture can lead to nutrient decrease. Folke and Kautsky (1989) have suggested that eutrophication as a result of cage farming can be avoided by integrating it with mussel farming who utilize the excess nutrients produced in the cage culture systems.

The species combination in polyculture of Chinese and Indian major carps in pond farms generally includes at least one scavenging species that feeds on faecal matter of the other species. This serves to reduce the load of faecal matter and thereby of soluble nutrients. The whole concept of polyculture based on full utilization of fish pond resources is geared to the reduction of wastes, and is, therefore, environmentally very sound.(Pillay, 1992).

Animals can also be modified to act as biosensors, detecting or monitoring hazardous materials in the environment (Jyoti and Sehgal, 1979; Sehgal, 1980). Researchers are now working to engineer zebra fish to act as biosensors for pollutants such as dioxin or PCBs.

Recirculation Technology

Traditional aquaculture production in ponds requires large quantities of water. Approximately, 1 million gallons (3.2 million litres) of water per acre are required to fill a pond and an equivalent volume is required to compensate for evaporation and seepage during the year (Losordo *et al.*, 1998). In many areas, traditional aquaculture in ponds is not possible because of limited water supplies or an absence of suitable land for pond construction. Recirculating aquaculture production systems offer an alternative to pond aquaculture technology. Through water treatment/remediation and reuse, recirculating systems use a fraction of the water required by ponds to produce similar yields. Because recirculating systems usually use tanks for aquaculture production, substantially less land is required.

Recirculation is the use of water systems to minimize or reduce dependence on water exchange and flushing in fish culture units. Water is typically recirculated when there is a specific need to minimize water replacement, maintain water quality conditions, which differ from the supply water, or to compensate for an insufficient water supply. It is used mainly to accomplish i) aeration, ii) removal of particulate matter, iii) biological filtration to remove waste ammonia and nitrite, and iv) buffering of pH.

From the view point of minimum environmental impact, recycling or recirculating systems of aquaculture would be ideal, as there will be only reduced discharges into open land or water bodies, thus making aquaculture an environment-friendly activity. A number of experimental and commercial projects using recycled water have been undertaken in industrialized countries. This technology is still not fully developed. Although, recirculating systems have generated much interest in recent years, little information is available to compare different systems under controlled laboratory conditions either on technical merits or for economics.

Recirculated systems are used to raise fish in a controlled environment. The fish are raised in tanks with continuously flowing water that is recirculated through a water treatment system and returned to the production tanks. The treatment may include mechanical filters to remove solids and biological filters to degrade the BOD and nitrify the ammonia, and oxygenation. Because construction of recirculating systems requires considerable capital investment, the fish produced in these systems are generally high valued species.

Recirculating systems are mechanically sophisticated, and biologically complex and intense where fish is usually reared intensively (1.1 Kg/3.2 litres or greater). To provide a suitable environment for intensive fish production, recirculating systems must maintain uniform flow rates (water and air/oxygen), fixed water levels, and uninterrupted operation.

Recirculatory system design : Recirculation production technology is most often used in tank systems because sufficient water is not available on site to "wash" fish wastes out of production tanks in a flow-through configuration or production system that uses water only once. By recirculating tank water through a water treatment system that "removes" ammonia and other waste products, the same effect is achieved as with the flow-through configuration.

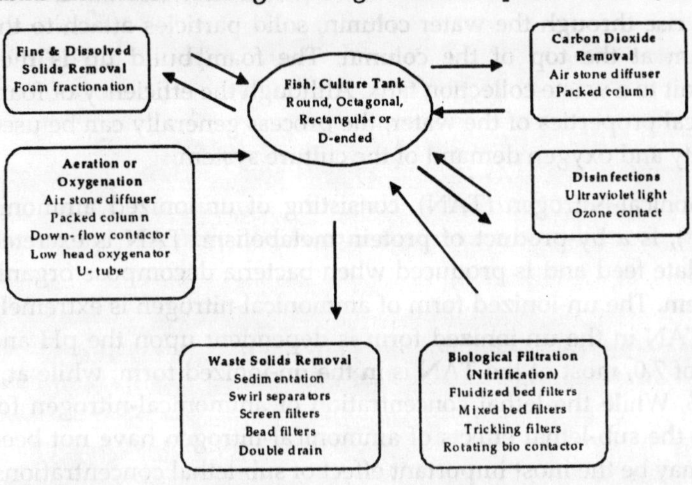

Fig. 3 : Required unit Processes and some typical components used in recirculating aquaculture production systems

A Key to successful recirculating production systems is the use of cost-effective water treatment system components. All recirculating production systems remove waste solids, oxidize ammonia and nitrite-nitrogen, remove carbon dioxide, and aerate or oxygenate the water before returning it to the fish tank (Fig. 3). More intensive systems or systems culturing sensitive species may require additional treatment processes such as fine solids removal, dissolved organics removal, or some form of disinfections.

Removal of solid wastes : Recirculating production systems are entirely feed based systems where high quality, water stable pelleted feeds are used to feed the aqua crop. Since feeds contain protein, carbohydrates, fat, minerals, and vitamins, the portion of feed not assimilated by the culture organism will be excreted as a highly organic waste (faecal solids). When broken down by bacteria within the system, faecal solids and uneaten feed will consume dissolved oxygen and generate ammonical-nitrogen. For this reason, waste solids should be removed from the system as quickly as possible. Solid wastes can be classified into four categories: settle able, suspended, floating and dissolved solids.

Settleable solids are the easiest of the four categories of solids to remove. These can be removed as they accumulate on the tank bottom through proper placement of drains, or they can be kept in suspension with continuous agitation and removed with a sedimentation tank (clarifier), mechanical filter (granular or screen), or swirl separator. Suspended solids don't settle to the bottom of fish culture tank and if not removed, these can significantly limit the amount of fish that can be grown in the system, and can irritate the gills of fish. The most popular treatment method for removing suspended solids generally involves some form of mechanical filtration. Fine suspended solids have been shown to contribute more than 50% of the total suspended solids in a recirculating system. Fine suspended solids increase the oxygen demand of the system and cause gill irritation and damage in finfish. Dissolved organic solids (protein) can contribute significantly to the oxygen demand of the total system.

Fine and dissolved solids cannot be easily or economically removed by sedimentation or mechanical filtration technology. Foam fractionation, as employed in aquaculture, is a process of introducing air bubbles at the bottom of a closed column of water that creates foam at the top air/water interface. As the bubbles rise through the water column, solid particles attach to the bubbles' surfaces, forming the foam at the top of the column. The foam build up is then channeled out of the fractionation unit to a waste collection tank. Although the efficiency of foam fractionation is subject to the chemical properties of the water, the process generally can be used to significantly reduce water turbidity and oxygen demand of the culture system.

Removal of nitrogen : Total ammonical-Nitrogen (TAN), consisting of un-ionized ammonia (NH_3) and ionized ammonia (NH_4^+), is a by-product of protein metabolism. TAN is excreted from the gills of fish as they assimilate feed and is produced when bacteria decompose organic waste solids within the culture system. The un-ionized form of ammonical-nitrogen is extremely toxic to most fish. The fraction of TAN in the un-ionized form is dependent upon the pH and temperature of the water. At a pH of 7.0, most of the TAN is in the un-ionized form; while at a pH of 8.75 it is reduced up to 30%, While the lethal concentration of ammonical-nitrogen for many species has been established, the sub-lethal effects of ammonical-nitrogen have not been well defined. Reduction in growth may be the most important effect of sub-lethal concentrations. In general, the concentration of un-ionized ammonical-nitrogen in tanks should not exceed 0.05 mg/l.

Nitrite-nitrogen (NO_2^-) is a product of the oxidation of ammonical- nitrogen. Nitrifying bacteria (*Nitrosomonas*) in the production system utilize ammonical-nitrogen as an energy source for growth and produce nitrite- nitrogen as a by-product. These bacteria are the basis for biological filtration. The nitrifying bacteria grow on the surface of the biofilter substrate, although all tank production system components will have nitrifying bacteria present to some extent. While nitrite-nitrogen is not as toxic as ammonia-nitrogen, it is harmful to aquatic species and must be controlled within the tank.

Nitrite-nitrogen binds with hemoglobin to produce methemoglobin, which is not capable of binding and transporting oxygen. The affected fish become starved of oxygen. The toxicity of nitrite-nitrogen is species specific. However, a common practice for reducing nitrite-nitrogen toxicity is to increase the chloride concentration of the culture water. Maintaining a chloride to nitrite-nitrogen ratio of 10:1 generally will protect against methemoglobin build-up and nitrite-nitrogen toxicity. Fortunately, *Nitrobacter* bacteria, which also are present in most biological filters, utilize nitrite-nitrogen as an energy source and produce nitrate as a by-product. In a recirculating system with a mature biofilter, nitrite-nitrogen concentrations should not exceed 10mg/l for long periods of time and in most cases should remain below 1mg/l.

Nitrates are not generally of great concern to an aquaculturist. Studies have shown that aquatic species can tolerate extremely high levels (>200mg/l) of nitrate-nitrogen in production systems. Nitrate-nitrogen concentrations do not generally reach such high levels in recirculating systems. Nitrate-nitrogen is either flushed from a system during system maintenance operations (such as settled solids removal or filter backwashing), or denitrification occurs within a treatment system component such as a settling tank. Denitrification occurs when anaerobic bacteria metabolize nitrate-nitrogen to produce nitrogen gas that is released to the atmosphere during the aeration process.

Role of biofiltration : Controlling the concentration of un-ionized ammonical-nitrogen (NH_3) in the culture tank is a primary objective of recirculating treatment system design. Ammonical-nitrogen must be "removed" from the culture tank at a rate equal to the rate of production to maintain a safe concentration. While there are a number of technologies available for removing ammonical- nitrogen from water, biological filtration or biofiltration is the most widely used. In biofiltration, there is a substrate with a large surface area where nitrifying bacteria can attach and grow. As mentioned earlier, *Nitrosomonas* and *Nitrobacter* bacteria oxidize ammonia and nitrite-nitrogen in the recycle stream to nitrite and nitrate-nitrogen, respectively. Gravel, sand, plastic beads, plastic rings, plastic tubes, and plastic plates are common biofiltration substrates. The configuration of the substrate and the manner in which it comes into contact with wastewater define the water treatment characteristics of the biological filtration unit. The most common configuration for biological filters include rotating biological contactors (RBC), fixed film reactors, expandable media filters and mixed bed reactors.

pH and alkalinity control : The measure of the hydrogen ion (H^+) concentration, or pH, in water indicates the degree to which water is either acidic or basic. The pH of water affects many other water quality parameters and the rates of many biological and chemical processes. Thus, pH is considered an important parameter to be monitored and controlled in recirculating aquaculture systems. Alkalinity is a measure of the water's capacity to neutralize acidity (hydrogen ions). Bicarbonate (HCO_3^-) and carbonate (CO_3^{--}) are the predominant bases or sources of alkalinity in most waters. Highly alkaline waters are more strongly buffered against pH change than less alkaline waters.

Nitrification is an acid-producing process. As ammonical-nitrogen is transformed to nitrate-nitrogen by nitrifying bacteria, hydrogen ions are produced. As hydrogen ions combine with bases such as hydroxide (OH^-), carbonate and bicarbonate, alkalinity is consumed and the pH decreases. Levels of pH below 7.0 will reduce the activity of nitrifying bacteria. If the source water for a recirculating system is low in alkalinity, then pH and alkalinity should be monitored and alkalinity must be maintained with additions of bases. Some bases commonly used include hydrated lime, Ca $(OH)_2$, quick lime, CaO, and sodium bicarbonate, $NaHCO_3$.

Concentration of Dissolved gases : Although ammonical-nitrogen build-up can severely limit a recirculating system's carrying capacity, maintaining adequate dissolved oxygen (DO) concentrations in the culture tank and filter system also is of critical importance. To maintain adequate DO levels in the culture tank, oxygen must be added to the tank at a rate equal to that of the rate of consumption by fish and bacteria.

Carbon dioxide (CO_2) is a by-product of fish and bacterial respiration and can accumulate within recirculating systems. Elevated carbon dioxide concentrations in water are not highly toxic to fish when sufficient dissolved oxygen is present. However, for most species, free carbon dioxide concentrations in the culture tank should be maintained at less than 20mg/l to maintain good growing conditions.

The build-up of dissolved nitrogen gas is rarely a problem in warm water aquaculture systems. However, caution is advised when pressurized aeration or oxygenation systems are used because atmospheric nitrogen can become supersaturated in water if air is entrained into the pressurized flow stream. Aquatic organisms subjected to elevated concentrations of dissolved nitrogen gas can develop "gas bubbles" in their circulatory systems and die.

Maintaining adequate dissolved oxygen levels and minimizing carbon dioxide concentrations in the culture tank cannot be overlooked in recirculating system design. In a typical intensively loaded recirculating system, aeration or oxygenation system failure can lead to a total loss of the fish crop in half an hour or even less.

Role of aeration and degassing : Addition of atmospheric oxygen to water or the release of excess carbon dioxide from water can be accomplished in recirculating systems through a variety of devices such as air diffusers, surface agitators, and pressurized or non-pressurized packed columns. System aeration is commonly carried out in the culture tanks, although this is not a particularly good place to add dissolved oxygen. This is because the oxygen transfer efficiency of aerators drops as the concentration of dissolved oxygen increases to near saturation levels in the tank water. Since, saturated conditions are desirable in the culture tank, aeration in this location is extremely inefficient.

In recirculating systems, a better place to aerate and degas water is in the recycled flow-stream just prior to re-entry into the culture tank. At this location, in systems using submerged biofilters, the concentration of dissolved oxygen should be at it's lowest and carbon dioxide concentration will be at it's highest. Packed column aerators (PCAs) are an effective and simple means of aerating water that is already in a flow-stream. A number of recirculating system designs use air-lift pumps (vertical pipes with air injection) to recycle water through treatment processes and back to the culture tank. Airlifts agitate water with air bubbles in the process and remove CO_2 and add dissolved oxygen.

Pure oxygen injection : In intensive production systems, the rate of oxygen consumption by the fish and bacteria may exceed the capabilities of typical aeration equipment to diffuse atmospheric oxygen into the water. In these cases, pure gaseous oxygen diffusion is used to increase the rate of oxygen addition and allow for a higher oxygen utilization rate. Properly designed oxygen diffusion devices can produce an oxygen absorption efficiency of more than 90%. However, as with tank aeration (with air), the culture tank is not the best location for oxygen diffusion with common "air stone" diffusers. Because of the short contact time of bubbles rising through a shallow (<6 feet) water column in tanks, air stone diffusers have oxygen absorption efficiencies of not greater than 40 percent. Efficient oxygen injection systems are designed to maximize both the oxygen/ water contact area and time. This can be achieved through the use of a counter-current contact column, a closed packed- column, a closed packed-column contact unit, a u-tube column or a down flow bubble contactor. Pure oxygen injections have also been used in lake restoration programmes to restore hyper-eutrophic lakes for the production of fish (Sehgal and Welch, 1991).

Summary

In this century, basic and applied research in genomics, biotechnology and genetic engineering is sparking a new scientific revolution. An important aspect of biotechnology that has received a wider attention in recent years is its multifaceted role in meeting National goals for innovation, environmental protection and sustainable development.

Modern aquaculture, though traditional in origin, is a relatively new science, and the industry based on aquaculture is not very old. Increase in fish production is being achieved by the expansion of areas of land and water under culture and use of more intensive and modern farming technologies, which involve higher usage of inputs such as water, feeds, fertilizers and chemicals. Higher inputs give rise to sedimentation adversely affecting water quality and fish production. These pose health risks to culture organisms and farm workers from water-borne diseases. Additionally, increased waste discharges degrade the environment.

Therefore, to ensure sustainable aquaculture yields, it is in the long-term interest of aquaculture to follow and apply the developments in environmental biotechnology. Application of Environmental Biotechnology holds a great promise for the future. The load of wastes either in the culture waters or in the discharges from aquaculture farms can be treated to some extent using biotechnologies, so that the adverse effects on the production organisms as well as on the environment can be considerably reduced.

Bioremediation technology uses living processes of microorganisms or plants or animals to clean up the environment. It has a great scope for improving aquaculture environment and is one of the most cost-effective technologies for dealing with environmental remediation problems. Areas of bioremediation, which are important from the viewpoint of improving aquaculture environment, are biodegradation, biosorption, phytoreme-diation, and zooremediation.

Recirculation technology uses water systems to minimize or reduce dependence on water exchange and flushing in fish culture units. It is ideal from the viewpoint of minimum environmental impact. This technology recirculates fish culture tank water through water treatment system, consisting of special devices for removing metabolic wastes and other waste products, maintaining adequate oxygen levels, and releasing excess carbon dioxide from the water.

References

Abdullah, A.R., Bajet, C.M., Matin, M.A., Nhan, D.D. and Sulaiman, A.H. (1997). Exotoxicology of pesticides in the tropical paddy field system. *Environ. Toxicol. Chem.* **16**: 59-70.

Adler, P.R. (1998). Phytoremediation of aquaculture effluents. *Aquaponics J.* **4**(4): 10-15.

Angadi, S.B., Hiremath, S. and Pujari, S. (1996). Toxicity of copper, nickel, manganese and cadmium on cyanobacteria *Hapalosiphon stuhlmannii*. *J. Environ. Biol.* **17** (2): 107-113.

Ayyappan, S., and Jana, J.K. (1999). Recent trends in global fish production with particular reference to aquaculture. *Fishing Chimes* 19 (3): 11-16.

Bowler, G., Montagu, M.V. and Inze, D. (1992). Superoxide dismutase and stress tolerance. *Ann. Rev. Plant Physiol Plant Mol. Biol.* **43**: 83-116.

Canini, A., Civitareate, P., Marini, S., Caiola, M.G. and Rotillo, G. (1992). Purification of iron superoxide dismutase from the cyanobacterium *Anabaena cylindrica* and localization of the enzyme in heterocysts by immunogold labeling. *Planta Berl.* **187**: 438- 444.

Chadd, H.E., Newman, J., Man, N.H. and Carr, N.G. (1996).. Identification of iron superoxide dismutase and a copper/zinc superoxide enzyme activity within marine cyanobacterium *Synechococcus sp.* WH 7803. *FEMS Microbiol. Lett.* **138**: 161-165.

Dash, A.K. and Mishra, P.C. (1999) Role of cyanobacteria in water pollution abatement.. *In:* Tasneem Fatma (ed.) *Cyanobacterial and Algal Metabolism and Environmental Biotechnology;* Narosa Publishing House, New Delhi, India. p 196- 207

Farm Chemical Handbook. (1993). 79th ed. Meister publishing Co., Willoughby, OH.

Felix, S. (2003). Halophyte crops as bio-filters for shrimp aquaculture treatment. Fishing Chimes, 23 (6): 35-37.

Folke,C.,and Kautsky, N. (1989). The role of ecosystems for sustainable development of aquaculture. *Ambio* **18** (4): 234-243.

Foster, P.L. (1977) Copper exclusion as mechanism of heavy metal tolerance in green algae, *Nature* **269**: 322-323.

Goolsby, D.A., Thurman, E.M., Pomes, M.L., Meyer, M.T. and Battaglin, W.A. (1997). Herbicides and their metabolites in rainfall: origin, transport, and deposition patterns across the Midwestern and Northeastern United States, 1990-1991, *Environ. Sci. Technol.* **31**: 1325-1333.

Jensen, J.E., Baxter, M., Rachlin, J.W. and Jani, V. (1982). Uptake of heavy metals by *Plectonema boryanum* (Cyanophyceae) into cellular components, especially polyphosphate bodies. An X-ray energy dispersive study, *Env. Poll. Ser.* **A 27**, 119-127.

Jyoti, M.K. and Sehgal, H. (1979). Ecology of rotifers of Surinsar, a sub-tropical freshwater lake in Jammu (J&K), India. *Hydrobiologia* **65**(1): 23-32.

Korzeniewski, K, and Korzeniewska, J. (1982). Changes in the composition and physiological properties of the bacterial flora of water and bottom sediments in Lake Lotowo caused by intensive trout culture. *Polskie Arch. Hydrobiol* **29** (3-4): 671-82.

Kuritz, T. (1999) Degradation of chlorinated organic compounds by cyanobacteria. p 168-176. *In: Cyanobacterial and Algal Metabolism and Environmental Biotechnology;* ed Tasneem Fatma; Narosa Publishing House, New Delhi, India.

Kuritz, and Wolk, C.P. (1995) Use of filamentous cyanobacteria for biode-gradation of organic pollutants. *Appl. Environ. Microbiol.* **61**: 234-238.

Losordo, T.M., Masser, M.P., Rakocy, J. (1998) Recirculating aquaculture tank production systems- an overview of critical considerations. *SRAC Publication No. 451.*

Mac Lean, F.L., Lucis, J.O., Shakh, Z.A. and Jansz,E..R. (1972). The uptake and subcellular distribution of Cd and Zn in microorganisms.*Proc. Fedn. Am. Soc. Ep. Biol.* **31**: 699.

Malhotra, Y.R., Jyoti, M.K. and Sehgal, H. (1979). Causes of size and sex restricted kill of *Puntius conchonius* in a subtropical lake in Jammu, . *Indian J. Experimental Biol.* **17**(8): 836-837.

Mallick, N., Pandey, S., and Rai, L.C. (1994). Involvement of a cadmium induced low molecular weight protein in regulating cadmium toxicity in the diazotrophic cyanobacterium *Anabaena doliolum. Biometals* **7**: 299- 304.

Manohar, C. and Subramaniam, G. (1992). Sewage- Cyanobacteria interaction- A case study. *Indian J. Environ. Protect.* **12**(4): 251-258.

Manohar, C. and Subramaniam, G. (1993). Feasibility studies using cyanobacteria in ossein effluent treatment. *Indian J. Environ. Hlth.* **35**(2): 88-96.

Majewski, M.S. and Capel, P.D. (1995). Pesticides in the Atmosphere. *Ann Arbor Press. Ann. Arbor, MI.*

Pandey, P.K., Pradeep, B. and Ayyappan, S. (2003) Mechanism of biofilm formation and it's relevance in fisheries. *Fishing Chimes* **23** (5): 63-64.

Pinas, F.F., Mateo, P. and Bonila, I. (1995). Cadmium toxicity in *Nostoc UAM 208* protection by calcium. *New Phytol.* **131**: 403-407.

Pillay, T.V.R. (1992). *Aquaculture and the Environment.*, Fishing News Book, Oxford, England: p189.

Proulx, D. and None de la J. (1988). Removal of macronutrients from waste water by immobilized microalgae. *In:* M. Moo *(ed.). Bioreactor Immobilized Enzymes and Cell fundamental and Applications,* Young. Elsevier Applied Science: 301-310.

Robinson, N.J. (1989). Algal metallothioneins: Secondary metabolite and proteins. *J. Appl. Phycol.* **1**: 5-18.

Sandhu, J.S., Toor, H.S. and Sehgal, H.S. (1985). Zooplankton community in manured fish tanks. *Annal. Biol.* **1(1):** 94-97.

Sehgal, H.S., Brar, C.S., and Sehdev, R.S. (1981). A note on large-scale mortality of *Cyprinus carpio var. communis* (Linn.) and the related ecological factors. *Indian J. Ecol.* **8 (2):** 310-312.

Sehgal, H.S., Kaur, K., and Sehgal, G.K. (1991). Phytoplankton response to biogas slurry. *Bioresour. Technol.* **37 (3):** 229-234.

Sehgal, H.S., Kaur, K., and Sehgal, G.K. (1992). Zoooplankton response to biogas slurry. *Bioresour. Technol.* **41** (**2**): 111-116.

Sehgal, H.S. and Sehgal, G.K. (2002). Fish diseases in Punjab: present status and challenges ahead. *Fishing Chimes.* **21** (10&11): 80-81.

Sehgal, H.S. and Welch, E.B. (1991). A case of unusually high oxygen demand in a eutrophic lake in Washington, U.S.A. *Hydrobiologia* **209**: 235-243

Singh, R., Toor, H.S. and Sehgal, H.S. (1992). Effect of some manures on phytoplankton production in fish tanks. *Zoologica Orientalis,* **8-9 (1&2)**: 14-18.

Toor, H.S., Sehgal, H.S. and Sehdev, R.S. (1983). A case study of acute fish diseases in tanks treated with high organic load. *Aquaculture* **35:** 277-282.

Uma, L. and Subramanian, G. (1990). Effective use of cyanobacteria in effluent treatment, *Proc. Nat. Symp. On Cyanobacteria in nitrogen fixation, IARI, New Delhi,* 437-443

Xu, Y., Zhang, Y., Oxynos, K. Schmitzer, J. and Kettrup, A. (1994). Hexachlorocyclohexane (HCH) residues in Ya-Er Lake area, China. *Intern. J. Enviorn. Anal. Chim.* **57**: 53-61.

Mallick, N., Pandey, S. and Rai, L.C. (1994). Involvement of a cadmium-induced low molecular weight protein in regulating cadmium toxicity in the diazotrophic cyanobacterium *Anabaena doliolum*. Biometals 7: 299-304.

Manohar, C. and Subramanian, G. (1992). Sewage-Cyanobacteria interaction- A case study. *Indian J. Environ. Protect.* 12(4): 251-258.

Manohar, C. and Subramanian, G. (1993). Feasibility studies using cyanobacteria in ossein effluent treatment. *Indian J. Environ. Hlth.* 35(2): 88-96.

Majewski, M.S. and Capel, P.D. (1995). *Pesticides in the Atmosphere*. Ann Arbor Press, Ann Arbor, MI.

Pandey, P.K., Pradeep, B. and Ayyappan, S. (2003). Mechanism of biofilm formation and its relevance in fisheries. Fishing Chimes 23 (5): 63-64.

Pinas, F.F., Mateo, P. and Bonilla, I. (1995). Cadmium toxicity in *Nostoc UAM 208* protection by calcium. *New Phytol.* 131: 403-407.

Pillay, T.V.R. (1992). *Aquaculture and the Environment*. Fishing News Book, Oxford, England, p189.

Proulx, D. and Noue de la J. (1988). Removal of macronutrients from waste water by immobilized microalgae. In: M. Moo (ed.): *Bioreactor Immobilized Enzymes and Cells: Fundamental and Applications*, Young, Elsevier Applied Science: 301-310.

Robinson, N.J. (1989). Algal metallothioneins: Secondary metabolite and proteins. *J. Appl. Phycol.* 1: 5-18.

Sandhu, I.S., Toor, H.S. and Sehgal, H.S. (1985). Zooplankton community in manured fish tanks. *Annal. Biol.* 1(1): 94-97.

Sehgal, H.S., Brar, C.S. and Sehdev, R.S. (1981). A note on large scale mortality of *Cyprinus carpio var. communis* (Linn.) and the related ecological factors. *Indian J. Ecol.* 8(2): 310-312.

Sehgal, H.S., Kaur, K. and Sehgal, G.K. (1991). Phytoplankton response to biogas slurry. *Bioresour. Technol.* 37(3): 229-234.

Sehgal, H.S., Kaur, K. and Sehgal, G.K. (1992). Zooplankton response to biogas slurry. *Bioresour. Technol.* 41(2): 111-120.

Sehgal, H.S. and Sehgal, G.K. (2002). Fish diseases in Punjab: present status and challenges ahead. Fishing Chimes. 21 (10&11): 80-81.

Sehgal, H.S. and Welch, E.B. (1991). A case of unusually high oxygen demand in a eutrophic lake in Washington, U.S.A. *Hydrobiologia* 209: 235-243.

Singh, R., Toor, H.S. and Sehgal, H.S. (1992). Effect of some manures on phytoplankton production in fish tanks. *Zoologica Orientalis* 8-9 (1&2): 14-18.

Toor, H.S., Sehgal, H.S. and Sehdev, R.S. (1983). A case study of acute fish diseases in tanks treated with high organic load. *Aquaculture* 35: 277-282.

Uma, L. and Subramanian, G. (1990). Effective use of cyanobacteria in effluent treatment. *Proc. Nat. Symp. On Cyanobacteria in nitrogen fixation*, IARI, New Delhi, 437-443.

Zhu, Y., Zhang, Y., Oxynos, K., Schmitzer, J. and Kettrup, A. (1994). Hexachlorocyclohexane (HCH) residues in Ya-Er Lake area, China. *Intern. J. Environ. Anal. Chem.* 57: 53-55.

29

Microbial Bioremediation of Heavy Metals

P. W. Ramteke

Introduction

In the last two decades numerous studies have shown that microorganisms can interact with ions such as, heavy metals or radionuclides. Many kinds of phenomena can then take place: namely, biosorption (Volesky, 1990), bioaccumulation (Gadd and White, 1993), direct or indirect use in the microbial metabolism (Lovely, 1995) etc. The schematic diagram (Fig.1) depicts the possible interaction(s) between metals and microbes. Such mechanisms are important components of natural biogeochemical cycles for metals and metalloids as well as there are potential application for bioremediation (White *et al.*, 1995).

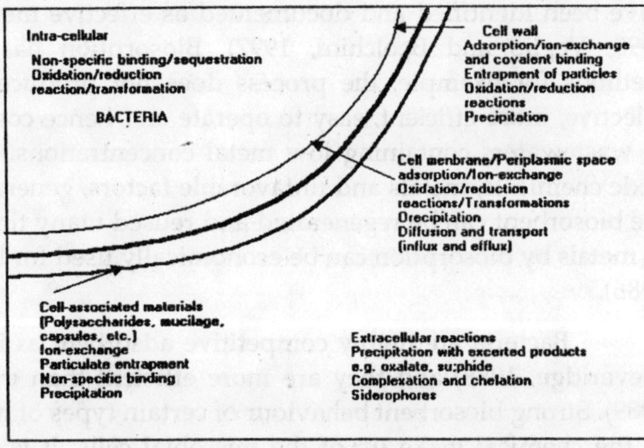

Fig. 1 The processes contributing to microbial uptake and detoxification of toxic heavy metals

Biosorption

The work of Adams and Holmes (1935) represented an early attempt on biosorption. They described the removal of Ca^{2+} and Mg^{2+} ions by tannin resin, black wattle bark (*Acacia mollissima*), which were treated directly so that the condensation product of *meta*-phenylenediamine and formaldihyde was used for anion sorption. Another early reference to the biosorption was cited by Rothstein *et al.* (1948). They presented the evidence that uranium acts at the yeast cell surface "by complexion with unknown groups associated with glucose metabolism". The complex was reported to have a one-to-one ratio with certain active 'groups' on the cell surface. A second complex was later postulated to be "groups associated with glucose mechanism".

The biosorption concept was used in 1949 by Ruchhoft, who obtained removal of ^{239}Pu from water by activated sludge. This decontamination process was referred to a process "having gelatinous matrices with tremendous suface area that are capable of absorbing radioactive materials". Polikarpov in 1966 pointed that radionuclides present in an aquatic (sea) environment are taken up by the marine organisms through direct absorption from the water. He pointed out that the above property is independent of the life functions of the microbial cells, since dead cells exhibited this property as well or better than the live cells. Tezuka (1968) suggested that the reversible flocculation of activated sludge bacteria, with the help of bivalent cations like Ca^{2+} or Mg^{2+}, is the result of ionic bond bridge formed among the negatively charged cell surfaces and cations in the solution (Volesky, 1990). Although, most of the past and current research was obtained towards the removal of cationic heavy metals, the biosorption of anions, like chromate, by biomass has become a growing concern in the field of biosorption (Kratochvil and Volesky, 1998).

Metal removal processes based on biosorbent properties of microbial biomass that can be used for metal removal from wastewaters have been well documented and reviewed (Volesky and Holan, 1995; Gadd, 1988). Mechanisms such as complexiation, ion exchange, coordination, adsorption, chelation and microprecipitation may be synergistically or independently involved in metal biosorption (Hu *et al.*, 1996). Various biomass materials including microbial biomass have been identified and documented as effective metal removing agents (Volesky and Holan, 1995; Veglio and Beolchini, 1997). Biosorption has distinct advantages over conventional methods for example, the process does not produce chemical sludges and it can be highly selective, more efficient, easy to operate and hence cost effective for treatment of large volumes of wastewaters containing low metal concentrations. Secondly, the biomass is not affected by toxic chemicals, metals and unfavorable factors, generally associated with the effluents. Thirdly, the biosorbent can be regenerated and reused many times (Gadd, 1990a). Moreover, the removal of metals by biosorption can be economically used for high-volume and low-tenor (Brierley *et al.*, 1986).

Bacteria are highly competitive adsorbent as it has a high surface area-to-volume ratio (Beveridge, 1988) and they are more efficient than minerals (Kurek *et al.*, 1982; Walker *et al.*, 1989). Strong biosorbent behaviour of certain types of microbes toward metallic ions is a function of the chemical make up of the microbial cells. It is necessary to emphasize that this type of active biomass consists of dead and metabolically inactive cells. This aspect is particularly important when it comes to the process application, whereby new biosorbents represent "chemicals" capable of sequestering a relatively large amount of the metal (Volesky, 1990). Localizing the metal deposition site within the biosorbent biomass and understanding the metal-sequestering mechanism, in combination with elucidation of the relevant metal solution chemistry and chemical structure of the metal deposition site, are all crucial aspects of the quest for an efficient biosorption process which should feature high metal selectivity and uptake.

Fungi and yeast have received attention in connection with metal biosorption particularly because waste fungal biomass arises as a by product from several industrial fermentations (Tsezos and Volesky,1981; Ehrlich and Brierley,1990; Volesky,1990; Tobin *et al*; 1984), while algae (including macroalgae) can be viewed as a renewable source of metal-sorbing biomass (Gadd, 1988; Volesky, 1990). Of related interest, is the removal of metal-containing particulates, e.g. zinc dust, magnetic and metal sulphides, from solution by fungal biomass including citric acid fermentation waste (*Aspergillus niger*) (Wainwright *et al.*,1990; Singleton *et al.*,1990). This process is independent of metabolism, but is favoured by growth.

Biosorption phenomenon is more or less like a chemical reaction and so several parameters are bound to affect the process. These are, nature and structure of microbial layers, the physiological state of the cell, pH of the medium, contact time, microbial growth conditions, temperature, ionic speciation, concentration of biomass and concerned metal and so on. Dwelling into these intricate features are reviewed in the succeeding paragraphs.

The pH of the metal solution is the most influential factor as it affects the surface properties of the biomass and metal speciation. The uptake of metallic cations by cell/biomass is reduced at pH below 2 and above 8; lowest at its isoelectric pH. Numerous investigators have shown that the pH range of 4.0 to 8.0 is optimal for uptake of cations by the negatively charged surface of bacteria (Tobin *et al.*, 1984; Volesky and Holan, 1995; Kratochvi and Volesky, 1998). The maximum biosorption of nickel by nickel tolerant *Pseudomonas stutzeri* was observed at pH 8.0 (Ramteke, 2000). However, the oxyanions like chromate are sorbed favorably under acidic conditions (Sharma and Forster, 1994a,b; Kratochvil *et al.*, 1998; Srinath *et al.*, 2002a,2003a). This very necessity of specific pH for recovery of Cr(VI) makes its selective recovery from a solution containing amalgam of heavy metal cations and anions (Volesky, 1990). It is suggested that chromate sorbs on biomass by ion exchange and subsequently oxidizes the biomass with an imminent requirement of protons for increase of equilibrium of metal uptake. The biosorption of metals is more or less ionic exchange, thus shift of solution pH was observed. This is attributed to the intrinsic ability of the microbial biomass to shift solution pH towards a favourable value. Thus, it is the final pH of the solution that determines the reaction equilibrium (Sharma and Forster, 1994a,b; Kratochvil *et al.*, 1998, Srinath *et al.*, 2002a). Thus, use of a buffered system or pH-conditioned biomass is promoted for factual analysis.

Mechanism of biosorption is metabolism independent and may seem to be little affected by the changes in temperature. But it seems it cannot be generalized to all the biosorbents. Specific biosorption of lead, cadmium and zinc by *Citrobactor* strain MCM B-181 (Puranik and Paknikar, 1999), uranium by *Pseudomanas* sp (Marques, 1991) was independent of temperature. Sorption of zinc by *Penicillium* sp (Mishra and Coudhury, 1996), copper and chromium (III) by *Aspergillus carbnarius* (Al-Asheh and Duvnjak, 1995) were inconsistent with temperature variation. Similarly, Cr(VI) biosorption by *Rhizopus arrhizus* (Sag and Kutsal 1996a,b), *Zoogloea ramigera* (Sag and Kutsal 1989) was affected by temperature and thus had a narrow optimum range. Similar observation was made for *Pseudomonas stutzeri* (Ramteke, 2000). However, biosorption of Cr(VI) by *Bacillus coagulans* was less affected and had a broad optimal range of temperature (Srinath *et al.*, 2003). Such observation is encountered when the chemical absorption plays the dominant role (Sag and Kutsal, 1989). The enhancement in metal sorption can be attributed to increase in the energy of the system that facilitates the metal attachment to the cell surface and the decrease in metal sorption to disorption of some sites of the cell surface available for metal adsorption (Al-Asheh and Duvnjak, 1995). It is also suggested that increase in metal uptake with increase in temperature is due to either affinity of sites for metal or an increase in binding sites on the relevant biomass (Marques *et al.*, 1991).

The cell surface is the major unit that performs the biosorption and its character can be altered by the constituents in the growth medium (Volesky and May-Phillips, 1995; Andres *et al.*, 1999; Puranik and Paknikar, 1999) and pretreatment with chemicals (Akthar *et al.*,1995; Puranik and Paknikar, 1999; Srinath *et al.*,2002a, 2003). Cell of *Ralstonia metallidurans* CH34 (Anders *et al.*,1999), *Citrobactor* strain MCM B-181 (Puranik and Paknikar, 1999) exhibited higher specific metal uptake when grown in complex medium instead of synthetic medium. This could be ascribed to anabolic metabolism where the readily available source of nutrients, e.g., peptone, tryptone and yeast extract in media which provide essential amino acids and growth factors that are incorporated in cell polymers binding to metals. Whereas, in synthetic medium the cells have to depend on catabolic metabolism. Simmons and Singleton (1996) reported that *L*-cysteine-supplemented media increased silver biosorption properties of *Saccharomyces cerevisiae* as compared to biomass from non-supplemented media. Pretreatment of cell surface may enhance biosorption character due to simple cleansing of site or complex actions like formation of electrostatic bonds, change in overall surface charge, modification of binding sites and masking/unmasking of binding sites. Distinct morphology of cell of bacteria is exhibited during the growth cycle. Thus, culture age can also affect the surface character and biosorption significantly. Chang and co-workers(1997) reported an increase in biosorption of lead and cadmium by *P. aeruginosa* with increasing culture age, however, this was not true in case of copper removal. However, in case of biosorption of nickel by *Pseudomonas stutzeri*, the age of the biomass did not influence uptake of the metal (Ramteke, 2000). Uranium removal by *S. cerevisiae* in logarithmic phase was 4.6 times more than the older-phase cells (Volesky and May-Phillips, 1995). The changes during the growth phase can be attributed to the changes in wall composition, ultrastructure and expression of exopolysaccharides (Scott and Palmer,1990). One interesting observation by Siegel *et al.*,(1986) was the biomass of a mycelium which lacked the Cr(VI) biosorption capability, showed this feature when the mycelium was pretreated with $Pb(NO_3)_2$. It suggests that Pb loaded on biomass reacted and precipitated as lead chromate.

Binding of metals to cell surface, functional groups, exopolymers and other metabolic products of bacteria is extensively studied (Rudd *et al.*, 1984; Greesey *et al.*, 1988; Schinner and Burgstaller, 1989; Francis and Dodge, 1990). Gram negative bacteria possess lipopolysacharides and phospholipids in their cell walls, with phosphoryl groups as the most abundant electronegative sites available for metal bindings (Coughlin *et al.*, 1983; Ferris, 1989). *Serratia marcescens*, a gram nigative bacteria, excretes a pigmented secondary metabolite called prodigiosin which has additional electronegative sites for metal bindings. Manifold biosorption of nickel was found in *S. marcescens* as compared to *Pseudomona stutzeri* (Kannan and Ramteke, 2002).

Contact time of biomass with the metal to attain equilibrium is relevant for know how of a process as well as it is a driving factor for economy of a process to be used for industrial purpose. Adsorption isotherms have been traditionally used to describe stoichiometric solution-solid interactions. These isotherms can provide insight into the nature of the physicochemical interactions involved in the extracellular immobilization of toxic metals. Biosorption is a rapid phenomenon, where equilibrium is attained in 30-60 min in the optimum conditions. The Cr(VI) removal by leaf mould (Sharma and Forster, 1994b) was of second order reaction, where solute and biosorbent concentrations play important roles in speed of adsorption reaction. Biosorption of Cr(VI) by *Rhizopus arrhizus* was rapid with 82% adsorption occurring in one hour and subsequently 92.5% after 8 hour at pH 2 (Merrin *et al.*, 1998). Subsequest incubation did not enhance the concentration of bound Cr(VI) concentration. Likewise, no enhancement in Cr(VI) biosorption was accounted for *B. coagulans* on further incubation (Srinath *et al.*, 2003). Nevertheless, it was not so in case of Cr(VI) biosorption by other different strains of *Rhizopus arrhizus* which showed a slower enhancement of biosorption level (Sag and Kutsal, 1996a,b; Prakasham *et al.*,1999). Probably, there was a very fast adsorption mechanism followed by a slower and different secondary metal binding mechanism (Kuyucak and Volesky,1989a,b; Ramteke, 2000).

Industrial effluents with presence of a single metal ion species can never occur. They are usually a blend of numerous metals and non-metals as cations and anions and thus they could impart some hindrance and /or assistance in binding of the concerned metal.

The ratio of metal ion to the biosorbent plays the decisive role during long run and in initial optimization. It has been seen that at a given equilibrium of metal concentration, biomass adsorbs more metal ions than at low cell densities as compared to high cell densities (Sharma and Forester, 1994a; Paknikar *et al.*, 1999; Srinath *et al.*, 2003). The changes in the biomass concentration affects the electrostatic interactions and interactions between binding sites. At lower concentrations of biomass, errors can be magnified consequently underestimating the specific metal uptake, as there will be many sites on the biomass still available. When a high initial concentration of metal is examined with a low biomass concentration, then after full saturation of the binding sites, the residual concentration will mislead the efficacy of a probable potent strain.

It is necessary to address here that there are several cations, anions of metals and non-metals as well as complexion agents, which can substantially reduce/enhance biosorption phenomenon. Puranik and Paknikar (1997) found that in multimetal solutions *Streptoverticillium cinnamoneum* exhibited preferential sorption of tested cationic metals. The order of the preference was also different. There was no observed enhancement in the biosorption in any metal. On the contrary to this observation, biosorption of gold by *Cladosporium cladosporioides* was enhanced by nickel and zinc with the observation that Zn was not taken up by the biomass (Pethkar and Patikar, 1998). Presence of non-metallic anionic compounds like carbonate, nitrate, sulfate and acetate was decreased to varying levels (Puranik and Patnikar,1999). Cell of *Chloral vulgaris* although biosorbed Cr(VI) and Cu optimally at different pH (2.0 and 4.0, respectively), biosorption of these metal was competitive (Aksu and Acikel, 1999). Biosorption of Cr(VI) and Fe(III) was found to be antagonistic for cells of *R. arrihzus* (Sag *et al.*,1998a) but was selective when a mixture of Cr(VI), Fe(III) and Cu(II) was exposed to the cells (Sag *et al.*,1998b). At pH 5.0, *Zoogloea* and zooglan biosorbed Cu, Cd and Zn selectively, but chromium was not biosorbed demonstrating that biosorption of a metal can be selective and can vary according to species and strain (Ahm *et al.*,1998). Even among different salts of same metal, selective biosorption of them was observed (Kannan and Ramteke, 2002). They observed that biosorption of nickel by *S. marcescens* was maximum for nickel nitrate and nickel chloride as compared to other nickel salts of ammonium, sulfate, oxide and carbonate.

Biosorption by immobilized living or dead biomass

Freely suspended microbial biomass has disadvantages that include small particle size, low mechanical strength and difficult biomass/effluent separation. Immobilized biomass particles in packed- or fluidized-bed reactors minimize these disadvantages (Macaskie and Dean, 1989; Gadd and White, 1990). Immobilized living biomass has mainly taken the form of biofilms on support made of a range of inert materials. These have been used in a variety of bioreactor configurations including rotating biological contractors, fixed bed reactors, trickle filters, fluidized bed and air-lift bioreactors (Gadd, 1988). Reactor type can be mixed, e.g. the use of biolfilm reactors in conjunction with a tandem stirred-tank reactor (Shumate and Strandberg, 1985). Living cell biofilms may provide an additional capacity for the removal of other pollutants including hydrocarbons, pesticides and nitrates. A large-scale commercial process (>5 x 10^6 gallons per day) treats effluents from gold mining and milling using rotating disc biofilm-contacting units for simultaneous degradation of cyanide, thiocyanite and ammonia as well as metal removal by biosorption (Hutchins *et al.*, 1986).

Bioflims, living or dead biomass of all groups has been immobilized by encapsulation or cross-linking (Brierley, 1990a; de Rome & Gadd, 1991; Hutchins *et al.*, 1986; Tobin *et al.*, 1994). Supports include agar, cellulose, alginates, cross-linked ethyl acrylate-ethylene glycoldimethylacrylate, polyacrylamide, silica gel and the cross-linking reagents tikuene disocyanate and glutaraldehyde. The biomass may be used in its "natural state" or modified, e.g. by alkali treatment, to improve biosorption efficiency (Hutchins *et al.*, 1986; Brierley, 1990c). Small –scale systems may be adequate for low volume waste containing valuable elements, e.g. Au. In order to use conventional reactor technology in larger systems, immobilized biomass particles should have similar properties to other commercial adsorbents, for example, in size (0.5-1.5 mm diameter), particle strength and chemical resistance. Diffusion into particles may present a problem, so they also require high porosity, hydrophilicity with a maximum amount of biomass and minimal amounts of binding agent (Tobin *et al.*, 1988; Tsezos *et al.*, 1989). Waste *Bacillus* sp. from industrial fermentations has been treated with alkali to enhance metal uptake, chemically cross-linked, pelleted by extrusion or milling and dried to provide an indefinite shelf life (Brierley, 1990c). This granulated Bacillus preparation is non-selective, removing Cd, Cr, Cu, Hg, Ni, Pb, U and Zn (singly or mixed) at a range of concentrations. Metal loadings are up to > 10% of the dry weight, giving a removal efficiency of >99% and effluents with total concentrations around 10-50 ppb. The biosorbent can be used in fixed-or fluidized-bed contactors depending on the scale of process required. After loading, metals are stripped from the biomass using sulphuric acid, sodium hydroxide or complexing agents and recovered using chemical methods. The granules are regenerated by alkali treatment for re-use (Brierley *et al.*, 1989).

Immobilized particles containing *Rhizopus arrhizus* biomass of 0.7-1.3 mm diameter contained 12-33% added polymer with improved uranium removal at lower polymer contents and lower particle diameters (Tsezos and Deutschmann, 1990). Complete uranium removal was possible from dilute uranium ore bioleaching solutions (<300 mg/l) with eluate concentrations after desorption being >5000 mg/l. The particles maintained full loading capacity (50mg U/g) over multiple biosorption- desorption cycles (Tsezos and Deutschmann, 1990).

Fluidized beds of alginate-and polyacrylamide-immobilized algae, e.g. *Chlorella vulgaris* and *Spirulina platensis* have been used to remove a variety of metals, including Cu, Pb, Zn, and Au, from mixtures and several schemes for selective recovery have been devised (Volesky, 1990; Harris and Ramelow, 1990). Alginate and polyacrylamide provide good resistance to hydrostatic pressure and mechanical degradation although it is thought that polyacrylamide is not strong enough for commercial applications (Tsezos and Deutschmann, 1990).

Immobilization in a cilica matrix is used in a batch or column system. Columns are slurry-packed with immobilized algal particles, 40-100-mesh size. Selective metal recovery is by treatment with appropriate reagents after which the regenerated biomass retains approximately 90% of the original metal uptake efficiency even after > 18 months of regular use. AlgaSORB™ has been successfully used for the removal of Ag, Al, Au, Co, Cu, Cr, Hg, Ni, Pb, Pt, U and Zn from contaminated effluents and process streams (Bedell and Darnall, 1990). Bio-Fix is a bisorbent using biomass from a number of origins including cyanobacteria (*Spirulina*), yeast, algae and plants (*Lemma* sp. and *Sphagnum* sp.) with xanthan and guar gums, blended to give a consistent product and immobilized as beads using polysulfone. The loading of Zn obtained is approximately four-fold that attained by an ion-exchange resin. The affinity series is discrimination. Metal ions are eluted using hydrochloride, or nitric acid and can be reused for more than 120 extraction-elution cycles (Brierley, 1990b).

Bioaccumulation

Most of the heavy metals have the ability to form complex compounds that may or may not be oxidation/reduction active. Thus, the heavy metals play an important role as "trace element" in sophisticated biochemical reactions for growth and function, including microorganisms (Huges and Poole, 1989). Out of 22 heavy metals, 17 have biological influence under physiological conditions (Nies, 1999). Among them Fe, Mo and Mn are important trace metals with low toxicity and Zn, Ni, Cu, V, Co, and Cr are toxic elements with high to moderate importance as trace elements. Other heavy metals, As, Ag, Sb, Cd, Hg, Pb and U have limited beneficial functions, but are considered as toxins. Bacteria are in the front line when it comes to interaction with metal in environment. Consequently, living organisms have evolved transport mechanisms for the active uptake and/or extrusion of these ions, thereby enabling cells to regulate their intercellular concentrations by a diverse family of metal ion transport proteins (Paulsen and Saier, 1997).

Wutrich (1892) is credited for the discovery of first systematic study of species uptake of copper by spores of some species of fungi in 19th century (Muraleedharan *et al.*, 1991). The first quantitive uptake of copper by spores of *Trilletia tritici* and *Ustilago crameri* was reported by Hecke (1902), whereas Pichler and Wobler (1922) observed better uptake for Cu by autoclaved spores than live spores. Goodman and Roberts (1971) reported first practical use of bioaccumulation as indicator for aerial metal levels. Role of activated sludge in uptake of metal was observed as early as 1940's and subsequently quantitive uptake of various metals by activated sludge was assessed. In case of chromium, its oxidation state was found to effect the removal by activated sludge (Muraleedharan *et al.*, 1991). While 70-90% of Cr(III) was removed, only 20% of Cr(VI) uptake was observed. Owing to the ionic charge and toxicity, there are very few reports on bioaccumulation of Cr(VI) by bacteria.

Microbial heavy metal accumulation often comprises of two stages (Failla, 1977; Trevors *et al.*, 1986; Belliveau *et al.*, 1987; Gadd, 1988, 1990b). An initial rapid and passive process involves physical adsorption or ion exchange at cell surface. Occasionally, a subsequent phase that is slower involves active metabolism-dependent transport into bacterial cell owing to its metabolism dependence. This second phase is inhibited by low temperatures, by absence of an energy source and by metabolic inhibitors and uncouplers. Rate of uptake is also influenced by the state of the cells and the composition of external medium (Gadd, 1990b). The initial concentration of metal is also influential on the rate of uptake (Srinath *et al.*, 2002a). Most mechanisms of metal transport rely on electrochemical proton gradient across the cell membrane that has a chemical component (pH gradient) and an electrochemical component (the membrane potential). Each of them can drive transport of ionized solutes across membranes. Specific uptake system does not exist for chromate, but in *Ralstonia* sp strain CH 34, chromate enters cells by the sulphate-uptake system (Nies, 1999). This is achieved only due to the similitude in their three-dimensional structure.

Since bioaccumulation is an active process, physiological state of cells affects bioaccumulation process. Starved cells (50 days) of a denitrifying bacterial consortium showed 10-15% higher amount of Cr(VI) bioaccumulation (Kong *et al.*,1992, 1993a). More than 100% of Cr(VI) was in the cell wall fraction of the starved cells in comparison to the fresh cells (Kong *et al.*,1992). This may be due to reduction of Cr(VI) to Cr(III) in the cell wall (Volesky, 1990). In fresh cells, maximum amount of chromium was present in the soluble fraction. Nevertheless, chromium content in the membrane fraction was comparable. Role of exopolymers is insignificant, owing to their anionic nature (Geesey and Jang, 1989). However, in presence of sulphate ions, Cr(VI) bioaccumulation was reduced to half in both fresh and starved cells (Kong *et al.*,1993a). Apart from other ions, pH and presence of CCl_4, as a toxicant, significantly affected bioaccumulation of Cr(VI) by these denitrifying cells (Guan *et al.*,1993a,b). Remobilization of the accumulated metal can be a significant problem during *in situ* bioremediation process. These cells notably remobilized less than 20% of chromium in 32 days (Kong *et al.*,1993b).

Cells of *Pseudomonas aeruginosa* G-1 were also found to contain highest amount of chromium in the soluble fraction (Horitsu *et al.*, 1983). Intracellular sequestration (compartmentalization), as found to occur with Cu(II) by *Pseudomonas syringae* (Cha and Cooksey, 1991), *Pseudomonas aeruginosa* (Kazy *et al.*,1999) has not been found to occur in case of Cr(VI). Extracellular precipitation of chromium has been reported in *Thiobacillus ferrooxidans* DSM 583 (Baillet *et al.*,1998) and a consortium of sulphate reducing bacteria, called SRB-III (Fude *et al.*,1994). The study of Cr(VI) tolerance was not possible in the medium containing ferrous iron as this chemolithic source was immediately reduced to ferric iron. Oxidation of ferrous compound came below 90% when the Cr(III) concentration was risen above 75 mM. Cells of *Thiobacillus ferrooxidans* bioaccumulated 501 Cr mg mg^{-1} dry weight at pH 2.4. The metal precipitated mainly at the outer surface owing to the positive surface charge of the cells (Kar and Dasgupta, 1996).

The consortium of sulphate reducing bacteria (SBR-III), isolated from an electroplating effluent tolerated up to 2,500 Cr(VI) mg 1^{-1} (Fude *et al.*,1994). They found that UV-killed dead cells did not accumulate significant amount of chromium, whereas, the uptake of Cr(VI) by living cells was significantly hampered in presence of molybdate. These observations ascertained that uptake of Cr(VI) was dependent on cellular metabolism and sulphate port was utilized for it. After 80 h incubation, ≤94% Cr(VI) removal was attained from concentrations ranging from 50-1000 mg 1^{-1}. In presence of Zn and U, 5.1% increase in chromium removal was found but at higher concentrations, precipitation of chromium was reduced significantly. The removal of Cr(VI) was credited as the H_2S produced during growth.

Role of eukaryotic microorganisms in bioaccumulation of Cr(VI) is now receiving attention. Analysis revealed that Cr(VI) reduces to Cr(III) prior to intercellular accumulation. This was the first report pertaining to Cr(VI) reduction by cyanobacteria. Role of heterocysts in it has been sought as *Synechococcus* PC6301 lacked it, when compared to *A. variabilis*. Cells of *Anabaena doliolum* and *Chrollea vulgaris* in immobilized form bioaccumulated 60-73% of Cr(VI) in comparison to 50% of free cells (Mallick and Rai, 1994).

Microbial mats, a self-organised laminated structure composed of heterotrophic and autotrophic organisms have come into foreplay of bioaccumulation of toxic metals. Such a constructed mat composed of cyanobacteria, purple bacteria (anoxygenic photosynthetic bacteria) and numerous heterotrophs were capable of bioaccumulating around 100% of Cr(VI) (Bender *et al.*,1995). Sulphate reducing bacteria (SRB) couples oxidation of an energy source to sulphate reduction during growth (Postgate, 1984). The sulphate is reduced to sulphide, which can interact with many toxic metals to precipitate into insoluble metal sulphides (Lovley and Coates, 1997; White and Gadd, 1997, 1998) and this serves as a basis for several bioremedial treatment methods (White *et al.*,1997, 1998). Removal of metals as precipitates by cells of *Desulfovibrio desulfuricans* was found to remove 86% of Cr(VI) in 24h when the initial concentration was 500 µM (Turick *et al.*,1998), while 83% removal the 960 µM of Cr(VI) was obtained in a 5 days period using a mixed consortium, SRB-III (Fude *et al.*,1994). Whereas another SRB in biofilm form of growth was found to be capable of removing 88% of Cr(VI) in 48h (Smith and Gadd, 2000). Biofilm formation and growth in that mode confers more tolerance to environmental stress factors, e.g. desiccation, and elevated levels of organic and inorganic pollutants in comparison to suspended form of growth (Costerton *et al.*, 1994). However, less than 10% of precipitated chromium was retained by these cells whereas, SRB-III retained most of the chromium as a precipitate around cells. Fude *et al.*, (1994) suggested that the chromate was indirectly reduced by H_2S and the precipitate was lodged in the cell wall. Absence of such a mechanisms would have resulted in insignificant amount of chromium accumulation by the SRB as studied by Smith and Gadd (2000).

Recently, services of metal accumulating bacteria as 'living factory' for the production of metal micro/nano size particles have been sought (Klaus-Joerger *et al.*, 2001). This led to carving of a new branch of biosciences called **biomimetics**. The biological material can be seen as a nanophase system in its own right and as the starting point for producing other novel nanophase systems. Magnetic-iron particles called magnetosomes of cubo-octahedral, elongated hexagonal prismatic and bullet-shaped morphologies are found in magnetotactic bacteria (Schueler and Frankel, 1999). Diatoms are known for active precipitation of carbonates (Kajander and Ciftcioglu, 1998; Rivadeneyra *et al.*, 1999) and formation of calcium pyrophosphate crystals (Keefe, 1976). The metal-peptide complexes forming cadmium sulfide micro-crystallites after active intracellular accumulation by *Schizosaccharomyces pombe* or formed on surface of *Klebsiella aerogenes* can fuction as quantum semiconductor crystallites and are ~1.8 mm in diameter. Such crystals have optical absorption, photosynthetic and electron transfer properties. One example of metal accumulating bacterium, *Pseudomonas stutzeri* AG259, can produce silver based single crystals in a range of few to 200 nm. Such characteristic production of nanoparticles by microorganisms has aspired adoption of bottom-up strategy as a counterpart to the present top-down-structuring in microelectronics.

Revoery of metal from cells/biomass

The cells/biomass used during biosorption as well as bioaccumulation are loaded with metal. If successful recovery were not achieved then the process would be nothing more than transfer of heavy metal from one component to another. Thus, recovery of metal is essential and to make the process economical, the subsequent regeneration of biomass can be reused for next round of biosorption. Elution of bound chromium from cells/biomass depends on its ionic state. It is normally found that, when chromium is bound in hexavalent state, its simple elution by acidic solutions is based on its reduction to trivalent state and then subsequently it releases into the eluent fraction. However, 100% elution of Cr(VI) as Cr(II) was not achieved from *Sargassum* biomass (Kratochvil *et al.*, 1998), probably due to slow reduction of Cr(VI) bound by biomass to Cr(III). They found higher amount of chromium recovery from biomass equilibrated to higher concentrations. This was attributed to the elution of chromium adsorbed to lower affinity sites, which bound to biomass after higher affinity sites were occupied. Akthar and Mohan (1995) found that chromium biosorbed on alkali treated biomass of *Aspergillus niger* could be eluted by 0.1 N HCl removing up to 98% of Cr(VI) from electroplating effluent in five cycles. The chromium elution efficiency also decreased with subsequent steps. In one study, biomass of *Myxococcus xanthus* and *S. cerevisiae* which biosorbed Cr(III), when desorption was done with 0.1 M sodium carbonate, not more than 20% could be recovered. Similarly, mild acidic solution (0.1 M HCl) could not elute biosorbed Cr(III) on *S. cerevisiae* biomass, but on increasing the concentration of acid, the yield marginally increased to 34% (Wilhelmi and Duncan, 1995).

Recently, Cot *et al.* (1999) has developed an innovative process where chromium present in solid tannery waste in Cr(III) form, is oxidized to Cr(VI) in alkaline conditions. This is achieved through action of alkaline peroxidase. The added advantage of this patented process is generation of chromium free gelatin in shorter time with better yield and superior quality. Thus generating highly added value for industry as gelatin is an essential constituent in medicine, pharmacy and cosmetics. The Cr(VI) generated is soluble in nature and thus nearly 100% recovery is possible. Similar oxidation of Cr(III) to Cr(VI) in effluent is achieved by addition of H_2O_2 with a combination of sodium hydroxide and sodium carbonate at pH between 11 and 13 (Celma *et al.*, 1999).

Bold initiatives have been taken to decontaminate heavy metal-laden soils and sediments. White and co-workers (1998) have described an integrated microbial process utilizing oxidizing and reducing bacteria to first mobilize and then precipitate toxic metal contamination in soil. Initially, metal contaminants (including Cd, Co, Cr, Cu, Mn, Ni and Zn) leached (66-99% efficiency) from artificially contaminated soil using sulfuric acid produced by sulpher-oxidising bacteria. Subsequently, the leachate was fed into another reactor in which the toxic metals were bio-precipitated as metal sulfide by a mixed culture of sulphate-reducing bacteria. Using this integrated approach, an industrially contaminated soil was leached of toxic metals with approximately 69% of the metals removed over 175 days. The procedure produced liquor effluent of sufficiently low metal concentration that could be safely discharged into the environment. The integrated procedure will probably be more commonly applied to *ex situ* applications (eg, slurry reactors), but its successful use in soil is a prime example of advances in this field.

Microbial metal transformation

Microorganisms can transform metal and metalloid species by oxidation, reduction, methylation and dealkylation (Gadd, 1993*b*). Contonuous cultures of Hg^{2+}-resistant bacteria, which can reduce Hg^{2+} to Hg^0 with mercuric reductase, volatilized Hg^0 from contaminated sewage at a rate of 2.5 mg l/h (98% removal) (Hansen *et al.*, 1984). Many bacteria, algae, fungi and yeasts can reduce Au(III) to elemental Au(O) and Ag(I) to elemental Ag(O) (Kierans *et al.*, 1991). Microbial transformations of arsenic and chromium species are also associated with a decrease in toxicity and may have relevance to wastewater treatment (Williams and Silver, 1984).

Organomercurials may be detoxified by organomercurial lyase, the resulting Hg^{2+} then being reduced to Hg^0 by mercuric reductase (Silver *et al.*, 1989; Gadd, 1990). Microbial dealkylation of organometallic compounds such as, organotins can result in the formation of ionic species, which could possibly be removed using a biosorptive process (Gadd, 1993 *b*). Biomethylated metal derivatives, although toxic, are often volatile and may be eliminated from a system by evaporation (Trevors, 1986).

A taxonomically diverse group of heterotrophic bacteria utilize metallic cations as terminal electron acceptors under anaerobic conditions. In the process, the metal is reduced to a lower valency state. There are several metallic elements that possess multiple valency which can potentially be utilized in this way by microorganisms. However, unlike sulphate-reduction, technology has not been developed to date, which employs these processes, despite the microbial reduction of elements such as, Fe and Mn being of great environmental significance (Lovely, 1993). Fe(III) and Mn(IV) appear to be the most commonly utilized metals as terminal electron acceptors in the biosphere and metal reduction organisms from many habitats frequently utilize both of these metals. However, since the solubility of both Fe and Mn increases by bacterial reduction, and neither metal posseses a major toxic threat, the most potentially important dissimilatory metal reduction reactions for biotechnological treatment of toxic metal pollution utilize other metals. Molybdenum (VI) was reduced to molybdenum blue by a strain of *Enterobacter cloacae,* which was isolated from a molybdate-polluted aquatic environment (Ghani *et al.*, 1993). Another stain of this organism, also isolated from a polluted habitat, was able to reduce Cr(VI) to Cr(III) under similar conditions (Wang *et al.*, 1989). The fact that these strains, isolated from polluted waters, may indicate that natural waters do not contain sufficient chromate or molybdate to support similar organisms and that this type of metabolism is an adaptation specifically exploiting environmental pollution by these oxyanions. Dissimilatory of Cr(VI) reduction was also carried out by a strain of *Escherichia coli,* which was unusual it apparently reduced Cr(VI) as a terminal electron acceptor under anaerobic conditions and also under aerobic conditions, albeit at a slower rate (Shan and Wang, 1993). A strain of *Shewanella (Alteromonas) putrefaciens* which reduced Fe(III) and Mn(IV) also reduced U(VI) to U(IV) (Lovely *et al.*, 1991). Due to the low solubility of U(VI), the reaction was accompanied by the formation of a black precipitate of U(IV) carbonate. When the organisms were contained by dialysis tubing, the precipitate was associated with the organisms, indicating that it was the result of an enzymatic reaction rather than reduction by reduced products (Gorby and Lovely, 1992). U(VI) was also reduced by the sulphate-reducing bacterium *Desulfovibrio desulfuricans* utilizing a mechanism that involved the electron transport chain although the organism could not utilize U(VI) as an electron acceptor for growth (Lovely and Phillips, 1992).

Reduction of Cr(VI) to the water insoluble and less toxic Cr(III) results in precipitation and offers a promising bioremediation strategy (Turick *et al.*,1998). Many bacteria belonging to genera *Pseudomanas, Aeromonas, Enterobacter, Escherichia, Bacillus, Streptomyces* can reduce Cr(VI) to Cr(III) (Cervantes and Silver, 1992). The reduction of Cr(VI) to Cr(III) was found to occur in at least in two steps, with Cr(V) as an intermediate. Complete and qualitative transformation of Cr(VI) to Cr(III) have been shown in *Escherichia coli* (Shen and Wang, 1993) and *Agrobacterium radiobacter* (Lovera *et al.*, 1993). Cosier and Dita (1996s), on comparing methods for reduction of Cr(VI) to Cr(III), found that the chemical method using ferrous sulphate for reduction of Cr(VI) is less efficient and more expensive than the microorganisms. The third, thermal method in comparison to the chemical and microbial methods was disadvantageous owing to much more energy consumption and thus, much more expensive, and the advantage of producing the least amount of final residues.

Significant reduction is generally achieved when cell density is very high. The rate of Cr (VI) reduction was increased with increasing cell density, as observed with *E. coli* ATCC 33456 (Shen and Wang, 1994a), *Pseudomonas fluorescens* LB300 (wang and Xiao, 1995), *Bacillus* sp (Wang and Xiao, 1995), *A. radiobacter* (Lovera *et.al.*, 1993) and *Enterobacter cloacae* H01 (Wang *et.al.*, 1991). However, the specific rate of Cr(VI) reduction by *E. coli* was higher at relatively lower cell density of 3×10^8 cell ml^{-1} (Shen and Wang, 1994a). Growth phase is not a stipulated requirement for Cr(VI) reduction to occur. Resting cells of *P. fluorescens* (Bopp and Ehrlich, 1988), *E. coli* (Sen and Wang, 1994a) and *A. radiobacter* (Lovera *et al.*, 1993) were reduced Cr(VI) at the same rate as in the growth media.

A variety of organic compounds as electron donors for chromium reduction is known, however, the majority of the known electron donors are natural aliphatic compounds, mainly low molecular weight carbohydrates, amino acids and fatty acids (Shen and Wang, 1995). Hydrogen serves as the electron donor in *Desulfovibrio vulgaris* (Lovely and Phillips, 1994). Chromium (VI) reduction is an energetic process involving respiratory chains of cells. Numerous environmental factors have been reported to impart notable effects. In aerobic conditions, dissolved oxygen itself caused uncompetitive inhabitation of Cr(VI) reduction in many strains (Wang and Shen, 1995). Rajwade *et al.* (1999) found that nitrite, nitrate and sulphate could reduce specific activity of the chromium reductase in *Pseudomonas mendicina* MCM B-180. They suggested that nitrate reducing enzymes of this strain was responsible for Cr(VI) reduction. In this strain, the Cr(VI) reducing capability conferred by plasmid (Dhakephalkar *et al.*, 1996) and when added to soil microcosms could immobilize 100 chromate µg mg^{-1} (2 mM) soil in 8 h by converting it into trivalent form (Salunkhe *et al.*, 1998). The chromate-contaminated soils, after microbiological treatment, supported growth of wheat seedlings without exerting any toxic effect. In *E. cloacae* H01, Cr(VI) reduction was inhibited slightly by sulphate (120 SO_4^{2-} mg 1^{-1}) or nitrate (150 N mg 1^{-1}). In comparison, nitrate (2,240 N mg 1^{-1}) had no effect on Cr(VI) reduction by *E. coli* (Shen and Wang, 1994a) whereas, no inhibition was observed at 4000, 5760 and 96 SO_4^{2-} mg 1^{-1} for *E. coli* (Shen and Wang, 1994a), *E. cloacae* (Komori *et al.*,1989, 1990) and *P. putida* (Ishibashi, *et al.*,1990), respectively. The inhibitory effect of nitrate on Cr(VI) reduction could be due to the competition for electron between Cr(VI) and nitrate. Whereas, sulphate inhibition could be due to competition in the sulphate transport pathway as there is chemical similarity between CrO_4^{2-} and SO_4^{2-} (Chen and Hao, 1996).

Influence of Cr(VI) concentration on its reduction seems to yield two opposite trends. One, the time required for complete reduction increased progressively as the initial concentration Cr(VI) increased in the cultures of *E. cloacae (*Wang *et al.*,1989), *E. coli* (Shen and Wang, 1994a), *P. fluroescens* (Wang and Xiao, 1995) and *Bacillus* sp (Wang and Xiao, 1995). Second trend was of a lower initial rates of Cr(VI) reduction at higher initial Cr(VI) concentrations (Komari *et al.*, 1989, 1990).

Biological reduction of Cr(VI) may occur both aerobically and anaerobically. In aerobic condition, Cr(VI) reduction is generally associated with soluble proteins with NADH as an electron donor either for requirement or for enhanced activity (Horitsu *et al.*,1987; Ishibashi *et al.*,1990; Shen and Wang, 1993; Wang and Xaio, 1995). In absence of added electron donors, endogenous reserves are used for reduction of Cr(VI) (Ishibashi *et al.*,1990; Shen and Wang, 1993; Wang and Xiao, 1995). Under anaerobic conditions, cytochrome c in *E. cloacae* (Wang *et al.*, 1989), cytochrome *b, d* in *E. coli* (Shen and Wang, 1993) and cytochrome c_3 in *D. vulgaris* (Lovely and Phillips, 1994) serve as the terminal electron acceptors. Although several carbohydrates, amino acids, fatty acids, hydrogen, NADH and endogenous reserves may be used as electron donor, their oxidation products during anaerobic Cr(VI) reduction is not yet known. Till date, there is no evidence to show that electron transport to Cr(VI) may yield energy to support anaerobic growth of Cr(VI) reducing strains. Cells of *A. radiobactor* EPS-916 (Lovera *et al.*, 1993) and *E. coli* ATCC33456 (Shen and Wang, 1993), *Aerococcus sp, Aeromonas* sp and *Micrococcus* sp (Srinath *et al.*, 2001b) are capable of reducing Cr(VI) in liquid media both aerobically and anaerobically, better in the latter condition. They isolated large population of bacteria capable of Cr-reduction from a CETP. It has recently been demonstrated (Kashefi and Lovely, 2000) that a diversity of hypothermophilic microorganisms have the ability to reduce Cr(VI) as well as Fe(III), U(VI), Tc(VII), Co(III) and Mn(IV) with hydrogen as the electron donor. The reduction of U(VI), Tc(VII), Cr(VI), Co(III) and Mn(IV) by *Pyrobaculum islandicum* and *Pyrobaculum aerophilum* at 100^0-C was dependent upon the presence of cells and hydrogen. The metalloid arsenate and selenate were not reduced. Poorly crystalline Fe(III) oxide for *P. islandicum* and *P. aerophilum* and Fe(III)-citrate, proton and yeast extract served as the electron acceptor and hydrogen as the electron acceptor for *P. islandicum* only. Reduction of toxic metals with hyperthermophilic microorganisms or their enzymes may be beneficial in remediation of metal-contaminated waters or waste streams in extreme temperatures.

The optimum pH ranged from 6.0 to 8.5 and the temperature from 30 to 37^0C for all the isolates mentioned above (Komori *et al.*, 1989, 1990; Wang and Shen, 1995; Wang and Xiao, 1995). Susceptibility of Cr(VI) reducing organisms to one or more heavy metals has been reported and unfortunately these heavy metals are concomitantly present in waste water of industrial process. Heavy metals like Cu(II) (Ohtake *et al.*,1990; Chen and Hao, 1996; Garbisu *et al.*,1997), Cd(II) (Chen and Hao, 1996; Garbisu *et al.*, 1997), Hg(II) (Ishibashi *et al.*,1990) Zn(II) (Ohteke *et al.*,1990; Garbisu *et al.*,1997) Ag(II) (Ishibashi *et al.*, 1990), Ni(II) (Chen and Hao, 1996; Garbisu *et al.*,1997), Mo(VI) (Lovera *et al.*,1993) and Fe(III) (Lovera *et al.*,1993) have varying patterns of inhibition depending on the species and strain of bacteria. Low levels of Pb (>2 mg 1^{-1}) have been found to have stimulatory effect on *E. coli* (Shen and Wang, 1994b) and enriched culture (Chen and Hao, 1996). It is probable that mineral chrocoite ($PbCrO_4$), a Cr(III) compound is the final product (Lebedeva and Lyalikova, 1979). *Pseudomonas fluorescens* ATCC13525 essentially required Fe(III) supplementation for detoxification of Cr(VI) (Appanna *et al.*, 1996).

In addition to heavy metals, phenolic compounds are generally present in chromimum containing wastewater and chromium contamination sites. Phenol and *para*-cresol at 5 mM and 2-chloromphenicol at 2 mM concentration severely inhibited both Cr(VI) reduction and cell growth of *P. fluorescens* LB300 (Wang and Xiao, 1995). Anaerobic cultures of *E. coli* ATCC33456 were susceptible to phenolic compounds than the aerobic cultures (Shen and Wang, 1993). Presence of biocides can also hamper Cr(VI) reduction capability (Bhide *et al.*, 1996).

Most of the studies have been carried out with simulated effluent. When strains are challenged with industrial effluent, numerous factors come into fray. Apart from heavy metals presence, need of additional and specific carbon sources occurs (Ohtake *et al.*, 1990; Bhide *et al.*, 1996; Chen and Hao, 1996; Ganguli and Tripathi, 1999). The *in situ* bio-reduction strategy has to be site- specific and requires proper designing depending on different parameters (Turick *et al.*, 1996). This also necessitates undertaking those microbes that are indigenous to that contaminated site and effluent (Gadd and White, 1993; Turick *et al.*, 1996). Another challenge is to overcome the Cr(VI) toxicity that may lead to cell inactivation and loss of Cr(VI) reduction capability (Yamamoto *et al.*, 1993; Shen and Wang, 1994a), necessitating the need for a two stage bioreactor to separate cell growth from the Cr(VI) contact phase (Shen and Wang, 1995). Sometimes, few percent of chromate may remain unreduced and not contemplating to the pollution norms (Smith and Gadd, 2000).

Nutrient amendment of Cr(VI) contaminated soil was found to induce CO_2 evolution and significant Cr(VI) reduction compared to uncontaminated soil (Turick *et al.*, 1998). Glucose (Turick *et al.*, 1998) and a more economically viable alternative molasses (Rege *et al.*, 1997; Turick *et al.*, 1997) were studied as the nutrient supplement. Mixed cultures of Cr(VI) resistant bacteria derived from diverse sources, when grown on sucrose, showed to reduce with very similar kinetics, suggesting that it may be possible to develop a near-universal engineering strategy to accelerate Cr(VI) reduction in soils (Turick *et al.*, 1998).

To augment the Cr(VI) reduction process by lowering the oxidation/reduction status of soil, irrigation by water was examined (Losi *et al.*, 1994). After 20 weeks, the removal percentage of Cr(VI) from the enriched water ranged from 51 to 98% and increased with increasing organic matter loading. Chromium concentrations in the drainage water were consistently < 50 µg 1^{-1} in the organic amended soil. Daily irrigation yielded lower drainage water Cr concentration than weekly irrigation (at a constant weekly volume) due to increased residence time of the water in the bioactive zone. Chromate adsorption accounted for < 1% of the total immobilized Cr and the amount taken up by the *Alfalfa* shoots was < 0.5% of the total added. This method shows promise as a cost-effective treatment for Cr-contaminated groundwater.

Krishnamurthy and Wilkens (1994) have postulated that Cr(VI) can be immobilized in aquifers in a solid phase forming barium chromate, a hexavalent chromium compound. Such solid phase cannot be easily solubilized and may cause problem in soil remediation by "Pump and Treat" technology by acting as a source of Cr(VI) contamination.

Microbial metalloid transformation

Two major transformation processes have been described for metalloids:

(i) reduction of metalloid oxyanions to elemental forms, eg. SeO_4^{2-} and SeO_3^{2-} to Se^0;

(ii) methylatioin of metalloids, metalloid oxyanions or organometalloids to methyl derivatives, eg. AsO_4^{3-} , AsO_2 - and methyllarsonic acid to $(CH_3)_3As$ (trimethylarsine).

Transformation processes have geological significance, since they may modify the mobility and toxicity of metalloids, as well as being of biotechnological potential in bioremediation (Lovely, 1993; Karlson and Frankenberger, 1993; Gadd, 1993*a,b*). Reduction of SeO_4^{2-} and SeO_3^{2-} (both toxic) to elemental selenium (Se^0) results in immobilization and detoxification (Oremland *et al.*, 1990). Methylation of arsenic and selenium compounds results in volatilization (Tamaki and Frankenberger, 1992; Karlson and Frankenberger, 1993).

Microbially-mediated reduction of selenate [Se(VI)] and selenite [Se(IV)] to elemental selenium is well known and numerous bacterial species are known to carry out this process (Lovely, 1993). Maiers *et al.*, (1988) described reduction of SeO_4^{2-} to elemental selenium, resulting in a red precipitate, by microbial populations isolated from water, sediment and soil from the selenium-rich Kesterson Resecoir in California. Microbial reduction of 100 mg Se/1 (1.3 mmol Se/1) was affected within 1 week of incubation. Up to 75 mg Se/1 (0.9 mmol Se/1) was reduced to Se^0. *Wolinella succinogenes,* adapted to grow in SeO_4^{2-} and SeO_3^{2-} at concentrations up to 10 mmol/1, was able to reduce these oxyanions to Se^0, but only during the stationary phase of growth (Tomei *et al.*, 1992). Bacterial cells treated with SeO_4^{2-} or SeO_3^{2-} contained electron dense granules in the cytoplasm, which were verified by energy dispersive X-ray analusis to consist of Se^0. Similar electron dense bodies, also consisting of Se^0, were found both inside and outside cells of *Pseudomonas maltophila* O-2, a strain isolated from a toxic waste site, following incubation in medium containing 40 mmol/1 SeO_3^{2-} (Blake *et al.*, 1993), although it was thought that the presence of extracellular crystals was due to lysis of cells. Reduction of SeO_3^{2-} was inhibited by the addition of buthionine sulfoximine (BSO), a well known inhibitor of reduced glutathione (GSH) biosynthesis, suggesting that GSH was involved in the reduction process. Steinberg and Oremland (1990) measured bacterial reduction of $^{75}SeO_4^{2-}$ to $^{75}Se^0$ in numerous sediment types. Potential rates of SeO_4^{2-} reduction range from 0.07 to 22 µmol SeO42- reduced/1/h., NO_3^-, NO_2^-, MoO42- and WO_4^{2-} inhibited SeO_4^{2-} reduction in some sediments, while SO_4^{2-} partially inhibited reducing activity in fresh water samples. Certain organisms are known to use SeO_4^{2-} as an electron acceptor to support growth. Oremland *et al.* (1989) reported oxidation of acetate, with concomitant reduction of SeO_4^{2-} to Se^0, by bacteria isolated from anoxic sediments. SeO_4^{2-} reduction was inhibited by WO_4^{2-} and CrO_4^{2-} but not by MoO_4^{2-}. The authors proposed the following mechanism for SeO_4^{2-} reduction (1):

$$4CH_3COO^- + 3\,SeO_4^{2-} \rightarrow 3\,Se^0 + 8\,CO_2 + 4\,H_2O + 4\,H^+ \qquad (1)$$

It was not clear, however, whether reduction occurred in a single step or via SeO_4^{2-}. Macy *et al.* (1989) reported that a *Pseudomonas* sp. was able to respire SeO_4^{2-} to SeO_3^{2-}, with oxidation of ^{14}C-labelled acetate to $^{14}CO^2$ by the mechanism given in equation (2)

$$4CH_3COO^- + H^+ + 4\,SeO_4^{2-} \rightarrow 2\,CO_2 + 4\,SeO_4^{2-} + 2\,H_2O \qquad (2)$$

Another isolate, a strict anaerobe, was able to reduce SeO_3^{2-} to Se^0, though it was not able to reduce SeO_4^{2-}. Co-culture of these organisms resulted in reduction of SeO_4^{2-} to Se^0. More recently, a novel species, *Thauera selenatis*, has benn isolated that is capable of anaerobic SeO_4^{2-} respiration to SeO_3^{2-} with concomitant reduction of NO_3^- (Macy *et al.*, 1993*a,b*; Macy and Lawson, 1993). SeO_3^{2-} formed from this reduction was further reduced to Se^0, though reduction of SeO_3^{2-} to Se^0 did not support growth. Mutants of *T. selenatis* lacking nitrite reductase activity were unable to reduce NO_2^- and SeO_3^{2-}, while those with higher nitrate reductase activity than the wild-type strain were capable of greater levels of NO_2^- and SeO_3^{2-} reduction. The authors suggested that the periplasmic nitrite reductase was responsible for SeO_3^{2-} reduction; cytochrome C551 may be a component of this enzyme, though levels of C551 in mutual and wild type cells were not greatly different (Demolldecker & Macy, 1993). It is generally believed that while SeO_4^{2-}

may act as a terminal electron acceptor to support growth of some organisms, SeO_3^{2-} reduction does not support growth and is more likely to be a detoxification mechanism (Lovley, 1993). Reduction of elemental selenium to selenide (Se^{2-}) has been noted in cultures of *Thiobacillus ferroxidans* (Bacon and Ingledew, 1989).

Reduction of TeO32- (also toxic) to Te^0 is also an apparent means of detoxification, as measured in bacteria. The genetics of tellurium resistance in Gram-negative bacteria has been received (Walter and Taylor, 1992). Uptake of TeO_3^{2-} by resistance cell is followed by reduction to Te^{0+}, which is deposited in or around cells, especially near the cytoplasmic membrane (Bradley *et al.*, 1988; Taylor *et al.*, 1988; Lloyd – Jones *et al.*, 1991, 1994). Blake *et al.*, (1993) described TeO_3^{2-} reduction to Te^0 by *P. maltophila* O-2. This process inhibited by BSO, as was SeO_3^{2-} reduction, indicating that GSH was involved in TeO_3^{2-} reduction. The photosynthetic bacterium *Rhodobacter sphaeroides*, grown photoheterotrophically, reduced TeO_3^{2-} at 100 μg/ml (0.4 mmol/1) to Te^0 (Moore and Kaplan, 1992, 1994). The latter was deposited intracellularly, mainly at the cytoplasmic membrane. SeO_3^{2-} is toxic to most microorganisms at 4 μmol/1 (Summers and Jacoby, 1977). The concomitant evolution of H_2 gas from photoheterotrophically-grown cells were observed with reduction of both SeO_3^{2-} and TeO_3^{2-}. Significant levels of $FADH_2$ – dependent TeO_3^{2-} -reducing activity were measured in the membrane fraction of *R. sphaeroides* cells (Moore and Kaplan, 1992).

In contrast to bacterial systems, fungal reduction of metalloids has received relatively little attention. Reduction of SeO_3^{2-} to Se^0 has been observed with *Fusarium* sp. (Ramadan *et al.*, 1988; Gharieb, 1993; Gharieb *et al.*, 1995), *Mortierella* sp. (Zieve *et al.*, 1985), *Saccharomyces cerevisiae, Candida albicans* (Falcone and Nickerson, 1963) and *Aspergillus funiculosus* (Gharieb, 1993; Gharieb *et al.*, 1995). Both extracellular and intracellular deposition of Se^0 has been demonstrated. Gharieb *et al.* (1995) showed that numerous filamentous and unicellular fungal species were capable of SeO_3^{2-} reduction to Se^0, resulting in a red colouration of coloinies. Smith (1974) showed that *Schizosaccharomyces pombe* reduced TeO_3^{2-} to Te^0, giving black or grey colonies.

Removal of selenium (as SeO_4^{2-}) from contaminated water and soil by bacterial and fungal species or populations has been demonstrated. Oremland *et al.* (1990, 1991) described the *in situ* removal of SeO_4^{2-}, by reduction to Se^0, by sediment bacteria in agricultural drainage regions of Nevada. Selenate removal from overlying water was rapid, with real removal rates estimated up to 155 μmol/m/day, depending on the ambient selenium oxyanion concentration. Experiments using $^{75}SeO_4^{2-}$ showed that 85% of *in situ* SeO_4^{2-} removal occurred in the upper 4-8 cm of sediments, while most SeO42- reducing activity was found at depth below 8 cm (Oremland *et al.*, 1990). Flooding of exposed sediments at Kesterson Reservoir with selenium-free water (in order to create anoxic conditions) resulted in reduction (and thereby immobilization) of large quantities of selenium that had been present in sediments (Long *et al.*, 1990). Anoxic conditions must be maintained, however, if selenium is to remain immobilized (Alemi *et al.*, 1988). Microbial methylation of selenium, resulting in its volatization, has also been used for *in situ* bioremediation of selenium-containing land and water, particularly at Kesterson Reservoir (Thompson-Eagle and Frankenberger, 1992). Selenium-contaminated agricultural drainage water was evaporated to dryness until the sediment selenium concentration approached 100 mg Se/Kg dry weight. Conditions such as, carbon source, moisture, temperature and aeration were then optimized for selenium volatilization and the process continued until selenium levels in sediments declined to acceptable levels. Atmospheric levels of selenium did not exceed safety limits, and prevailing weather conditions resulted in selenium being blown towards selenium-deficient areas.

The potential for *ex situ* treatment of selenium-contaminated water has also been demonstrated. SeO_4^{2-} in uranium-mine discharge waters was microbially reduced to Se^0 after passage through a soil column (Kauffman *et al.*, 1986). Removal of SeO_4^{2-} in an algal-bacteria system has also been suggested (Gerhardt *et al., 1991).* NO_3^-, an inhibitor of SeO_4^{2-} reduction, was removed during algal growth. The algal biomass also provides a source of electron donors for the removal of any residual NO_3^- denitrifying bacteria and for SeO_4^{2-} reduction, and thus precipitation, to Se^0. Macy *et al.* (1993*a*) described the simultaneous removal of SeO_4^{2-} and NO_3^- under anaerobic conditions from selenium contaminated drainage water by *Thauera selenatis* in recycled sludge blanket and fluidized bed reactors. SeO_4^{2-} reduction by *T. selenatis* was not inhibited by NO_3^- indeed both SeO_4^{2-} and NO_3^- were reduced concomitantly. SeO_4^{2-} levels were declined from 350 to 400 µg Se/1 to 4-5 µmol Se/1 (0.07 µmol Se/1). The final product of the reduction process was Se^0. NO_3^- in drainage water was reduced by 98%.

Methylation of metalloids

Microbial methylation of metalloids to yield volatile derivatives such as, dimethylselenide or trimethylarsine is a well known phenomenon (Karlson and Frankenberger, 1993; Gadd, 1993 *a,b*). Bacterial species known to produce methyl derivatives of selenium from SeO_4^{2-} and SeO_3^{2-} include *Aeromonas sp.* (Chau *et al.*, 1976), *Bacillus* sp. (Razak *et al.*, 1990) and *Pseudomonas* sp. (Chasteen *et al.*, 1990). Dimethylselenide [$(CH_3)_2Se$] is most common methylated product. There has been much research into selenium methylation by soil fungi (Thompson-Eagle *et al.*, 1991; Karlson and Frankenberger, 1993). *Alternaria alternata* methylated inorganic selenium (SeO_4^{2-}, SeO_3^{2-}) more rapidly than organic forms (Thompson-Eagle *et al.*, 1989). Both dimethylselenide and dimethyldiselenise [$(CH_3)_2Se$] have been detected as volatile products of fungal selenium methylation. While bacteria and fungi are thought to be important in volatilization of selenium from soils and sediments, bacteria are thought to play a more dominate role in selenium-contaminated waters (Thompson-Eagle *et al.*, 1989). The mechanism for selenium methylation appears to involve transfer of methyl groups as carbonium (CH_3^+) ions via the S-adenosyl methionine system (Gadd, 1993*b*). Much less work have been done on tellurium methylation by fungi (Karlson and Frankenberger, 1993). There is evidence of dimethyltelluride and dimethylditelluride production by *Penicillium* sp. (Huysmans and Frankenberger, 1991). However, it is not clear whether tellurium methylation is a detoxification process, since certain methyl derivatives of tellurium are no less toxic than TeO_3^{2-} (Karlson and Frankenbeerger, 1993). Several bacterial and fungal species have been shown to methylatearsenic compounds such as, arsenate [As(V), AsO_4^{3-}], arsenite [As(III), AsO_2^-] and methylarsonic acid ($CH_3H_2AsO_3$) to volatile dimethyl- [$(CH_3)_2HAs$] or trimethylarsine [$(CH_3)_3As$] (Tamaki & Frankenberger, 1992). Methyl derivatives of arsenic are less toxic to bacteria than organic forms. The machenism for anaerobic biomethylation of AsO_3^- by *Methanobacterium* sp. appeared to consist of reduction to AsO_3^- and subsequent methylation to methylarsonic acid, dimethyllarsenic acid and finally dimethyllarsine. Vitamin B_{12} was thought to be the methyl group donor (McBride and Wolfe, 1971). Under aerobic conditions, however, fungal methylation of arsenic compounds appears to involve Sadenosymethionine (Gadd, 1993*b*). Huysmans and Frankerberger (1991) demonstrated production of dimethyl- and trimethyllarsine from methylarsonic acid by *Penicillium* sp.

Sulphide precipitation

Hydrogen sulphide is produced by sulphate-reducing bacteria, eg. *Desulfovibrio* and/or *Desulfotomaculum* sp. The solubility products of most metal sulphides are extremely low and they are readily precipitated as sulphides, eg. ZnS, CdS, CuS and FeS. Sulphate-reducing activity can occur as a useful auxiliary metal-removing mechanism (Brierley, 1990*b*). A 9 m^3 stainless steel sludge-blanket reactor using sulphide-reducung bacteria has been piloted by Shell Research Ltd and Budelco B.V. (Barnes, 1990). This plant successfully removed toxic metals and sulphate from contaminated groundwater at a long –standing smelter site precipitation as metal sulphides, yielding out flow metal concentrations below the ppb range (Barnes *et al.*, 1991). A detailed analysis of this process including mass balance was also carried out (Barnes *et al.*, 1992). The process has since been expanded to a commercial pilot scale using an 1800 m^3 concrete reactor, built by Parques B.V. (Netherlands) at a cost of 33 million guilders for the entire installation. This plant is capable of treating 7000 m^3/day and has been in operation since 30 October 1992. In another pilot-plant study, carried out by the US Bureau of Mines, a 4500 m^3 reactor using spent mushroom compost as substrate and support for sulphide-reducing bacteria removed 95% of the metals and 20% of the sulphate from 19.3 m^3 of coal mine drainage waters (Dvorak *et al.*, 1992).

Cell wall components

Mcrobial exopolymers that have received most attention in relation to metal binding are those, which form capsules or slime layers. Most of thses are composed of polysaccharide, glycoproteins and lipopolysaccharide, which may be associated with protein. Potential interactions of metals with these have been reviewed (Geesey and Jang, 1990). Generally, a correlation exists between high anionic charge and metal complexing capacity. Additionally, there may be deposition of metal in a chemically-altered form (Beveridge and Doyle, 1989).

The cell walls of bacteria also have several metal-binding components, which contribute to the biosorption processes, already, described. The carboxyl groups of the peptidoglycan are the main binding sites in Gram-positive cell walls with the phosphate groups contributing significantly in Gram-negative organisms (McLean and Beveridge, 1990). Many fungi have a high chitin content in cell walls, and this polymer of *N*-acetyl glucosamine is an effective metal and radionuclide biosorbent (Macaskie and Dean, 1990;Tobin *et al.*, 1994). Actinide accumulation in intact biomass appears mainly to comprise metabolism-independent biosorption (Tobin *et al.*, 1987, 1988, 1994), the main site of uptake being the cell wall (Volesky, 1990), although permeabilization of cell walls with carbonates of detergents can increase uptake (Strandberg *et al.*, 1981; Tsezos and Volesky, 1982; Gadd and White, 1989 *a, b, c;* Tsezos, 1983). Chitosan and other chitin derivatives also have a significant biosorptive capability (Wales and Sagar, 1990). Insoluble chitosan-glucan complexes and glucans possessing amino- or sugur acid groups from *Aspergillus niger* exhibit biosorptive properties and effect efficient removal of transition metal ions from solution (Muzzarelli *et al.*, 1986). Melanins are fungal pigments, which enhance survival under environmental stress (Bell and Wheeler, 1986). Fungal melanins are located in and/or exterior to cell walls where they may appear as electron-dence deposit and granules.

Granules may be released into the external medium and be termed "extracellular melanin" although this is generally of identical composition to salt-associated melanin (Bell & Wheeler, 1986). Fungal phenolic polymers and melanins contain phenolic units. Peptides, carbohydrates, aliphatic hydrocarbons and fatty acids possess many potential metal binding sites (Senesi *et al.*, 1987; Sakaguchi and Nakajima, 1987). Oxygen-aontaining groups in these substances, including carboxyl, phenolic and alcoholic gydroxyl, carbonyl and methoxyl groups may be particularly important in metal binding (Gadd & de Rome, 1988). A variety of heavy metals can induce or accelerate melanin production in fungi and melanized cell forms. Chlamydospores can have high metal uptake capacities for metals and organometallic compounds like tributyltin chloride (Mowll and Gadd, 1984; Senesi *et al.*, 1987; Gadd and de Rome, 1988; Gadd *et al.*, 1990).

Metal binding proteins and peptides

Virtually all biological material has a high affinity for toxic metals and radionulides, and are induced by their presence. Specific metal-binding proteins and peptides have been recorded in all microbial groups examined, although most detailed work has been carried out with yeasts (Butt and Ecker, 1987; Dameron *et al.*, 1989; Mehruand& Winge, 1991).

Metallothioneins are small cysenin-rich polypeptides that can bind essential metals, e.g. Cu and Zn, as well as non-essential metals like Cd. Metallothioneins are low molecular weight 6-7 kDa), cysteine-rich protein found in animals, higher plants, eukaryotic microorganisms and some prokaryotes (Hamer, 1986). They form complexes with metal for their detoxification. They mediate copper resistance in *Saccharomyces cerevisiae* (Fogel *et al.*, 1988) and also bind other metals (Mehra and Winge, 1991). Metal ϒ-glutamyl peptides (Phytochelatins) are short peptides involved in heavy metal detoxification in algae, plants and some fungi and yeasts (Mehra and Winge, 1991; Gadd, 1993*a*). Both of these classes of metal-binding peptides may have an application for metal-removal in the future.

One approach that has evolved to enhance binding of metals on surface of bacterial cells is expression of metallothioneins. Metallothioneins of yeast (CUP1) and mammalians (HMTIA), when expressed as fusions to LanB, enhanced the metal binding capacity of the cells between 15 and 20 fold (Sousa *et al.*, 1998). When the *Neurospora crassa* metallothioneins was expressed in 1-12 repeat as a fusion with the maltose binding protein in the periplasm of *E. coli* and the Cd binding capacity was increased ten folds with one repeat and maximum 65 fold with nine repeats (Mauro and Pazirandeh, 2000). Novel binding peptides, rich in either histidines or cysteines, were engineered to lamb and expressed on the surface of *E. coli* (Kotrba *et al.*, 1999). Surface display of histidines and cysteines increased biosorption of heavy metals by four-fold and two-fold, respectively. Expressions of such motifs in diverse fusion proteins have enhanced biosorption of Hg^{2+} and Cd^{2+} by around 11 folds (Pazirandeh *et al.*, 1995; Sousa *et al.*, 1996). It is of interest to evaluate the biosorption capability of these peptides in more environmentally robust bacteria, such as *Pseudomonas,* for their potential in bioremediation (Mejare and Bulow, 2001).

Potential for future development

Several biotechnological approaches seem to be established as a means of combating toxic metal pollution from industrial and other sources, although none are yet in widespread use. Several design preferences are beginning to encourage within the field with the processes currently in our near to practical operation, mainly utilization biosorption and or bioprecipitation as these require the least modification of biological material and can be used in those process technologies which are already well established for ion-exchange or biological treatment of organic wastes. There has also been some attention given to the possibilities of integrating biological metal removal with sewage and other waste treatment (Cowling *et al.*, 1992). A technically simpler approach using artificial wetlands with undefined biota has been found useful for cost-effective treatment of high volume, low concentration wastes such as, mine runoff (Barierly, 1999*b*). In the longer term, the use of purified biopolymers as specific metal binding agents holds out considerable potential (Ashley & Roach, 1990). The application of both genetic and protein engineering could conceivably lead to peptides or other biopolymers with enhanced metal specificity, stability and other useful properties. It is clear that some microbiological methods for the treatment of metal/radionuclide-containing wastes offer potentially efficient and cost-effective alternatives or adjuncts to existing treatment technologies.

Summary

In recent years significant advances have been made in the understanding of microbe-metal interactions. Many of these microbe-metal interactions have potential applications for the remediation of metal contaminated environments and waste streams; however, potential bioremediation strategies for metal have yet to be significantly applied to large-scale environmental restoration efforts. Collaboration between microbiologists, environmental engineers and geochemists is required in order to bring this promising technology to fruition.

References

Adams, B.A. and Holmes, E.L. (1935) Adsorptive properties of synthetic resins. I, *J. Soc. Chem. Ind.*, 54:1-6

Ahm, D.-H., Chung, Y.-C. and Pak, D. (1998) Biosorption of heavy metal ions by immobilized *Zoogloea* and zooglan. *Appl. Biotechnol.*, 73:43-50.

Aksu, Z. and Acikel, U. (1999) A single-staged bioseparation process for simultaneous removal of copper (II) and chromium (VI) by using *C. vulgaris*. *Process Biochem.*, 34:589-599.

Akthar, M.N. and Mohan, P.M. (1995) Bioremediation of toxic metal ions from polluted lake waters and industrial effluents by fungal biosorbent. *Curr. Sci.*, 69, 1028-1030.

Al-Asheh, S. and Dvunjak, Z. (1995) Adsorption of copper and chromium by *Aspergillus carbonarius*. *Biotechnol. Prog.*, 11:638-642.

Appanna, V.D., Gazso, L.G., Huang, L. and Pierre, M. St. (1996) Mechanism of chromium detoxification in *Pseudomonas fluorescens* is dependent on iron. *Bull. Environ. Contam. Toxicol.*, 57, 875-880.

Ashley, N. V. and Roach, D.J.W. (1990). Review of biotechnology applications to nuclear waste treatment. *J. Chem. Technol. Biotechnol.*, 49, 381-394.

Bacon, M. and Ingledew, W. J. (1989). The reductive reactions of *Thiobacillus ferroxidans* on sulphur and selenium. *FEMS Microbiol. Lett.*, 58, 189-94.

Baillet, F., Margnin, J.-P., Cheruy, A. and Ozil, P. (1998) Chromium precipitation by the acidophilic bacterium *Thiobacillus ferrooxidans*. *Biotechnol. Lett.*, 20, 95-99.

Barnes, L. J., Janssen, F. J., Scheeren, P. J. M., J., Versteegh, J.H. and Koch, R. O. (1992). Simultaneous microbial removal of sulphate and heavy metals from wastewater. *Trans. Inst. Mining Metall.* C, 101, 183-189.

Barnes, L. J., Janssen, F. J., Sherren, J., Versteegh, J.H., Koch, R. O. and Scheeren, P. J. H. (1991). A new process for the microbial removal of sulphate and heavy metals from contaminated wasters extracted by a geohydrological control system. *Tra::s. Inst. Chem. Engrs*, 69, 184-186.

Bedell, G. W. and Darnall, D.W. (1990). Immobilization of nonviable, biosorbent algal biomass for the recovery of metal ions. In: *Biosorption* of Heavy Metals, ed. B. Volesky, CRC Press, Boca Raton, FL, pp. 313-326.

Bell, A. A. and Wheeler, M. H. (1986). Biosynthesis and functions of fungal melanins. *A. Rev. Phytopathol.*, 24, 411-451.

Belliveau, B.H., Starodub, M.E., Cotter, C. and Trevors, J.T. (1987) Metal resistance and accumulation in bacteria. *Biotechnol. Adv.*, 5, 101-127.

Bender, J., Lee, R.F. and Phillips, p. (1995) Uptake and transformation of metal and metalloids by microbial mats and their use in bioremediation. *J. Ind. Microbial.*, 14, 113-118.

Beveridge, T. J. and Doyle, R. J. (1989). *Metal Ions and Bacteria.* J Wiley, New York.

Beveridge, T.J. (1988) The bacterial surface: general consideration towards design and function. *Con. J. Microbial.*, 34: 363-372.

Bhide, J.V., Dhakephalkar, P.K. and Paknikar, K.M. (1996) Microbiological process for the removal of Cr(VI) from chromate-bearing cooling tower effluent. *Biotechnol. Lett.*, 18, 667-672.

Blake, R. C., Choate, D. M., Bardhan, S., Revis, N., Barton, L. L. and Zocco, T. G. (1993). Chemical transformation of toxic metals by a *Pseudomonas* strain from a toxic waste site. *Environ. Toxicol. Chem.*, 12, 1365-1376.

Bopp, L.H. and Ehrlich, H.L. (1988) Chromate resistance and reduction in *Pseudomonas fluorescens* stain LB300. Arch. Microbiol., 150, 426-431.

Bradley, D. E., Grewal, K. K., Taylor, D. E. and Whelan, J. (1988). Characteristics of RP4 tellurite-resistance transposon Tn521. *J. Gen. Microbiol.*, 134, 2009-2018.

Brierley, C. L. (1990a). Metal immobilization using bacteria. pp. 303-323, In: *Microbial Mineral Recovery*, eds H. L. Ehrlich and C. L. Brierley. McGraw-Hill, New York.

Brierley, C. L. (1990b). Bioremediation of metal-contaminated surface and groundwaters. *Geomicrobiol. J.*, 8, 201-223.

Brierley, J. A. (1990c). Production and application of a Bacillus-based product for use in metals biosorption. pp. 305-311, In: *Biosorption of Heavy Metals*, ed. B. Volesky. CRC Press, Boca Raton, FL.

Brierley, C. L., Brrierley, J. A. and Davidson, M. S. (1989). Applied microbial processes for metals recovery and removal from wastewater. pp. 359-382, In: *Metals Ions and Bacteria*, eds T. J. Beveridge and R. J. Doyle. J Wiley, New York.

Brierley, J.A., Goyak, G.M. and Brierley, C.L. (1986) Considerations for commercial use of natural products for metal recovery, pp. 105-117, In : *Immobilization of ions by biosorption*, Eccles, h. and Hunt, S, eds., Horwood, Chichester.

Butt, T. R. and Ecker, D. J. (1987). Yeast metallothionein and applications in biotechnology. *Microbiol. Rev.*, 51, 351-364.

Celma, P.J., Carrio, R., Choque, R., Cabeza, L.F. and Cot, J. (1999) Chrome recovery from sheepsking tanning baths by an ion exchange (anionic) process. Pilot scale triale and in tannery chrome reuse. *Proc. XXV IULTCS Congress*, Tata McGraw-Hill Publishers, New Delhi, pp. 458-461.

Cervantes, C. and Silver, S. (1992) Bacterial chromate resistance and reduction. *Plasmid*, 27, 65-71.

Cha, J.-S. and Cooksey, D. A. (1991) Copper resistance in Pseudomonas syringae mediated by periplasmic and outer membrane proteins. *Proc, Natl. Acad. Sci. USA*, 88, 8915-8919.

Chaney, R.L., Malik, M., Li, Y.M., Brown, S.L., Brewer, E.P., Angel, J.S. and Baker, A.J. (1997) Phytorremediation of soil metals. *Curr. Opin. Biotechnol.*, 8, 279-284.

Chang, J.S., Law, R. and chang, C.C. (1997) Biosorption of lead, copper and cadmium by biomass of *Pseudomonas aeruginosa PU21*. *Water Res.*, 31: 1651-1658.

Chasteen, T. G., Silver, G. M., Briks, J. W. and Fall, R. (1990). Fluorine-induced chemiluminescence detection of biologically methylated tellurium, selenium and sulfur compounds. *Chromatographia*, 30, 181-185.

Chau, Y. K., Wong, P. T. S., Silverberg, B. A., Luxin, P. L. and Bangert, G. A. (1976). Methylation of selenium in the aquatic environment. *Science*, 912, 1130-1131.

Chen, J.M. and Hao, O.J. (1996) Environmental factors and modeling in microbial chromium (VI) reduction. *Water Environ. Res.*, 68, 1156-1164.

Cosier, V. and Dita, M. (1996) Microbial reduction of hexavalent chromium in activated sludge. *Studia Universitatis Babes-Bolyai Biologia*, 41, 169-176.

Costerton, J.W., Lewandowski, Z., de Beer, D., Caldwell, D., Korber, D. and James, G. (1994) Biofilms: the customized microniche. *J. Bacteriol.*, 176, 2137-2142.

Cot, J., Manich, A.M., Marsal, A., Fort, M., Celma, P.J., Carrio, R., Choque, R. and Cabeza, L.F. (1999) Processing of collagenic residues. Isolation of gelarin by the action of peroxochromates, *Proc. XXV JULTCS Congress*, Tata McGraw-Hill Publishers, New Delhi, pp. 135-145.

Cowling, S. J., Gardner, M. L. and Hunt, D. T. E. (1992). Removel of heavy metals from sewage by sulphide precipitation: thermodynamic calculations and tests on a pilot-scale anaerobic reactor. *Environ. Technol.*, 13, 281-291.

Cunningham, S.D. and Berti, W.R. (1993) Remediation of contaminated soils with green plants: An overview. *In Vitro Cell Dev. Biol.*, 29, 207-212.

Cunningham, S.D. and Ow, D.W. (1996) Promises and prospects of phytoremediation. *Plant Physiol.*, 110, 715-719.

Dameerson, C. T., Reese, R. N., Mehra, R. K., Kortan, A. R., Carroll, D. J., Steigerwald, M. L., Brus, L. E. and Winge, D. R. (1989). Biosynthesis of cadmium sulphide quantum semiconductor crystallites. *Nature*, 338, 596-598.

De Rome, L. and Gadd, G. M. (1991). Use of pelleted and immobilized yeast and fungal biomass for heavy metal and radionuclide recovery. *J. Indust. Microbiol.*, 7, 97-104.

Demolldecker, H. and Macy, J. M. (1993). The periplasmic nitrite reductase of *Thauera selenatis* may catalyze the reduction of selenite to elemental selenium. *Arch. Microbiol.*, 160, 241-247.

Dhakephalkar, P.K., Bhide, J.V. and Paknikar, K.M. (1996) Plasmid mediated chromate resistance and reduction in *Pseudomonas mendocina* MCM B-180. *Biotechnol. Lett.*, 18, 1119-1122.

Dvorak, D. H., Hedin, R. S., Edonborn, H. S., and McIntire, P. E. (1992). Treatment of metal-contaminated water using bacterial sulfate-reduction: results from pilot-scale reactors. *Biotechnol. Bioengng*, 40, 609-616.

Ehrlich, H. L. and Brierley, C. L. (1990). *Microbial Mineral Recovery*. McGraw-Hill, New York.

Failla, M.L. (1977) Zinc: Function and transport in microorganisms. P. 151-214, In: Microorganisms and minerals, Weinberg, E.D., ed., Marcel Dekker.

Flatman, P.E. and Lanza, G.R. (1998) phytoremediation: Current views on an emerging green technology. *J. Soil Contamin.*, 7, 415-432.

Fogel, S., Welch, J. W. and Mahoney, D. H. (1988). The molecular genetics of copper resistance in *Saccharomyces cerevisiae* – a paradigm for non-conventional yeasts. *J. Basic Microbiol.*, 28, 147-60.

Fude, L., Harris, B., Urrutia, M.M. and Beveridge, T.J. (1994) Reduction of Cr(VI) by a consortium of sulfate-reducing bacteria (SRB-III) *Appl. Environ. Microbiol.*, 60, 1525-1531.

Gadd, G. M. (1993a). Interactions of fungi with toxi metals. *New Phytol.*, 124, 25-60.

Gadd, G. M. (1993b). Microbial formation and transformation of organometallic and organometalloid compounds. *FEMS Microbiol. Rev.*, 11, 297-316.

Gadd, G. M. and Rome, L. de (1988). Biosorption of copper by fungal melanin. *Appl. Microbiol. Biotechnol.*, 29, 610-617.

Gadd, G. M. and White, C. (1989a). Heavy metal and radionuclide accumulation and toxicity in fungi and yeasts. In: Metal – Microbe Interactions, eds R. K. Poole and G. M. Gadd, pp. 19-38. IRL Press, Oxford.

Gadd, G. M. and White, C. (1989b). The removal of thorium from simulated acid process streams by fungal biomass. *Biotechnol. Bioengng*, 33, 592-597.

Gadd, G. M. and White, C. (1989c). Uptake and intracellular distribution of thorium in *Saccharomyces cerevisiae. Environmental Pollution*, 61, 187-197.

Gadd, G. M. and White, C. (1990). Biosorption of radionuclides by yeast and fungal biomass. *J. Chem. Technol. Biotechnol.*, 49, 331-343.

Gadd, G.M. (1988) Accumulation of metals by microorganisms and algae, pp. 401-433, In: *Biotechnology- a comprehensive treatise*, Vol. 6b, Special microbial process, VCH Verlagsgesellschaft, Weinheim, Germany.

Gadd, G.M. (1990a) Fungi and yeasts for metal accumulation, pp. 249-275, In: *Microbial mineral recovery*, Ehrlich, H.L. and Brierley, C.L., eds., McGraw-hill, New York.

Gadd, G.M. (1990b) Heavy metal accumulation by bacteria and other microorganisms. *Experientia*, 46, 834-840.

Gadd, G.M. (1996) Influence of microorganisms on the environmental fate of radionuclides. *Endeavour*, 20, 150-156.

Gadd, G.M. and White, C. (1993) Microbial treatment of metal pollution – a working biotechnology? *Trends Biotechnol.*, 11, 353-359.

Ganguli, A. and Tripathi, A.K. (1999) Survival and chromate reducing ability of *Pseudomonas aeruginosa* in industrial effluents. *Lett. Appl. Microbiol.*, 28, 76-80.

Garbişu, C., LIama, M.J. and Serra, J.L. (1997) Effect of heavy metals on chromate reductiuon by *Bacillus subtilis*. *J. Gen. Appl. Microbiol.*, 43, 369-371.

Garnham, G.W. and Green, M. (1995) Chromate (VI) uptake by and interactions with cyanobacteria. *J. Ind. Microbial.*, 14, 247-251.

Geesey, G. and Jang, L. (1990). Extracellular polymers for metal binding. pp. 223-247, In: *Microbial Mineral Recovery*, eds H. L. Ehrlich and C. L. Brierley. McGraw-Hill, New York.

Geesey, G.G. and Jang, j. (1989) Interactions between metal ions and capsular polymers, . Pp. 325-357, In: *Metal ions and bacteria*, Beveridge, T.J. and Doyles, R.J., eds., John Wiley and Co., New York.

Gerhardty, M. B., Green, F. B., Newman, D., Lundquist, T. J., Tresan, R. and Oswald, W. J. (1991). Removal of selenium using a novel algal-bacterial process. *Res. J. Water Pollution Control Fed.*, 63, 799-805.

Ghani, B., Takai, M., Hishani, N. Z., Kishimoto, N., Ismail, A. K., Tano, T. and Sungio, T. (1993). Isolation and Characterisation of a Mo^{6+} -reducing bacterium. *Appl. Environ. Microbiol.*, 59, 1176-1180.

Gharieb, M. M. (1993). Selenium toxicity, accumulation and metabolism by fungi and influence of the fungicide dithane. Ph.D. Thesis, University of Dundee.

Gharieb, M. M., Wilkinson, S. C. and Gadd, G. M. (1995). Reduction of selenium oxyanions by unicellular, polymorphic and filamentous funfi: cellular location of reduced selenium and implications for tolerance. *J. Indust. Microbiol.*, 14, 300-311.

Goodman, G.T. and Roberts, T.M. (1971) Plants and soils as indicators of metals in air. *Nature*, 231, 287-292.

Gorby, Y. and Lovley, D. R. (1992). Enzymatic uranium precipitation. *Environ. Sci. Technol.*, 26, 205-207.

Guan, L., Petersen., J.N., Johnstone, D.L., yonge, D.R. and Brouns, T.M. (1993a) Equilibrium sorption of Cr^{6+} by a consortia of denitrifying bacteria. *Biotechnol. Lett.*, 15, 727-732.

Guan, L., Petersen., Johnstone, D.L., Yonge, D.R. Brouns, T.M., (1993b) Sorption rates for the uptake of Cr^{6+} by a consortia of denitrifying bacteria. *Biotechnol. Lett.*, 15, 733-736.

Hamer, D.H. (1986) Metallothionein. *Ann. Rev. Biochem.*, 55, 913-951.

Hansen, C. L., Zwolinski, G., Martin, D. and Williams, J. W. (1984). Bacterial removal of mercury from sewage. *Biotechnol. Bioengng.*, 26, 1330-1333.

Harris, P. A. and Ramelow, G. J. (1990). Binding of metal ions by particulate biomass derived from Chlorella vulgaris and Scenedesmus quadricauda. *Environ. Sci. Technol.*, 24, 220-228.

Horitsu, H., Futo, S., Miyazawa, Y., Ogai, S. and Kawai, K. (1987) Enzymatic reduction of hexavalent chromium by hexavalent chromium tolerant *Pseudomonas ambigua* G-1. *Agric. Biol. Chem.*, 51, 2417-2420.

Horitsu, H., Futo, S., ozawa, K. and Kawai, K. (1983) Comparison of characteristics of hexavalent chromium-tolerant bacterium, *Pseudomonas ambigua* G-1, and its hexavalent chromium-sensitive mutant. *Agric. Biol. Chem.*, 47, 2907-2908.

Hu, M.Z.-C., Norman, J.M., Faison, B.D. and Reeves, M.F. (1996) Biosorption of uranium by *Pseudomonas aeruginosa* strain CSU: characterization and comparison studies. *Biotechnol. Bioeng.*, 51, 237-247.

Huges, M.N. and Poole, R.K. (1989) *Metals and Microorganisms*, Chapman and Hall, London.

Hutchins, S. R., Davidson, M. S., Brierley, J. A. and Brierley, C. L. (1986). Microorganisms in reclamation of metals. *A. Rev. Microbiol.*, 40, 311-336.

Huysmans, K. D. and Frankenberger, W. T. (1991). Evolution of trimethylarsine by a *Penicillium* sp. isolated from agricultural evaporation pond water. *Sci. Total Environ.*, 105, 13-28.

Ishibashi, Y., Cervantes, C. and Silver, S. (1990) Chromium reduction in *Pseudomonas putida. Appl. Environ. Microbiol.*, 56, 2268-2270.

Kajander, E.O. and Ciftcioglu, N. (1998) Nanobacteria: an alternative mechanism for pathogenic intra-cellular calcification and stone formation. *Proc. Natl. Acad. Sci. U.S.A.*, 95, 8274-8279.

Kannan, A, and Ramteke, P.W. (2002) Uptake of Nikel (H) by *Serratia marcescens. J. Cnviron Biol.* 23: 57-59.

Kannan, A. and Ramteke, P.W. (2000) uptake of nickel (II) by Serratia marcescens. *J. Envmon. Biol.* 23:57-59.

Kar, N.S. and Dasgupta, A.K. (1996) The possible role of Surface Change in membrane organisation in an acidophile Ind. *J. Biochem. Biophys.* 33 398-402.

Kar, N.S. and Lovley, D.R. (1996) The possible role of surface charge in membrane organization in an acidophilie. *Ind. J. Biochem. Biophys.*, 33, 398-402.

Karlson, U. and Frankenberger, W. T. (1993). Biological alkylation of selenium and tellurium. pp. 185-227, In: *Metal Ions in Biological Systems*, Vol. 29, *Biological Properties of Metal Alkyl Derivatives*, eds H. Sigel and A. Sigel. Marcel Dekker, New York.

Kashefi, K. and Lovley, D. R. (2000) Reduction of Fe(III), Mn(IV), and toxic metals at 100 °C by *Pyrobaculum islandicum, Appl. Environ. Microbiol.*, 1050-1056.

Kauffman, J. W., Laughlin, W. C. and Baldwin, R. A. (1986). Microbiological treatment of uranium mine waters. *Environ. Sci. Technol.*, 20, 243-248.

Kazy, S. K., Sar, P., Asthana. R. K. and Singh, S. P. (1999) Copper uptake and its compartmentalization in *Pseudomonas aeruginosa* strains: chemical nature of cellular metal *World J. Microbiol.*, 15, 599-605.

Keeffe, W.E. (1976) Formation of crystalline deposits by several genera of the family Enterobacteriaceae. *Infect. Imminol.*, 14, 590-592.

Kierans, M., Staines, A. M., Bennett, H. and Gadd, G. M. (1991). Silver tolerance and accumulation in yeasts. *Biol. Metals*, 4, 100-106.

Klaus-Joerger, T., Joerger, R., Olsson, E. and Granqvist, C.-G. (2001) Bacteria as workers in the living factory: metal-accumulating bacteria and their potential for materials science. *Trends Biotechnol.*, 19, 15-20.

Komori, K., Rivas, A., Toda, K. and Ohtake, H. (1990) Biological removal of toxic chromium using an *Enterobacter cloacae* strain that reduces chromate under anaerobic conditions. *Biotechnol. Bioeng.*, 35, 951-954.

Komori, K., Wang, P.C., Toda, K. and Ohtake, H. (1989) Factors affecting chromate reduction in *Enterobacter cloacae* strain H01. *Appl. Microbiol. Biotechnol.*, 31, 567-570.

Kong, S., Johnstone, D.L., Yonge, D.R., Petersen, J.N. and Brouns, T.M. (1993b) Remobilization of chromium from starved and fresh subsurface bacterial consortium. *Biotechnol. Lett.*, 15, 1081-1084.

Kong, S., Yonge, D.R., Johnstone, D.L., and Brouns, T.M. (1993a) Competing ion effect on chromium adsorption with fresh and starved subsurface bacterial consortium. *Biotechnol. Lett.*, 15, 87-92.

Kong, S., Yonge, D.R., Johnstone, D.L., Petersen, J.N. and Brouns, T.M. (1992) Chromium distribution in subcellular components between fresh and starved subsurface bacterial consortium. *Biotechnol. Lett.*, 14, 521-524.

Kotrba, P., Doleckova, L., de Loranzo and Ruml, T. (1999) Enhanced bioaccumulation of heavy metal ions by bacterial cells due to surface display of short metal binding peptides. *Appl. Environ. Microbiol.*, 65, 1092-1098.

Kratochvil, D. and Volesky, B. (1998) Advances in the biosorption of heavy metals. *Trends Biotechnol.*, 16, 291-300.

Krishnamurthy, S. and Wilkens, M.M. (1994) Environmental chemistry of chromium. *Northeastern Geol.*, 16, 14-17.

Kurek, E., Czaban, J. and Bollag, J.-M. (1982) Sorption of cadmium by microorganisms in competition with other soil constituents. *Appl. Environ. Microbiol.* 43, 1011-1015.

Kuyucak, N. and Volesky, B. (1989) The elution of gold sequestered on a natural biosorbent. *Biorecovery*, 1, 205-218.

Lebedeva, E.V. and Lyalikova, N.N. (1979) Reduction of crocoite by *Pseudomonas chromatophila* sp. nov. *Microbiologiya*, 48, 517-522.

Lloyd-Jones, G., Osborn, A. M., Ritchie, D. A., Strike, P., Hobman, J. L., Brown, N. L. and Rouch, D. A. (1994). Accumulation and intracellular fate of tellurite in tellurite-resistant *Escherichia coli*: a model for the mechanism of resistance. *FEMS Microbiol. Lett.*, 118, 113-120.

Lloyd-Jones, G., Ritchie, D. A. and Strike, P. (1991). Biochemical and biophysical analysis of plasmid pMJ600-encoded tellurite [$TeO^2{}_3{}^-$] resistance. *FEMS Microbiol. Lett.*, 81, 19-24.

Long, R. H. B., Benson, S. M., Tokunaga, T. K. and Yee, A. (1990). Selenium immobilization in a pond sediment at Kesterson Reservoir. *J. Environ. Qual.*, 19, 302-311.

Losi, M.E., Amrhein, C. and Frankenberger Jr., W.T. (1994) Bioremediation of chromate-contaminated groundwater by reduction and precipitation in surface soils. *J. Environ. Qual.*, 23, 1141-1151.

Lovely, D.R. and Phillips, E.J.P. (1994) Reduction of chromate by Desulfovibrio vulgaris and its c_3 cytochrome *Appl. Environ. Microbiol.* 60: 726-728.

Lovera, S.L., Bonet, R., Simon-Pujol, M.D. and Congregado, F. (1993) Effect of culture medium ions on chromate reduction by resting cells of *Agrobacterium radiobacter*. *Appl. Microbiol. Biotechnol.*, 39, 424-426.

Lovley, D. R. (1993). Dissimilatory metal reduction. *Ann. Rev. Microbiol.*, 47, 263-290.

Lovley, D. R. and Phillips, E. J. P. (1992). Reduction of uranium by *Desulfovibrio desulfuricans*. *Appl. Environ. Microbiol.*, 58, 850-856.

Lovley, D. R., Phillips, E. J. P., Gorby, Y. A. and Landa, E. R. (1991). Microbial reduction of uranium. *Nature*, 350, 413-416.

Lovley, D.R. (1995) Bioremediation of organic and metal contaminants with dissimilatory metal reduction. *J. Ind. Microbiol.*, 14, 85-93.

Lovley, D.R. and Coates, J.D. (1997) Bioremediation of metal contamination: *Curr. Opin. Biotechnol.*, 8, 285-289.

Macaskie, L. E. and Dean, A. C. R. (1989). Microbial metabolism, Desolubilization and deposition of heavy metals : uptake by immobilized cells and application to the treatment of liquid waetes. pp. 150-201, In *Biojogical Waste Treatment*, ed. A. Mizrahi. Liss, New York.

Macaskie, L. E. and Dean, A. C. R. (1990). Metal-sequestering biochemicals. pp. 199-248, In: *Biosorption of Heavy Metals*, ed. B. Volesky. CRC Press, Boca Raton, FL.

Macy, J. M. and Lawson, S. (1993). Can yield (Y(M)) of Thauera selenatis grown anaerobically with acetate plus selenate or nitrate. *Arch. Microbiol.*, 160, 295-298.

Macy, J. M., Lawson, S. and Demolldecker, H. (1993b). Bioremediation of selenium oxyanions in San Joaquin drainage water using Thauera selenatis in a biological reactor system. *Appl. Microbiol. Biotechnol.*, 40, 588-594.

Macy, J. M., Michel, T. A. and Kirsch, D. G. (1989). Selenate reduction by a Pseudomonas species: a new mode of anaerobic respiration. *FEMS Microbiol. Lett.*, 61, 195-198.

Maiers, D. T., Wichlacz, P. L., Thompson, D. L. and Bruhn, D. F. (1988). Selenate reduction by bacteria from a sclenium-rich environment. *Appl. Environ. Microbiol.*, 54, 2591-2593.

Mallick, N. and Rai, L.C. (1994) Removal of inorganic ions from wastewaters by immobilized microalgae. *World J. Biotechnol.*, 10, 439-443.

Marques, A.M., Roca, X., Simon-Pujol, M.D., Fuste, M.C. and Congregado, F. (1991), Uranium accumulation by Pseudomonas sp. EPS-5028. *Appl. Microbiol. Biotechnol.*, 35, 406-410.

Mauro, J.M. and Pazirandeh, M. (2000) Construction and expression of functional multidomain polypeptides in *Escherichia coli*: expression of the *Neurospora crassa* metallothionein gene. *Lett. Appl. Microbiol.*, 30, 161-166.

McBride, B. C. and Wolfe, R. S. (1971). Biosysnthesis of dimethylarsine by *Methanobacterium. Biochemistry*, 10, 4312-4317.

McLean, R. J. C. and Beveridge, T. J. (1990). Metal-binding capacity of bacterial surfaces and their ability to form mineralized aggregates. pp. 185-222, In: *Microbial Mineral Recovery*, eds H. L. Ehrlich and C. L. Brierley. McGraw-Hill, New York.

Mehra, R. K. and Winge, D. R. (1991). Metal ion residence in fungi: molecular mechanisms and their related expression. *J. Celluar Biochem.*, 45, 30-40.

Mejare, M. and Bulow, L. (2001) Metal-binding proteins and peptides in bioremediation and phytoremediation of heavy metals *Trends Biotechnol.*, 19, 67-73.

Merrin, J.S., Sheela, R., Saswathi, N., Prakasham, R.S. and Ramakrishna, S.V. (1998) Biosorption of chromium VI using *Rhizopus arrhizus*. *Indian J. Exp. Biol.*, 36, 1052-1055.

Mishra, S.P. and Chaudhary, G.R. (1996) Kinetics of Zn^{2+} adsorption by *Penicillium* sp. *Hydrometallurgy*, 40, 11-23.

Money, C.A. (1999) Clean technology challenges, *Proc. XXY IULTCS Congress*, Tata McGraw-Hill Publishers, New Delhi, pp. 284-294.

Moore, M. D. and Kaplan, S. (1992). Identification of intrinsic high-level residence to rare-earth oxides and oxyanions in members of the class Proteobacteria: characterization of tellurite, selenite and rhodium sesquioxide reduction in *Rhodobacter sphaeroides*. *J. Bacteriol.*, 174, 1505-1514.

Moore, M. D. and Kaplan, S. (1994). Members of the family Rhodospirillaceae reduce heavy-metal oxyanions to maintain redox poise during photo-synthetic growth. *Am. Soc. Microbiol. News*, 60, 17-23.

Mowll, J. L. and Gadd, G. M. (1984). Cadmium uptake by *Aureobasidium pullulans*. *J. Gen. Microbiol.*, 130, 279-284.

Muraleedharan, T.R., Iyengar, L. and Venkobachar, C. (1991) Biopsorption: an attractive alternative for metal removal and recovery. *Curr. Sci.*, 61, 379-385.

Muzzarelli, R. A. A., Bregani, F. and Sigon, F. (1986). Chelating abilities of amino acid glucans and sugar acid glucans derived from chitosan. pp. 173-182, In: *immobilization of Ions by Biosorption*, eds H. Eccles and S. Hunt, Ellis Horwood, Chichester.

Nies, D. H., (1999) microbial heavy-metal resistance, *Appl. Microbiol. Biotechnol*. 51, 730-750.

Ohtake, H., Fujii, E. and Toda, T. (1990) Reduction of toxic chromate in an industrial effluent by use of a chromate-reducing strain of *Enterobacter cloacae*. *Environ. Technol.*, 11, 663-668.

Oremland, R. S., Hollibaugh, J. T., Maest, A. S., Presser, T., Miller, L. G. and Culbertson, C. W. (1989). Selenate reduction to elemental selsnium by anaerobic bacteria in sediments and culture: biogeochemical significance of a novel, sulfate-independent respiration. *Appl. Environ. Microbiol.*, 55, 2333-2343.

Oremland, R. S., Steinberg, N. A., Maest, A. S., Miller, L. G. and Hollibaugh, J. T. (1990). Measurements of *in situ* rates of selenate removal by dissimilatory bacterial reduction in sediments. *Environ. Sci. Technol.*, 24, 1157-1164.

Oremland, R. S., Steinberg, N. A., Presser, T. S. and Miller, L. G. (1991). *In situ* bacterial selenate reduction in the agricultural drainage system of Western Nevada. *Appl. Environ. Microbiol.*, 57, 615-617.

Paknikar, K.M., Puranik, P.R. and Pethkar, A.V. (1999) Development of microbial biosorbents – need for standardizations of experimental protocols. pp. 363-373, In: *Biohydrometallurgy and the environment toward the mining of 21st century. Part B: Molecular biology, biosorption, bioremediation*. Amils, R. and Ballester, A., eds., Elsevier, Amsterdam.

Paulsen, I.T. and Saier Jr., M.H. (1997) A novel family of ubiquitous heavy metal ion transport proteins. *J. Membrane Biol.*, 156, 99-103.

Pazirandeh, M., L.A., Mauro, J.M., Campbell, J.R. and Gaber, B.P. (1995) Expression of the *Neurospora crassa* metallothionein gene in *Escherichia coli* and its effect on heavy-metal uptake. *Appl. Microbiol. Biotechnol.*, 43, 1112-1117.

Pethkar, A.V. and Paknikar, K.M. (1998) Recovery of gold from solutions using *Cladosporium cladosporiodies* biomass beads. *J. Biotechnol.*, 63, 121-136.

Polikarpov, C.G. (1966) *Radioecology of aquatic organisms*, North-Holland Publishing Co., Netherlands.

Postgate, J. (1984) *The sulfate-Reducing Bacteria*, Cambridge university press, Cambridge, pp. 56-100.

Prakasham, R,S., Merrie, J.S., Sheela, R., Saswathi, N. and Ramakrishna, S.V. (1999) Biosorption of chromium VI by free and immobilized *Rhizopus arrhizus*. *Environ. Pollut.*, 104, 421-427.

Puranik, P.R. and Paknikar, K.M. (1997) Biosorption of lead and zinc from solutions using *Streptoverticillium Cinnamoneum* waste Biomass. *J. Biotechnol.*, 55, 113-124.

Puranik, P.R. and Paknikar, K.M. (1999) Biosorption of lead, cadmium, and zinc by *Citrobacter* strain MCM B-181: Characterization studies. *Biotechnol. Prog.*, 15, 228-237.

Rajwade, J.M., Saluke, P.B. and Paknikar, K.M. (1999) Biochemical basis of chromate reduction by *Psrudomonas mendocina*, p. 105-114, In: *Biohydrometallurgy and the environment toward the mining of the 21st century*, Amils, R. and Ballester, A., eds., Elsevier, Amsterdam.

Ramadan, S. E., Razak, A. A., Yousseff, Y. A. and Sedky, N. M. (1988). Selenium metabolism in a strain of *Fusarium*. *Biol. Trace Element Res.*, 18, 161-170.

Ramteke, P.W. (2000) Biosorption of nickel (II) by *Pseudomonas Stutzeri*. J. *Environ. Biol.* 21: 219-221.

Ramteke, P.W. and Kannan, A (2003) Toxicity of nickel (II) in microbial test Systems. *Asian J. Microbiol. Biotech. Env. Sci.* 5: 77-80.

Razak, A. A., Ramadan, S. E. and Zawahry, K. E. L. (1990). Mobilization of selenium by a selenium-dependent bacterium. *Boil. Trace Element Res.*, 25, 193-199.

Rege, M.A., Petersen, J.N., Johnstone, D.L., Turick, C.E., Yonge, D.R. and Apel, W.A. (1997) Bacterial reduction of hexavalent chromium by *Enterobacter cloacae* strain H01 grown on sucrose. *Biotechnol. Lett.*, 19, 691-694.

Rivadeneyra, M.-A., Delgado, G., Soriano, M., Ramos-Cormenzana, A. and Delgado, R. (1999) Biomineralization of carbonates by *Marinococcus albus* and *Marinococcus halophilus* isolated from the Salar de Atacama (chile). *Curr. Microbiol.*, 39, 53-57.

Rothstein, A., Frenket, A. and Larrabee, C. (1948) The relation of the cell surface to metabolism, III. Certain characteristics of the uranium complex with cell surface groups of yeast. *J. Cell. Comp. Physiol.*, 32, 261.

Ruchhoft, C.C. (1949) The possibilities of disposal of radioactive waster by biological treatment methiods. *Sewage Works J.*, 21, 877.

Rugh, C.L., Senecoff, J.F., Meagher, R.B. and Merkle, S.A. (1998) Development of transgenic yellow popular for mercury phytoremediation. *Nature Biotechnol.*, 16, 925-928.

Sag, Y. and Kutsal, T. (1989) The use of Zoogloea ramigera in waste water treatment containing Cr(VI) and Cd(II) ions. *Biotechnol. Lett.*, 11, 145-148.

Sag, Y. and Kutsal, T. (1996a), The selective biosorption of chromium (VI) and copper (III) ions from binary metal mixtures by R. arrhizus. *Process Biochem.*, 31, 561-572.

Sag, Y. and Kutsal, T. (1996b) Fully competitive biosorption of chromium (VI) and iron (III) ions from binary metal mixtures by *R. arrhizus*. Use of the competitive Langmuir model. *Process Biochem.*, 31, 573-585.

Sag, Y., Acikel, U., Aksu, Z. and Kutsal, T. (1998a) A comparative study of the simultaneous biosorption of Cr(VI) and Fe(III) on *C. vulgaris* and *R. arrhizus*: application of the competitive adsorption models. *Process Biochem.*, 33, 273-281.

Sag, Y., Acikel, U., Aksu, Z. and Kutsal, T. (1998b) comparative biosorption of chromium(VI), iron (III) and copper(II) ions from binary metal mixtures by *R. arrhizus* and *C. vulgaris*. *Turkish J. Eng. Environ. Lett.*, 11, 145-154.

Sakaguchi, T. and Nakajima, A. (1987). Accumulation of uranium by biopigments. *J. Chem. Technol. Biotechnol.*, 40, 133-141.

Salt, D.E., Blaylock, M., Kumar, N.P.B.A., Dushenkov, V., Ensely, B.D., Chet, I. and Raskin, I. (1995) Phytoremediation: A novel strategy for the removal of toxic metals from the environment using plants. *Bio/Technology*, 13,468-474.

Salunkhe, P.B., Dhakephalkar, P.K. and Paknikar, K.M. (1999) Bioremediation of hexavalent chromium in soil microcosms. *Biotechnol. Lett.*, 20, 749-751.

Schueler, D. and Frankel, R.B. (1999) Bacterial magnetosomes: microbiology, biomineralization and biotechnological applications. *Appl. Microbiol. Biotechnol.*, 52, 64-473.

Scott, J.A. and Karanjkar, A.M. (1990) Sites of cadmium uptake in bacteria used for biosorption. *Appl. Microbiol. Biotechnol.*; 33, 221-225.

Sen, H. and Wang, Y.-T. (1993) Characterization of enzymatic reduction of hexavalent chromium by *Escherichia coli* ATCC33456. *Appl. Environ. Microbiol.*, 59, 3771-3777.

Sen, H. and Wang, Y.-T. (1994a) Modling hexavalent chromium reduction in *Escherichia coli* ATCC33456. *Biotechnol. Bioeng.*, 43, 293-300.

Senesi, N., Sposito, G. and Martin, J. P. (1987). Copper (II) and iron (III) complexation by humic acid-like polymers (melanins) from soil fungi. *Sci. Total Environ.*, 62, 241-252.

Sharma, D.C. and Forster, C.F. (1994a) The treatment of chromium wastewaters using the sorptive potential of leaf mould. *Bioresour. Technol.*, 49, 31-40.

Sharma, D.C. and Forster, C.F. (1994b), A preliminary examination into the adsorption of hexavalent chromium using low-cost adsorbents. *Bioresour. Technol.* 47, 257-264.

Shen, H. and Wang, Y.-T. (1994b) Biological reduction of chromium by *E. coli*. *J. Environ. Eng.*, 120, 560-572.

Shen, H. and Wang, Y.-T. (1995) Simultaneous chromium reduction and phenol degradation in a coculture of *Escherichia coli* ATCC33456 and *Pseudomonas putida* DMP-1. *Appl. Environ. Microbiol.*, 61, 2754-2758.

Shooner, F. and Tyagi, R.D. (1996) Thermophilion of microbial leaching of heavy metals from municipal sludge using indigenous sulphur-oxidizing microbiota. *Appl. Microbial. Biotechnol.*, 45, 440-446.

Shumate, S.E. and Strandberg, G.W. 1985 Accumulation of metals by microbial cells PP235-247, In: comprehensive biotechnology, vol. 4, Moo-Young, M., Robinson, C.W. and Howell, J.A., eds., Pergamon press, New Yprk.

Silver, S., Laddaga, R. A. and Misra, T. K. (1989). Plasmid-determined resitance to metal ions. pp. 49-63, In: Metal – Microbe Interactions, eds R. K. Poole and G. M. Gadd. IRL Press, Oxford.

Simmons, P. and Singletion, I. (1996) A. method to increase silver biosorption by an industrial strain of *Saccharomyces cerevisiae. Appl. Microbiol., Biotechnol.*, 45, 278-285.

Singleton, I., Wainwright, M. and Edyvean, R. G. J. (1990). Some factors influencing the adsorption of particulates by fungal mycelium. *Biorecovery*, 1, 271-289.

Smith, D. G. (1974). Tellurite reduction in *Schizosaccharomyces pombe. J. Gen. Microbiol.*, 83, 389-392.

Smith, W.L. and Gadd, G.M. (2000) Reduction and precipitation of chromate by mixed culture sulphate-reducing bacterial biofilms. *J. Appl. Microbiol.*, 88, 983-991.

Sousa, C., Cebolla, A. and de Lorenzo, V. (1996) Enhanced metalloadsorption of bacterial cells displaying poly-His peptides. *Nature Bio/Technology*, 14, 1017-1020.

Sousa, C., Kotrba, P., Ruml, T., Cebolla, A. and de Lorenzo, V. (1998) Metalloadsorption by *Escherichia coli* cells displaying yeast and mammalian metallothioneins anchored to the outer membrane protein LamB. *J. Bacteriol.*, 180, 2280-2284.

Srinath, T., Garg, S.K. and Ramteke, P.W. (2003) Biosorption and elution of Cr(VI) : A detoxification strategy In New Horizones in Biotechnology Proceedings of the International Biotechnology Conference (Ed. Roussos, S., Soccd, C.R., Panday, A. and Augur, C.

Srinath, T., Garg, S.K. and Ramteke, P.W. (2002a) Chromium (VI) accumulation by *Bacillus Circulans*: Effect of growth conditions. *Indian J. Microbiol.* 42: 141-146.

Srinath, T., Verma, T., Ramteke, P.W. and Garg, S.K. (2002b) Chromium (VI) biosorption and bioaccumulation by chromate resistant bacteria. *Chemosphere*, 48:427-435.

Steinbnerg, N. A. and Oremland, R. S. (1990). Dissimilatory selenate reduction potentials in a diversity of sediment types. *Appl. Environ. Microbiol.*, 56, 426-428.

Strandberg, G. W., Shumate, S. E. and Parrott, J. R. (1981). Microbial cell as biosorbents for heavy metals: accumulation of uranium by *Saccharomyces cerevisiae* and *Pseudomonas aeruginosa. Appl. Environ. Microbiol.*, 41, 227-245.

Summers, A. O. and Jacoby, G. A. (1977). Plasmid-determined resistance to tellurium compounds. *J. Bacteriol.*, 129, 276-281.

Tamaki, S. and Frankenberger, W. T. (1992). Environmental biochemistry of arsenic. *Rev. Environ. Contam. Toxicol.*, 124, 79-110.

Taylor, D. E., Walter, E. G., Sherburne, R. and Bazett-Jones, D. P. (1988). Structure and Iocation of tellurium deposited in *Escherichia coli* cells harboring tellurite resistance plasmids. *J. Ultrastructure Molec. Structure Res.*, 99, 18-26.

Tesezos, M. and Volesky, B. (1981). Biosorption of uranium and thorium. *Biotechnol. Bioengng*, 23, 583-604.

Tesezos, M. and Volesky, B. (1982). The mechanism of uranium biosorption by *Rhizopus arrhizus. Biotechnol. Bioengng*, 24, 385-401.

Thompson-Eagle, E. T. and Frankenberger, W. T. (1992). Bioremediation of soils contaminated with selenium. pp. 261-309, In: *Advances in Soil Science*, Vol. 17, eds R. Lal and B. A. Stewart. Springer, New York.

Thompson-Eagle, E. T., Frankenberger, W. T. and Karlson, U. (1989). Volatilization of selenium by *Alternaria alternate*. *Appl. Environ. Microbiol.*, 55, 1406-1413.

Tobin, J. M., Cooper, D. G. and Neufeld, R. J. (1987). Influence of anions on metal adsorption by *Rhizopus arrhizus* biomass. *Biotechnol. Bioengng*, 30, 882-886.

Tobin, J. M., Copper, D. G. and Neufeld, R. J. (1988). The effects of cation competition on metal biosorption by *Rhizopus arrhizus* biomass. *Biotechnol. Bioengng*, 31, 282-286.

Tobin, J., White, C. and Gadd, G. M. (1994). Fungal accumulation of toxic metals and application to environmental technology. *J. Indust. Microbiol.*, 13, 126-130.

Tobin, J.M., Cooper, D.G. and Neufeld, R.J. (1984) Uptake of metal ions by *Rhizopus arrhizus* biomass. *Appl. Environ. Microbiol.*, 47, 821-824.

Tomei, F. A., Barton, L. A., Lemanski, C. L. and Zocco, T. G. (1992). Reduction of selenate and selenite to elemental selenium by *Wolinella succinogenes*. *Can. J. Microbiol.*, 38, 1328-1333.

Trevors, J.T. and Stratton, G.W. and Gadd, G.M. (1986) Cadmium transport, resistance, and toxicity in bacteria, algae, and fungi. Can. *J. Microbial.*, 32, 447-464.

Tsezos, M. (1983). The role of chitin in uranium adsorption by *Rhizopus arrhizus*. *Biotechnol. Bioengng*, 25, 2025-2040.

Tsezos, M. and Deutschmann, A. A. (1990). An investigation of engineering parameters for the use of immobilized biomass particles in biosorption. *J. Chem. Technol. Biotechnol.*, 48, 29-39.

Tsezos, M., McCready, R. G. L. and Bell, J. P. (1989). The continuous recovery of uranium from biologically leached solutions using immobilized biomass. *Biotechnol. Bioengng*, 34, 10-17.

Turick, C.E. Apel, W.A. and Carmiol, N.S. (1996) Isolation of hexavalent chromium-reducing anaerobes from hexavalent-chromium-contaminated and noncontaminated and noncontaminated environments. *Appl. Microbiol. Biotechnol.*, 44, 683-688.

Turick, C.E., Camp, C.E. and Apel, W.A. (1997) Reduction of Cr(VI) to Cr(VIII) in a packed-bed reactor. *Appl. Biochem. Biotechnol.*, 63-65, 871-877.

Turick, C.E., Graves, C. and Apel, W.A. (1998) Bioremediation postential of Cr(VI)-contaminated soil using indigenous microorganisms. *Bioremediation J.*, 2, 1-6.

Veglio, F. and Beolchini,, F. (1997) Removal of metals by biosorption: a review. *Hydrometallurgy*, 44, 301-316.

Volesky, B. (1990) Biosorption of heavy metals, CRC press, Boca Raton, F.L.

Volesky, B. and Holan, Z.R. (1995) Biosorption of heavy metals. *Biotechnol. Prog.*, 11, 235-250.

Volesky, B. and May-Phillips, H.A. (1995) Biosorption of heavy metals by Saccharomyces cerevisiae. *Appl. Microbiol.*, 42, 797-806.

Wainwright, M., Singleton, I. and Edyvean, R. G. J. (1990). Magnetite adsorption as a means of making fungal biomass susceptible to a magnetic field. *Biorecovery*, 2, 37-53.

Wales, D. S. and Sagar, B. F. (1990). Recovery of metal ions by microfungal filters. *J. Chem. Technol. Biotechnol.*, 49, 345-355.

Walker, S.G., Flemming, C.A., Ferris, F.G., Beveridge, T.J. and Bailey, G.W. (1989) Physicohemical interaction of *Escherichia coli* cell envelopes and *Bacillus subtilis* cell walls with two clays and the ability of the composite to immobilize heavy metals from solution. *Appl. Environ. Microbiol.*, 55, 2976-2984.

Walter, E. G. and Taylor, D. E. (1992). Plasmid-mediated resistance to tellurite: expressed and cryptic. *Plasmid*, 27, 52-64.

Wang, P.C., Mori, T., Komori, K., Sasatu, M., Toda, K. and Ohtake, H. (1989) Isolation and characterization of an Enterobacter cloacae strain that reduces hexavalent chromium under anaerobic conditions. *Appl. Environ, Microbiol.*, 55, 1665-1669.

Wang, P.C., Toda, K., ohtake, H., Kusaka, I. And Yabe, I. (1991) Membrane-bound respiratory system of *Enterobacter cloacae* strain H01 grown anaerobically with chromate. *FEMS Microbiol. Lett.*, 78, 11-16.

Wang, Y.-T. and Shen, H. (1995) Bacterial reduction of hexavalent chromium. *J. Ind. Microbiol.*, 14, 159-163.

Wang, Y.-T. and Xiao, C. (1995) Factors affecting hexavalent chromium reduction in pure cultures of bacteria. *Water Res.*, 29, 2467-2474.

White, C. and Gadd, G.M. (1997) An internal sedimentation bioreactor for laboratory-scale removal of toxic chemicals from soil leachates using biogenic sulphide precipitation. *J. Ind. Microbiol. Biotechnol.*, 18, 414-421.

White, C. and Gadd, G.M. (1998) Accumulation and effects of cadmium on sulfate-reducing bacterial films. *Microbiology*, 144, 1407-1415.

White, C., Sayer, J.A. and Gadd, G.M. (1997) Microbial solubilization and immobilization of toxic metals: key biogeochemical processes for treatment of contamination. *FEMS Microbiol. Rev.*, 20, 503-516.

White, C., Sharma, A. K. and Gadd, G.M. (1998) An integrated microbial process for the bioremediation of soil contaminated with toxic metals. *Nature Biotechnol.*, 16, 572-575.

White, C., Wilkinson, S.C. and Gadd, G.M. (1995) the role of microorganisms in biosorption of toxic metals radionuclides. *Intl. Biodet. And Biodegrad.*, 35, 17-40.

Wilhemi, B.S. and Duncan, J.R. (1995) Metal recovery from *Saccharomyces cerevisiae* biosorption columns. *Biotechnol. Lett.*, 17, 1007-1012.

Williams, J. W. and Silve,r S. (1984). Bacterial resistance and detoxification of heavy metals. *Enzyme Microb. Technol.*, 6, 530-537.

Yamamoto, K., Kato, J., Yano, T. and Ohtake, H. (1993) Kinetics and modeling of hexavalent chromium in *Enterobacter cloacae. Biotechnol. Bioeng.*, 41, 129-133.

30

Genotoxicity Risk Assessment of Hazardous Chemicals and Wastes

K. Krishnamurthi, S. Saravana Devi and T. Chakrabarti

Introduction

The benefits to mankind provided by Industrialization are enormous, though wastes and effluents are the undesirable consequences of industrialization. Industrialization results in the formation of waste products, which are released into the environment in the form of wastewater, gaseous emissions, and solid residues, often untreated or partially treated, leading to environmental pollution and degradation. Such a situation poses a threat to the flora and fauna and the ecosystem as well especially when the pollutants are persistent in nature (Chakrabarti et al.,1988). The persistence of xenobiotics in the environment poses a new challenge to the environment as its long residence time in the environment could result in manifestations such as toxicity, bioaccumulation and biomagnifications.

Hazardous wastes - International Scenario

Globally more than 400 million metric tons of hazardous wastes are generated annually by various industries. In addition, enormous quantities of industrial effluents are discharged daily into rivers, streams, harbors, and municipal sewage systems throughout the world (USEPA, 1988).

Resource Conservation and Recovery Act (RCRA) of USA defines hazardous waste as, "a solid waste, or a combination of solid wastes which, because of its (their) quantity, concentration, or physical, chemical or infectious characteristics may cause, or significantly contribute to an increase in mortality or an increase in serious irreversible, or incapacitating reversible, illness or pose a substantial or potential hazard to human health or to the environment when improperly treated, stored, transported, or disposed of, or otherwise managed.

As per RCRA, a solid waste can become a hazardous waste by virtue of any of the following conditions:

- Being on one of four lists of hazardous wastes;
- Being a mixture of solid waste and a hazardous waste on one of the lists;
- Being derived from the treatment, storage, or disposal of a hazardous waste on one of the lists;
- Exhibiting one or more of the four characteristics of a hazardous waste namely corrosivity, reactivity, ignitability and toxicity (CRIT)

The above-mentioned four lists are

- Hazardous waste from non-specific sources,
- Hazardous wastes from specific sources,

- Acute hazardous list,
- Toxic list.

Initially the RCRA considered the criteria of corrosivity, reactivity, ignitability, infectiousness, radioactivity and toxicity but now a hazardous waste may have any one or more of the four characteristics viz. (CRIT)

Hazardous wastes - Indian Scenario

In India, wastes are classified as hazardous if the wastes possess one or more of hazardous physical characteristics (eg, ignitability) or if they comprise certain industrial process wastes or contain specifically regulated chemicals at levels more than the stipulated/ regulated limits. Approximately, 8.0 million metric tons of wastes generated in India are categorized as hazardous based on the Schedules 1 and 2 of the Hazardous Wastes (Management and Handling) Rules, 1989 was amended in January 2000 (Tables 1 and 2) (MOEF, 2000).

Table 1 : Schedule 1: Industry Specific Sources of Hazardous Wastes Categories of Hazardous Wastes (Rules, 1989)

Sr. No.	Industry/process		Wastes Streams	Waste Category
1	Petrochemical processes and pyrolytic operations	1.1	Oven debris	12
		1.2	Oil-water sluge mixture	10
		1.3	Oil-containing bleaching earth	10
		4.4	Acid tar	11
		1.5	Sulphur-containing residue from Sulphur removal	10
		1.6	Oil-water mixture	10
		1.7	Oil-containing sludge	10
		1.8	Oil emulsion	10
		1.9	Oil-containing acid	10
		1.10	Tar-residue made with coal tar	11
		1.11	Sludge from waste water purification	12
		1.12	residual liquid and paste-like organic substances made with aromatic, aliphatic and naphenic hydrocarbons	10
		1.12	residual liquid and paste-like organicsustances made with aromatic, aliphatic and naphenic hydrocarbons	13,15
		1.13	Residue from alkali wash of fuels	16
2	Natural gas production	2.1	Mercury-containing sludge	4
		2.2	Mercury-containing filter material	4
		2.3	Sulphur-containing residues	
3	Production or use of zinc	3.1	Zinc ashes	3
		3.2	Zinc slags	3
		3.3	Jarosite	3
4	Production or use of lead	4.1	lead ashes	3
		4.2	lead slags	3
		4.3	lead-containing filter material	3
5	Production or use of cadmium	5.1	cadmium-containing filter material	4
6	Production or use of arsenic	6.1	Arsenic-containing filter material	4

Guidelines for characterizing the biological toxicity of hazardous wastes have not yet been formulated; however, knowledge of the potential biological toxicity of hazardous waste could supplement the available chemical characterization and provide relevant information regarding potential health effects. As hazardous wastes contain a wide range of chemical mutagens and genotoxicants and because no single bioassay will detect all chemical classes of mutagens, several short-term assays with different genetic end points may be

Sr. No.	Industry/process		Wastes Streams	Waste Category
7	Production of cast iron	7.1	Cupola oven dust	3
8	Production of crude iron and steel with oxy-steel converters or Electro-ovens	8.1	Blast-furnace gas dust	12
		8.2	Filtered material	12
		8.4	Other exhaust dust	12
9	Production of aluminium (primary or secondary production)	9.1	Filtered material	12
		9.2	Cathode residues	12
		9.3	Overn debris	12
10	Non-ferro metallurgical process	10.1	Heavy metal-containing oven debris chalk	12
11	Hardening of steel	11.1	Cyanide, nitrate, or nitrite-containing sludge	1
		11.2	Hardening salt	1
12	Production or use of asbestos or asbestos-containing materials	12.1	Asbestos-containing residue	14
13	Production of chlorine by means of the electrolyses process	13.1	Asbestos-containing residue	14
14	Phenol production	14.1	Phenol mixture	13
		14.2	Phenol-water mixture	13
15	Matalworking	15.1	Selenium-containing metal waste	3
			berylium-containing metal waste	3&4
			coolant made with oil products	10
			oil-water sludge mixture	10
16	Metal surface treatment, such as etching, staining, polishing, galvanising, cleaning, degreasing and hot dip glavanising	16.1	Acid, acid residue or acid mixture	16
		16.2	Alkali, alkali resudue or alkali	16
		16.3	Mixture, glavanic bath and (half) concentrate made with sulphide chromium (IV), cyanide, copper zinc, cadmium, nickel or tin	1,2,3&4
		16.4	Copper etching fluid	2
		16.5	Halogen-free sludge from a bath which used organic solvents	6
		16.6	Halogen-containing sludge from a bath with organic solvents	5
		16.7	Phosphating sludge	2
		16.8	Halogen-containing organic degreasing bath	6
17	Treatment of galvanis and similar with baths and wter purification in metal surface treatment	17.1	Metal hydroxide sludge chromium, copper, zinc, nickel or silver	3
		17.2	Heavy metal-containing eluate from ion exchangers	2
		17.3	Heavy metal containing half-concentrates from memberance systems	2
18	Production of acids or ammonia	18.1	Acid containing residues	16
		18.2	Ammonia-containing residues	16
19	Production or use of solvents	19.1	Contaminated halogen-free aromatic or naphtenic solvents	5
		19.2	Contaminated halogen-free solvent made with phenols, ketones, ethers, acetates, alcohols of glycols	5

required to be performed for determining genotoxic potentials of hazardous wastes. Consequently, combinations of short-term assays have been proposed for screening hazardous wastes for their risk potential (DeMarini et al., 1987). Nonetheless, most investigations pertaining to genotoxicity assessment of hazardous wastes have been based on the *Salmonella* assay though mammalian cells and other eucaryotic cells have been used to a limited extent in bioassays (DeMarini et al., 1989).

Risk assessment

When improperly handled and disposed, industrial wastes imperil both human health and the environment. Human exposures (occupational and non-occupational)

Sr. No.	Industry/process		Wastes Streams	Waste Category
19	Production or use of solvents	19.1	Contaminated halogen-free aromatic or naphtenic solvents	5
		19.2	Contaminated halogen-free solvent made with phenols, ketones, ethers, acetates, alcohols of glycols	5
		19.3	Contaminated halogen-containing aliphiatic or naphtenic solvents	5
		19.4	Contaminated halogen-containing solvents made with phenols	6
		19.5	Containing solvents or mixture of solvents made with organic nitrogen-containing aromatics, naphtenes of aliphatics	9
		19.6	Contaminated solvents or mixture of solvents made with organic sulphur compounds	15
		19.7	Distillation residue	11
20	Removal of coatings from ships, bridges and locks electricity pylons	20.1	Blasting material contaminated with coating residues	17
21	Production or use of coating paints, lacquers, varnished and plastics, and of inks	21.1	Residuces of coatings of such as inks completely hardened	7
		21.2	Sludge from waste water purification in production processes	12
22	Production or use of glues, cements, adhesive and resings	22.1	Glue, cement of adhesive residue (not made with regetable or animal materials) if not completely dried out	7
		22.2	Sludge from waste waster purification in production processes	10
23	Production or use of latex	23.1	Resin oil residue	10
24	Production or use of paint removers	24.1	Paint remover residue	7
25	Printing and copying with liquid toner	25.1	Printing ink residue	7
		25.2	Silkscreen printing ink residue	7
		25.3	Lacquer residue	7
		25.4	Liquid toner residue	7
		25.5	Residue of cleaning agents made with solvents	6
		25.6	Etching Fluid residue	6&7
		25.7	Dispereive oil residue	6
		25.8	Oxidising agent residue	5
26	Production or use of photo-chemicals	26.1	Developing residue	7
		26.2	Fixer residue	6&7
		26.3	Bleachign fixer residue	6&7
27	Production or use of organic peroxides	27.1	Organics peroxide residue	5&6

to industrial wastes have led to toxic effects ranging from acute, sub-acute or chronic, teratogenicity, mutagenicity, genotoxicity and carcinogenicity (Grisham, 1986, Buffler *et. al.*, 1985). Evidence of genotoxic health effects, such as cancer, birth defects, and reproductive anomalies (Griffith, *et.al.*, 1989, Najem *et al.*, 1985), have been found in populations living near hazardous waste dump sites.

According to the U.S. Department of Health Human Services (DHHS, 1985), risk assessment is "the use of available information to evaluate and estimate exposure to substance(s) and its consequent adverse health effects". The USEPA, (1991) notes that risk assessment involves analyzing past exposures, determining the types and amounts of adverse effects, and predicting the outcome from subsequent exposure. In short,

Sr. No.	Industry/process		Wastes Streams	Waste Category
28	Production or use of halogen-containing hydrocarbons or of aromatics, alphatic or naphtenic hydrocarbons	28.1	Residue of fluid or pasty organic materials made with halogen-containing hydrocarbons	15
		28.2	Residue of luid or pasty organic materials made with aromatic, alphatic or naphtenic hydrocarbons	15
29	Production or use of organic nitrogen compounds or organis	29.1	Resiude of fluid or pasty organic materials made with oxygen compounds nitrogen or oxygen compunds (other than vegetable or animal carbonhydrates proteins fats and fatty acids)	15
30	Production or use of materials made with (excluding cements)	30.1	Silicone oil residue	12
		30.2	Silicone-containing residues	12
31	Production or canvas and textiles	31.1	Textile chemical residue	12
32	Production or use of plastics or raw materials for them	32.1	Halogen-free residue of additives for plastics (e.g. dyestuffs, stabiliser, or flame retardenes)	12
		32.2	Halogen-free residue of plasticiers for plastics	6
		32.3	Halogen-free residue of plasticiers for plastics	5
		32.4	Halogen containing residue of plasticiers for plastics	6
		32.5	Residue from the preparation of vinylchloride monomer	6
		32.6	Residue from the preparation of acrylonitrile monomer	6
		32.7	Residue of liquid or pastry rubber emulsion or rubber solution if not polymerised	6
		32.8	Sludge from waste water purification from rubber production if not polymerized	12
		32.9	PVC-containing residues if not polymerized	6
33	Production of cosmeties	33.1	Residue of chemical raw materials and additives (other than vegetable and animal carbohydrates, proteins, fats and fatty acids)	18
34	Production of pharmaceuticals	34.1	Residue from the production of medicines (other than vegetable and animal carbohydrates, proteins, fats and fatty acids)	5,6
35	Production, formulation of pesticides	35.1	Production or formulation use of pesticides residue	15

Risk = Hazard x Dose

The baseline risk assessment is a critical factor in the ranking and remediation of contaminated waste sites. Standard methodology for risk assessment requires the selection of principal hazardous constituents, an exposure assessment, and a toxicity assessment. This information is then combined to form the basis of the standard chemical-based risk assessment procedure and to calculate incremental carcinogenic risk or hazard index (for systemic effects) for specific principal hazardous constituents at reasonable maximum exposures. Risk assessment also enables delineation of risk based remediation goals.

Toxicology

Toxicology is defined as the study of the toxic properties, or adverse health effects, of agents or substances. In short, modern toxicology encompasses two facets: qualitative evaluation, and quantitative assessment

Sr. No.	Industry/process		Wastes Streams	Waste Category
		35.2	Sludge from waste water treatment	12
		35.3	Hexa or hexa containing residue made with hexa chlorocyclohexane or hexachlorobenzene	6
		35.4	Residues from the use of pesticides	15
36	Production, formulation or use of wood preservatives	36.1	Production and formulation residue	15
		36.2	Sludge from the waste water purification	12
		36.3	Residue from the use of wood preservatives wood alkali bath	6
37	Cleaning, emptying and maintenance of tanks and separators of vessels vehicles and of mobile and stationary storage ranks, washing water	37.1	Oil containing cargo residue, washing water and sludge	10
		37.2	Chemical containing cargo residue	9
		37.3	Oil water sludge mixture and oil containing air filter from oil, sludge or petrol separation	10
38	Cleaning of barrels which have held chemicals substances	38.1	Chemical containing residue from barrel cleaning	18
		38.2	Sludge from waste water purification	12
39	Purification procession for air and water	39.1	Sludge from waste water treatment from artificial fertilizer production	12
		39.2	Sludge from the treatment of waste water containing hydrofluoric acid	12
		39.3	Heavy metal containing residue from used ion exchange material in the water purification	3
		39.4	Flue gas cleaning residue	12
40	Purification procession for organic water	40.1	Filters and filter material which have organic liquids on them e.g. minieral oil, synthetic oil and organic chlorine compounds	12
41	Waste treatment processes, e.g. incineration distillation and separation and concentration techniques	41.1	Sludge from the incineration of exclusively chemical waste	12
		41.2	Fly ash from incineration of hazardous waste, except exclusively communal sewage sludge, flue gas cleaning residue	12

of toxicity. Qualitative evaluation here is the study of an agent, either chemical or physical, that can cause or have the potential to cause an adverse or harmful effect in living organisms (Randerath *et al.*,1994).

Acute toxicity

Acute toxicity evaluates the short-term (less than 30 days) adverse biological effects of a single exposure (or a small number of exposures over a week or less) to a material or physical agent. Acute toxicity tests provide information on the harmful effects associated with short-term exposure of chemicals or physical agents on living organisms. Acute toxicity assessments are generally performed first, and the information from theses studies is subsequently used to design protocols for longer-term studies.

Sub-acute/chronic toxicity

Chronic toxicity tests are designed to examine the effects of a

Sr. No.	Industry/process	Wastes Streams		Waste Category
		41.3	Residue from the burning of cables	12
		41.4	Battery acid	16
		41.5	Distillation residue from the work-up of	12
		41.6	Contaminated halogen free organic solvents distillation residue from the work-up of contaminated halogen containing organic solvents	5&6
42	The shredding of object made chiefly of plastic of rubber	42.1	Shredder waste	16
43	Performance of maintenance and repair work on vehicles and	43.1	Oil containing sludge and oil emulsion	5
		43.2	Filter and filter material which have organic liquids on them, e.g. mineral oil, synthetic oil and organic chlorine compounds	6&12
44	Every action relating to and every use of lubricating and system oil	44.1	Spent oil	5&6
		44.2	Other spent lubricating and system oil	5&6

substance following prolonged and repeated exposure (Alridge 1993). Many chemicals are not acutely toxic at the concentrations found in the environment. However, latent and cumulative effects can become apparent through chronic toxic studies which are designed to reveal long-term dose-response relationships.

Sub-chronic toxicity

The extended exposure period and more extensive observations made during repeated sub-chronic dose experiments permit the study of cumulative effects arising from bioaccumulation of the toxicant and, also, enable clearer identification of specific target organs where the manifestations maximally or specifically occur. Data from sub-chronic studies are used to choose dose levels for general chronic toxicity or carcinogenesis studies (such as the maximum tolerated dose, MTD).

Teratogenicity

Chemicals on exposure to pregnant mothers may induce hereditary birth defects in the progeny or child. Such chemicals are designated as teratogens.

Mutagenicity

Chemical or physical agents on exposure to organisms cause a sudden change in their characteristic. These agents are called as mutagens.

Carcinogenicity

Physical or chemical agents on exposure to organism, cause uncontrolled cellular growth or neoplasm (cancer) and such agents are known as carcinogens.

Table 2: Schedule 2 of Hazardous Waste Rules
[See rule 3(I) (b)]
List of waste substances with concentration limits

Classes	
Class A	
Concentration limit: 50 mg/kg	
A1	Antimony and antimony compounds
A2	Arsenic and arsenic compounds
A3	beryllium and cadmium compounds
A4	Cadmium and beryllium compounds
A5	Chromium (VI) compounds
A6	Mercury and mercury compounds
A7	Selenium and selenium compounds
A8	Tellurium and tellurium compounds
A9	Thallium and thallium compounds
A10	Inorganic cyanide compounds (cyanides)
A11	Metal carbonyls
A12	Naphthalene
A13	Anthracene
A14	Phenanthrene
A15	Chrysene, benzo(a) anthracene, fluoranthene, benzo(a) pyrene, benzo(K) fluoranthene, indeno (1,2,3-ed) pyrene and benzo(ghi) perylene
A16	Halogenated fused aromatic rings, e.g. polychlorobiphenyls plus derivatives
A17	Halogenated aromatic compounds
A18	Benzene
A19	Dieldrin, aldrin and endrin
A20	Organotin compounds
Class B	
Concentration limit : 5,000 mg/kg	
B1	Chromium (III) compounds
B2	Cobalt compounds
B3	Copper compounds
B4	Lead and lead compounds
B5	Molybdenum compounds
B6	Nickel compounds
B7	Tin compounds
B8	Vanadium compounds
B9	Tungsten compounds
B10	Silver compounds
B11	Organic halogen compounds
B12	Organic phosphorus compounds
B13	Organic peroxides
B14	Organic nitro- and nitroso- compounds
B15	Organic azo-and azo-oxy compounds
B16	Nitriles
B17	Amines
B18	(Iso-and thio-) cyanates
B19	Phenol and phenolic compounds
B20	Merceptens
B21	Asbestos
B22	Drilling, cutting, grinding and rolling oil or emulsions thereof
B23	Halogen-silanes
B24	Hydrazine(s)
B25	Fluorine
B26	Chlorine
B27	Bromine
B28	White phosphorus
B29	Ferro-silicon and alloys
B30	Manganese- silicon
B31	Halogen-containing substances which produce acidic vapours on contact with damp air or water, e.g. silicon tetrachloride, aluminium chloride, titanium tetrachloride
Class C	
Concentration limit : 20,000 mg/kg	
C1	Ammonia and ammonium compounds
C2	Inorganic peroxides
C3	Barium compounds, except barium sulphate
C4	Fluorine compounds

C5	Phosphorus compounds, except the phosphates of aluminium, calcium and iron	D3	Metal bisulphates
C6	Bromates, (hypo) bromites	D4	Oxides and hydroxides except those of : hydrogen, carbon, silicon, iron, Aluminum, titanium, manganese, magnesium, calcium
C7	Chlorates, (hypo) chlorites		
C8	Aromatic compounds		
C9	Organic silicon compounds	D5	Aliphatic and napthenic hydrocarbons
C10	Organic sulphur compounds		
C11	Iodates	D6	Organic oxygen compounds
C12	Nitrates, nitrites	D7	Organic oxygen compounds
C13	Sulphides	D8	Nitrides
C14	Zinc compounds	D9	Hydrides
C15	Salts of per acids		
C16	Acid halides, acid amides	Class E	
C17	Acid anhydrides	Regardless of concentration limit	
		E1	Highly flammable substances
Class D		E2	Substances which generate dangerous quantities of highly flammbale gases on contact with water or damp air.
Concentration limit : 50,000 mg/kg			
D1	Sulphur		
D2	Inorganic acids		

Note: All on dry weight basis.

In situ studies

Studies on plants and animals, indigenous to environment and impacted by hazardous wastes or industrial effluents, provide additional evidence of genotoxic effects. The production of sister chromatid exchange (SCE), micronucleus (MN), chromosomal aberration (CA) in Chinese hamster ovary (CHO) cells is a relatively common clastogenesis test for wastes and effluents (Houk 1992). It has been reported that chemical manufacturing effluents, petroleum refining wastes, pulp and paper mill effluents and coke plant discharges could induce CA, MN, SCE and mutation in *Salmonella* assay **(Table 3)**. Another test system used to detect the clastogenesis of industrial wastes and effluents is the whole plant species such as onion, fern and soybean. Klekowski and Levin (1979) studied fern growing in a river heavily polluted with the effluent from a pulp and paper mill and reported a statistically significant increase in chromosome mutations in the plants collected along the contaminated river when compared to the plants growing at an uncontaminated site.

Genetic toxicology/Genotoxicity

Genetic toxicology is a branch of general toxicology which addresses the problems of toxicity to the DNA and the study of damages to the genes by chemical or physical agents. Damages to the genes (i.e., to DNA), if not repaired timely and correctly, change the DNA sequence and cause mutations (Doll & Peto 1981). Genotoxicity is thus customarily defined as the ability to damage DNA and to change DNA sequence. DNA sequence changes can be single nucleotide changes that result in point mutations, or multiple nucleotide changes that result in visible chromosomal aberrations (Cohen & Elwein 1990).

Table 3 : Distribution of the Mutagenic Potencies of Industrial Wastes and Effluents Based on Activity per Unit Mass (Revertants per milligram [rev/mg][a])

Potency (rev/mg)		Waste
	10^7	
Extreme		Plastics tar Composited organic wastes from a hazardous waste incineration facility Petrochemical waste oil (coke plant waste and waste paints) crude waste
	10^6	
High	10^5	
		Chemical manufacturing (industrial organics) Condensate blend of wastewater from munitions production; herbicide manufacturing (acetone/water)
	10^4	
		Effluent from the manufacture of dyes and epoxy resins; chlorination stage effluent; pulp and paper mill effluent, biotreated; coke plant effluent; refinery waste (API oil-water separator sludge), basic fraction
		Pulp and paper waste; wastewaters from the manufacture of hair dyes and herbicides; used crankcase oil
Moderate	10^3	Paper dyeing and manufacture; plasticizer waste; chemical (peroxides) manufacturing effluent ; wood preserving waste, acid fraction; textile dyeing; pharmaceutical manufacturing; petrochemical plant waste (API oil-water separator sludge), acid fraction; combined API/slop-oil emulsions; neutral fraction; storm water runoff, neutral fraction; foundry effluent; vinyl chloride production waste (EDC tar); resins manufacturing effluent; paper dyeing; wood-preserving bottom sediment, crude waste; petroleum refinery effluent (2 independent studies)
	10^2	Kraft pulp mill fiber waste (water-soluble compounds); acetonitrile bottom stream, basic fraction
Low		
	Below Detection	Pharmaceutical wastewater; electroplating effluent; resin manufacturing; surfactants manufacture; PVC wastewater

Rationale in selection of genotoxicity studies

The demonstration that industrial waste and wastewater can induce genotoxic effects in aquatic and terrestrial species (including humans) has stimulated research in this area. In the past decade, sizeable scientific data have been generated. Progress in this field is largely due to the development and widespread use of short-term genetic bioassays. Their simplicity and sensitivity to genetic damage, the relative speed and low cost of experimentation, and the small amount of sample required for analysis have permitted research in genetic toxicology to flourish (Alridge, 1993).

Results from genetic bioassays are relevant to human health because the toxicological target is DNA, which exists in all-cellular life forms. Thus, it can be stated with reasonable degree of accuracy that compounds shown to be reactive with DNA in one species have the potential to produce similar effects in other species. In general, perturbations of genetic material is deleterious to the organism in case the damage is not repaired and can lead to severe and irreversible health consequences (Scott et al. 1991). A total of 25 biological endpoints (both genotoxic and epigenetic) that are related to genotoxicity and carcinogenicity have been documented (Lai et al. 1996). Mutagenicity assays, which exploits organisms such as bacteria, yeast, fungi and cultured mammalian cells to detect the mutagenic and genotoxic potential of a sample, have been used extensively to evaluate risks associated with industrial wastes (DeMarini et al. 1984).

In vitro and _In vivo_ short term genetic bioassays

Genetic damage is most accurately defined in terms of the specific structural lesions that occur along the DNA. Three major classifications of damage can be described which are mutagenesis, clastogenesis and aneuploidy. Mutagenesis refers to gene (or point) mutations which involve changes in the DNA sequence within a gene. Clastogenesis refers to changes in chromosome structure, usually resulting in a gain or loss, or rearrangement of chromosome pieces (segments) and aneuploidy refers to gain or loss of intact chromosomes.

More than 200 short-term tests, involving *in vitro/in vivo* systems, have been developed over the last 20-25 years to aid in the identification of agents that pose genetic hazards to humans (Waters et al. 1988). The tests can be separated into major groups on the basis of the biological system employed and the genetic endpoint detected. Prokaryotic (bacterial) bioassays detect agents that induce gene mutations and primary DNA damage. Included in this category is the well-known *Salmonella*, or Ames mutagenicity assay. Eucaryotic assays (fungal tests, plant assays, and assays involving mammalian cells) detect a broader range of genetic damage, including gene mutations, DNA damage and repair, chromosomal damage, and aneuploidy.

In vivo – short term tests

Short-term *in vivo* tests have been used to evaluate the genotoxic potential of industrial wastes and effluents. In these types of tests, the waste or waste extract is administered to a whole animal. Tissues or body fluids are collected and examined for genetic damage. Bioindicators of exposure include determination of urinary metabolites, the formation DNA adducts, and the detection of micronuclei. These techniques are more expensive and labor intensive than most *in vitro* assays. Therefore, their application in the risk assessment study involving industrial wastes has been limited (DeMarini et al. 1984).

In vitro tests

The rationale behind genotoxic/carcinogen screening programs is to have quick, efficient and reliable systems for identifying compounds that have genotoxic and carcinogenic potential. Though *in vivo* tests yield reliable results, the procedures involved utilize long time in completing the study. Further, *in vivo* testings require large infrastructure, expenditure, space, extrapolation of data from surrogate species and animals thus making their choice limited.

To replace the *in vivo* assays, a number of *in vitro* tests have been developed and are popular as these tests require shorter time, reduced cost, and reduced space to conduct the assay. Further, the ability of *in vitro* tests to utilize human and other mammalian cells and smaller quantities of testing agents, have resulted their inclusion in toxicity assessment protocols worldwide. *In vitro* toxicological models now play a vital role in elucidating cellular and molecular mechanisms involved in apoptotic and necrotic cell death and in carcinogenesis and have been extensively used to establish the role of mediators, such as free radicals and oncogenes in oncogenesis (Walum et al. 1990).

Ame's test of the numerous genetic bioassays available, the *Salmonella* mutagenicity assay (Ames test) is selected most often to evaluate the genotoxicity of industrial wastes, effluents and discharges. For quiet sometime, this assay has been used extensively. To increase the susceptibility of the *Salmonella* tester strains to mutagens, various modifications to the wild-type strains have been made which include the removal of the bacteria's DNA excision-repair system resulting in

- a decrease in the ability of the bacteria to correct DNA damage,
- the partial loss of the lipopolysaccharide structure of the bacterial cell walls permitting large mutagens (such as polycyclic aromatic hydrocarbons) to penetrate the cell,
- incorrect repair of certain types of DNA damage leading to the expression of a mutation in plasmid containing strains.

There are certain classes of chemicals for which the bacterial *Salmonella* assay has proven relatively insensitive. These mainly include chlorinated organics, solvents, organophosphorus pesticides and some metals. Benzene, for instance, is not positive in the *Salmonella* assay, but it is a potent clastogen and proven carcinogen by some other test system. As these compounds are common constituents of many industrial wastes, it is suggested to include other test systems for genotoxicity studies of industrial wastes and effluents (Woo et al. 1996).

Cytotoxicity : The assessment of overt toxicity in cultured cells and other cell types (e.g., hepatocytes) usually involves one of several methods for monitoring cell membrane integrity. These include methods for detecting enzyme leakage from cells and those that are based upon the uptake of dye compounds into the cells. Assessment of viability is based on uptake of dye, such as trypan blue, by non-viable cells and its active exclusion by viable cells. This method requires visual examination of cells by microscopy and scoring of cells for % survival. Another method for assessment of cytotoxicity involves the uptake of dye by viable cells (ex. in attachment cultures) and quantitation of incorporated dye by spectrophotometry.

DNA adduct assay : The determination of DNA adduct has been performed on a number of occasions as a part of human biomonitoring for exposure to carcinogens. The techniques encompass immunochemical assays, synchronous fluorescence and ^{32}P-post labeling assay. The ^{32}P-post labeling assay for adducts is widely used and is capable of achieving extreme sensitivity in the order of one adduct in 10^{10} nucleotides (Randerath & Randerath 1993). DNA adducts are viewed as a more relevant biological parameter for human exposure to carcinogens than environmental monitoring of chemicals or analyzing for chemicals in blood or urine, since adduct levels represent that part of the dose that escapes detoxification. However, this assay usage on wastes and effluents is very low (at a level of around 0.3%).

DNA single strand breaks (SSB) : The relatively high sensitivity of the SSB response to xenobiotic exposure, as well as the numerous toxicological sequelae associated with the SSB response, including clastogenesis, heritable mutations and cancer have attributed specific significance to SSB as one of the end points of genotoxicity monitoring.

Methods of quantitation of SSB are generally based upon exposure of the DNA strand to alkaline conditions (pH>12.5) such that unwinding of the helix occurs at the SSB sites. If an appropriate fixed period of unwinding is used, the formation of single stranded (SS) DNA will be proportional to the number of strand breaks, procedures exist for facilitating unwinding and for quantifying the SS and double stranded (DS) DNA fractions. One of the simplest procedures involves an alkaline unwinding step followed by DNA quantification using the fluorescent DNA–binding stain, Hoechst 33258/ Ethidium bromide, in a sample which contain both SS and DS fractions (Birnboim & Jevcak, 1981, Chow et al. 1989).

Micronucleus assay (MN) : Micronucleus (MN) formation, as its name suggests, is the appearance of structures in a cell that has undergone cell division and that are morphologically identical to, but smaller than the main nuclei. Micronuclei can be formed both from chromosome breaks causing acentric fragments (clastogenic), and from lagging chromosomes resulting in whole chromosome loss (aneugenic) when the chromosomes recondense. The micronuclei assay has traditionally been sensitive to damage from ionizing radiation but relatively insensitive to adduct forming chemical agents (Vnenchak et al. 1996). The scoring step in MN assay can be tedious and repetitive. Micronucleus determinations are very low, at a level of usage of 0.3%, as compared to chromosome aberration which accounts for 9.9% of tests on wastes and effluents.

Sister chromatid exchange (SCE) : Sister chromatid exchange involves a symmetrical exchange between chromatid within a chromosome and, therefore, involves identical base sequences in DNA. SCEs are known to occur as a normal feature of cell division in mammalian systems, SCEs are now believed to involve the interchange of DNA replication products at homologus loci caused by DNA breakage and reunion. The exact mechanisms of SCE formation, however, are still unknown. These assays are of low sensitivity and do not quantitatively predict exposure levels of wastes and effluents causing adverse health effects (Tucker et al. 1993).

Chromosomal aberration : The purpose of chromosomal aberration test is to identify agents that cause structural chromosomal aberrations in cultured mammalian cells. Structural aberrations may involve chromosome or chromatid. Majority of chemical mutagens induce aberrations of the chromatid type though chromosome type aberrations are not uncommon. Chromosomal aberrations are commonly scored in metaphase and anaphase cells. Chromosomal aberrations appearing in anaphase can more easily be registered than those in metaphase; they also provide a fast means of estimating chromosome damage. In contrast, metaphase analyses give more information on the types of induced aberration. Metaphase is the stage in which the various types of aberrations could be distinguished and a detailed analysis of the effect can be made. Chromosome mutations and related events are the cause of many human genetic diseases and there are substantial evidences to show that chromosome mutations and related events could cause alterations in oncogenes and tumour suppressor genes of somatic cells resulting in cancer induction in humans and experimental animals (Preston et al. 1987).

Oncogenes, tumor suppressor genes and environmental cancer

The connection between carcinogenesis (cancer generation) and mutagenesis has been established with three types of cancer agents: chemical carcinogens, which cause simple local changes in DNA sequence; ionizing radiation, which can break chromosomes and cause translocations; and viruses, which introduce foreign DNA into the cell (White 1994). It is observed that the cell has mechanisms to prevent the damaged DNA from being passed on to its daughter cells. This has led researchers to investigate tumor, or oncogenes/growth suppressor genes. Among these, p^{53} gene plays a major role as the most frequently encountered genetic events in human malignancy is the alterations of the p^{53} gene and its encoded protein (Colman et al. 2000).

Role of p^{53} gene and its protein

From epidemiological and experimental studies, it has been observed that more than 50% of cancers in humans are due to environmental exposure to chemicals and mutation of p^{53} gene (Colman et al. 2000). p^{53} is located on the short arm of the human chromosome 17, band 13 and is approximately 20-kilo bases in length. This gene yields a 2.8 kb mRNA transcript and encodes for a 53 kilo Dalton nuclear phosphoprotein which contains 393 amino acids (Greenblat et al. 1994, Lane 1992). Upon DNA damage, p^{53} gene gets activated and stops the cell cycle; it also induces cell to repair the damaged DNA. If the repair is successful, it allows the cell for further progress to S-phase. Otherwise, the p^{53} gene triggers the cell to undergo programmed cell death called apoptosis. By this way, the gene prevents its damaged DNA from being passed on to its daughter cells.

Phosphorylation of p^{53}

The biochemical function of p^{53} is regulated by reversible serine phosphorylation (Meek 1994) p^{53} protein can be phosphorylated in vitro by at least seven different kinases, including cdc2, casein kinase II (Meek et al. 1990). DNA-dependent protein kinase 1, a casein kinase I-like kinase, protein kinase C, mitogen-activated protein kinase, and JNK1 (Milne et al. 1995). Although the precise role(s) of these kinases in regulating p^{53} function is not understood, recent studies are providing clues. For example, mutation in the casein kinase II phosphorylation site at serine 392 can reduce the antiproliferative activity of p^{53}. The S and G2/M cyclin-dependent kinase complexes, cdk2-cyclin A and cdk2-cyclin B, phosphorylate serine 315 of p^{53} and stimulate its sequence-specific DNA binding to p^{21Waf1} and GADD45 sites preferentially (Wang 1995). The G1 cyclin-dependent kinase complexes, cyclin E/cdk2 and cyclin Dl/cdk6, do not phosphorylate p^{53}, which is consistent with results indicating that p^{53} is underphosphorylated at serine 315 in the G1 phase of the cell cycle. These data also are consistent with a hypothetic negative feedback loop in which increased cdk2 activity would generate a transcriptionally activated p^{53} that would increase p^{21Waf1} expression, an inhibitor of cdk2, and consequently reduce cdk2 phosphorylation of p^{53} and, thus, reduce p^{53} transcription-transactivation function. Protein kinase C also is likely to regulate p^{53} function as a transcription transactivator (Hupp 1992). *In vitro* protein kinase C phosphorylates serine residue at position 378 in the carboxyl-terminal region of p^{53} that contains the epitope recognized by the monoclonal antibody (Pab421) and is the domain for negative regulation of p^{53} transcription transactivation. Phosphorylation of this domain blocks Pab421 binding to p^{53} and Pab421 nonreactivity with p^{53} which correlates well with growth arrest. Activation of protein kinase C by phorbol 12-myristate 13-acetate also induces growth arrest at the G1 checkpoint. Both cdk2 and protein kinase C can participate in apoptosis (Skouv et al. 1994).

Molecular archaeology of p^{53} mutations

Mutations can arise by either endogenous mutagenic mechanisms or exogenous mutagenic agents and are archived in the spectrum of p^{53} mutations found in human cancer. Errors introduced during DNA replication, RNA splicing, DNA repair, and DNA deamination are examples of endogenous mutagenic mechanisms. The DNA sequence context is an important factor determining these events. Almost all short deletions and insertions occur at monotonic runs of two or more identical bases or at repeats of 2- to 8-base-pair DNA motifs, either in tandem or separated by a short intervening sequence (Greenbalt et al. 1996). The mechanism that has been extensively studied is called slipped mispairing, a misalignment of the template DNA strands during replication that leads to either deletion, if the nucleotides excluded from pairing are on the template strand, or insertion, if they are on the primer strand. When direct repeat sequences mispair with a complementary motif nearby, the intervening oligonucleotide sequence may form a loop between the two repeat motifs and be deleted (Krawczak & Cooper 1991). More lengthy runs and sequence repeats are more likely to generate frameshift mutations. The deletions and insertions in the p^{53} gene found in human tumors also may be biologically selected from the broad array of such mutations occurring in human cells. When compared with the distribution of missense mutations, these types of mutations occur more frequently in exons 2-4 (54%) and 9-11 (77%) than in exons 5-8 (20%). The N-terminus of the p^{53} protein (encoded by exons 2-4), has an abundance of acidic amino acids that are involved in transcriptional function of p^{53}, it binds to transcription factors such as TBP in the basal transcription multiprotein complex, TFIID (Mack et al. 1993). Experimental studies have shown that multiple point mutations are required to inactivate its transcription-transactivation function. The carboxy-terminus (encoded by exons 9-11) of the p^{53} protein is enriched in basic amino acids that are important in the oligomerization and nuclear localization of the p^{53} protein, recognition of DNA damage and induction of apoptosis (Bakalkin et al. (1994). Multiple point mutations are infrequently found in the p^{53} gene, which is consistent with the target theory (i.e., exogenous mutagens target the p^{53} gene within the context of the entire human genome) (Dimitrios et al. 2000). Therefore, deletions and insertions would be a more efficient mutagenic mechanism than single-point mutations in disrupting these N-terminal and C-terminal functional domains.

The interest in the assessment of human risk to carcinogenic compounds at contaminated sites appears to be justified, since epidemiological data over a number of years have shown that most human cancers are due to environmental factors. In addition, available experimental data have confirmed the role of environmental chemicals in multistep carcinogenesis. Though there has been considerable progress in recent times in the development of short-term genetic assays (Wang et al. 1996), the National Centre for Toxicological Research (NCTR) is involved in the development and validation of short-term assays for characterizing mutagenic, genotoxic and carcinogenic effects of chemicals on cellular systems.

Recommendation of IARC on battery of tests

The cytogenetic findings for the IARC on proven, probable and possible human carcinogens are presented in **Table 4**. It could be seen from this Table that though many compounds and substances induce SCE, MN and CA, at times, all the end points could not been induced by one compound, indicating that multiple endpoints need to be monitored in genotoxic investigations of contaminated sites and hazardous wastes (Vnenchak et al. 1996)

The genotoxicity of hazardous chemicals and wastes originating from petroleum refineries, petrochemical industries and coke ovens of steel industry as well as wastewaters from the above industries, pesticide industries, metal bearing wastes etc. have been carried out using multiple end points (battery of assays) such as DNA damage/stand break, chromosomal aberration (macro-lesion of genetic material), p^{53} protein induction (repair mechanism) and apoptosis (repair as well as cell integrity and selection of healthy population). Genotoxicity determination could provide valuable information, if used as such or in conjunction with current risk assessment protocols available elsewhere (Woo et al.1996).

Table 4 : Cytogenetic data for the chemicals, mixtures and processes in IARC categories

Agent/exposure	Cytogenetic finding*					
	Humans		Animals			
	CA	SCE	ML	DAP	BOE	MN
Human carcinogens (IARC group 1)	+	+	--	--	+	?
Alcoholic beverages	--	--	--	--	--	--
Aluminum production	?	?	--	+	-	+
Arsenic and arsenic compounds	--	?	--	--	--	--
Asbestos	--	?	--	--	--	--
Azothiophine	?	--	--	+	--	+
Benzene	+	--	--	--	--	+
Betelquid with tobacco	--	+	+	--	--	+
Bis(chloromethyl)ether and chloromethyl ether (technical grade)	(+)	--	--	--	--	--
1,4-buanediol dimethane sulfonate (Myleran)	(+)	--	--	+	--	+
Chlorambucil	?	+	--	+	--	--
Ciclosporin	(+)	--	--	--	--	--
Coal tars	+	--	--	--	--	--
Coke production	+					
Combined oral contraceptives	+					
Cyclophosphamide	+	+	--	+	+	+
Hgexavalent chromium compounds	+	+	--	+	+	+
Melphalan	+	+	--	+	--	--
8-methoxypsoralen plus ultraviolet A radiation	+	--	--	--	--	--
Mineral oils untreated and modly treated	+	--	--	--	--	--
nickel compounds	+	--	--	?	--	--
Painter, occupational exposure as	--	--	--	--	--	--
Radon	+	--	--	--	--	--
Rubber industry	?	--	--	--	--	--
Tobacco products smokeless	--	--	--	--	--	--
Tobacco smoke	+	--	--	--	--	--
Ins(1 azindiny) phosphine sulfate (1 biotepa)	--	--	+	--	--	+
Vinyl chloride	+	?	--	+	+	+
Probable human carciongens (group 2A)	--	--	--	--	--	--
Acrylonitrile	--	--	--	--	--	--
Adriamycin	+	+	--	+	+	+
Cadmium and Cadmium compounds	--	(-)	--	--	--	--

Reference:

Aldridge, W. N. (1993) The biochemical Principles of Toxicology, Cited in : Experimental toxicology second edition D. Anderson and D.M. conning (Eds.) Royal Society of Chemistry Publishers, Cambridge, USA.

Bakalkin G, Yakovleva T, Selivanova G, Magnusson KP, Szekely L and Kisaleva E. (1994) p^{53} binds single-stranded DNA ends and catalyzes DNA renaturation and strand transfer. *Proc. Natl. Acad. Sci. USA* 91; 413-417

Birnboim and J.J. Jevcak (1981) Fluorometric Method for Rapid Detection of DNA strand breaks in Human white blood cells produced by low Doses of Radiation, Cancer Res., 41: 1889 - 1892.

Buffler, P.A., M. Crane and M.M. Key (1985) Possibilities of detecting health effects by studies of populations exposed to chemicals from waste disposal sites, *Environ. Health Perspect. 62, 423-456.*

Chakrabarti, P.V.R. Subramanyam and B.B. Sundaresan (1988) Biodegradation of Recalcitrant industrial wastes in: Bitreatment Systems *Vol.II CRC Press, Donald L. Wise (Ed.) pp. 171-234.*

Chow, S.C., McConkey, D.J., Orrenius, S and Jondal, M (1989) Quantitation DNA fragmentation using fiberglass filters. *Anal. Biochem.* 183; 42-45

Cohen, S.M. & Elwein, L. B. (1990) cell proliferation in carcinogenesis, *Science, 249; 1007-1011*

Colman, M.S., Apshari, C.A. and Barrett, J.C. (2000) Regulation of p^{53} stability and activity in response to genotoxic stress. *Mutat. Res.* 462 (2-3), 179-188.

DeMarini, D.M., Brimer, P.A. and Hsie, A.W (1984) Cytotoxicity and mutagenicity of coal oils in the CHO HGPRT assay, *Environ. Mutagen.* 6, 517-527

DeMarini, D.M., D.J. Brussick and J. Lewtas (1987) Use of limited protocols to evaluate the genotoxicity of hazardous waste in mammalian cell assays : comparison to Salmonella, *J. Toxicol. Environ. Health, 22, 225-239.*

DeMarini, D.M., J. E. Gallagher, V.S. Houk and E. Simmons (1989) toxicological evaluation of complex industrial wastes; implications of exposure assessment, Toxicol. Letters, 49, 199-214.

DHHS, (1985). Risk assessment and risk management of toxic substances: A Report to the Secretary, Department of Health Human Services from the Executive committee, DHHS committee to Coordinate Environmental and Related Programs (CCERP).

Dimitrios Assimakopoulos, Evangelos Kolettas, Nektaria Zagarianakow, Angelos Evangelou, Antonious Skevas and Nisi J. Agnantis (2000) Prognostic significance of p^{53} in the cancer of the larynx. *Anticancer Research* 20; 3555-3564.

Doll & peto, R. (1981). The causes of cancer: quantitative estimates of avoidable risks of cancer in the United States. *J. Natn. Cancer. Inst. 66; pp. 1191 –1308*

Greenbalt MS, Grollman AP, and Harris CC (1996) Deletions and insertions in the p^{53} tumor suppressor gene in human cancers: confirmation of the DNA polymerase slippage/misalignment model. *Cancer Res.* 56; 2130-2136.

Greenblat MS, Bennett WP, Hollstein M and Harris CC (1994) Mutations in the p^{53} tumor suppressor gene: clues to cancer etiology and molecular pathogenesis. *Cancer Research* 54: 4855-4878.

Griffith, J., R.C. Duncan, W.B. Riggan and A.C. pellom (1989) Cancer mortality in U.S. counties with hazardous waste sites and ground water pollution, *Arch. Env. Health, 44, 69-74.*

Grisham, J.W., Ed. (1986) Health aspects of the Disposal of Waste Chemicals. *Pergamon Press, New York, 454 pp.*

Houk V.S. (1992) Waste treatment technologies and their effect on the genotoxicity of industrial wastes, effluents, *Mutat. Res. 277; 91-138.*

Hupp TR, Meek DW, Midgley CA and Lane DP (1992) Regulation of the specific DNA binding function of p^{53}, *Cell* 71; 875-886.

Klekowski, E. and D.E. Lewin (1979) Mutagens in a river heavily polluted with paper recycling wastes: Results of field and laboratory mutagen assays, *Environ. Mutagen. 1, 209-219.*

Krawczak M and Cooper DN (1991) Gene deletions causing human genetic disease: mechanisms of mutagenesis and the role of the local DNA sequence environment. *Hum. Genet.* 86; 425-441.

Lai, D.Y. Yin-tak Woo, Mary F. Argus and Joseph C. Arcos (1996) Carcinogenic potential of organic peroxides: Prediction based on Structure-Activity Relationships (SAR) and Mechanism based short-term tests. *Environ. Carcino and Ecotox. Revs.* C14 (1); 63-80.

Lane DP (1992) Cancer, p53, guardian of genome, *Nature*, 358,15-16.

Mack DH, Vartikar J, Pipas JM and Laimins LA (1993) Specific repression of TATA-mediated but not initiator-mediated transcription by wild-type p^{53}. *Nature* 363; 281-283.

Meek DW, (1994) post-translational modification of p^{53}. *Semin Cancer Biol.* 5;203-210.

Milne DM, Campbell LE, Campbell DG and Meek DW. (1995) p^{53} is phosphorylated in vitro and in vivo by an ultraviolet radiation-induced protein kinase characteristic of the c-Jun kinase, JNK 1. *J Biol. Chem.* 270; 5511-5518.

MOEF, (2000), Management and Handling of hazardous wastes Rules, MOEF, 2000.

Najem, G.R., D.B. Louria, M.A. Lavenhar and M. Feuerman (1985) Clusters of cancer mortality in New Jercy municipalities; with special reference to chemical toxic waste disposal sites and per capita income, *Intl. J. Epid., 14, 528-537*

Preston, R.J., San Sebastian, J.R. and McFee, A.F (1987) The *in vitro* human lymphocyte assay for assessing the clastogenicity of chemical agents. *Mutat. Res.* 189; 143-188.

Randerath K, Zhou G.D, Donnelly KC, Safe SH and Randerath, E (1994) DNA damage induced by wood preserving waste extracts in vitro without metabolic activation as assayed by P32-post labeling. Cancer Letters 83; 123-128.

Randerath, K and Randerath, (1993) In post labelling Methods for detection of DNA adducts, edited by D.H. Philops, m. Castegnaro and H. Bartesh, P.B. IARC, Scientific Publications No. 124, International Agency for Research on Cancer, Lyon.

Scott, D., Galloway S.M., Marshall, R.R. Ishidate, M., Jr. and Brussick, D. J. (1991) Genotoxicity under extreme culture conditions, A report from ICPEMC ask Group, *Mutat, Res. 257; 147-204.*

Skouv J, Jensen PO, Forchammer J, Larsen JK and Lund LR (1994) Tumor-promoting phorbol ester transiently down-modulates the p^{53} level and blocks the cell cycle. *Cell Growth Differ.* 5; 329-340.

Tucker, J.D., Auletta, A., Cimio, M.C., Dearfield, K.M. Jacobson0Kram, D., Tice, R and Carrano, A.V. (1993) *Mutat. Res.* 297; 101.

U.S. Environmental Protection Agency (1991) Technical support document for water quality-based toxics control. *EPA 505/2-90-001. Technical Report. Washington, DC.*

U.S. Environmental protection Agency (USEPA) (1988) environmental progress and challenges; EPA's Update. Office of policy planning and evaluation, *EPA-230-07-88-033, Washington, DC, August 1988, pp.140*

Vnenchak Magdalena Kosz, Katarzyna Rokosz, Anna Szuster and Roza Kubiak (1996) Use of micronucleus test for screening the potential genotoxicity of polluted water. *CEJOEM*, 2 (4); 329-334.

Walum, E., Stenberg, K. and Jenssen, D. (1990) Understanding cell toxicology principles and practices, Ellis Horwood, New York.

Wang XW, Vermeulen W, Coursen JD, Gibson M, Lupold SE and Forrester K, (1996) The XPB and XPD helicases are components of the p^{53}-mediated apoptosis pathway. *Genes Dev.* 10; 1219-1232.

Wang Y, Prives C (1995) Increased and altered DNA binding of human p^{53} by S and G2/M but not G1 cyclin-dependent kinases. *Nature*, 376; 88-91.

Waters, M.D., H.F. Stack, A.L. Brady, P.H.M. Lohman, I. Haroun and H. vainio (1988) Use of computerized data listings and activity profiles of genetic and related effects in the review of 195 compounds, Mutation Research, 205, 295-312.

White E. (1994) Tumor biology. P^{53}, guardian of Rb. *Nature* 371; 21-22.

Woo Yin-tak, David Y. Lai, Mary F. Argus and Joseph C. Arcos (1996) Carcinogenicity of organophosphorus pesticides/ compounds: An analysis of their Structure-Activity Relationships. Environ. Carcino. & Ecotox. Revs. C14 (1); 1-42.

WASTE MANAGEMENT

Development of New Materials based on Agro and Industrial Wastes Towards Ecofriendly Society

Kestur Gundappa Satyanarayana, Luiz Pereira Ramos and Fernando Wypych*

Introduction

The last century has seen one of the largest urbanization with its population growing by three times between 1950-1990 with developing countries contributing to its fourfold increase. It is expected to reach about 3.8 billion by 2025 in the developing countries (Dowall 1997). Hence there is a greater thrust now as never before for the use of eco friendly and energy saving materials and processes along with the reduction in the use of expensive but dwindling wood resource, and non-renewable petroleum and the mineral resources. It is reported that current rate of consumption of petroleum is reported to be about 10^5 faster than nature can create (Netravali and Chabba 2003). These have enormous impact on the environmental sustainability and in safeguarding ecosystems. It is worth mentioning here that the environment has become a growing concern of society and industry worldwide even as early as 1962, as mentioned in the book "Silent Spring" by Racel Carson with the threat posed from pollution to the global eco-system followed by focus on the global effect of pollution-global warming and deforestation in the nineties. Then the focus was primarily centered around the dominant and visible end-source polluters namely industry. However, catchwords coined by the United Nations in the mid 90s such as "Sustainable Growth", the "Business Council for Sustainable Development" and "eco-efficiency", the attention was drawn to the general environmental effect of per capita consumption and population growth. Thus a new dimension of "clean consumption" was added to "clean production" to the environmental agenda. Then the concept of "Industrial Ecology" was defined by various world agencies of which the one stated by the National Academy of Sciences, USA in 1992 seems to be more appropriate. It states "A new approach to the industrial design of products and processes and the implementation of sustainable manufacturing strategies is a concept in which an industrial system is viewed, not in isolation from its surrounding system, but in concert with them. Industrial ecology seeks to optimize the total material cycle from virgin material to finished material, to component, to product, to waste product and to ultimate disposal" (Marstrander 1998).

In view of ever growing environmental consciousness all over the world, several governments have passed laws for the use of natural materials in many applications in view of their eco-friendly attributes, availability and cost. Many developing countries such as Brazil and India possess abundantly available agro and minerals resources. But presently many of these are under utilized thus leading to non-exploitation of their potential resulting in low value products and unemployment. Also, a number of by-products are produced in all industries including agricultural activities, which are mostly termed as "wastes". These may pose threat to the environment, but possibly have range of uses giving excellent opportunities for scientists to discover methods for their better utilization. This is because the wastes may be considered to be as valuable, or may be more valuable, than the main products. Table 1 lists some of the byproducts from different sources/industries and their possible applications. It is also well known that agricultural processing yields many byproducts that have significant value and in developing countries these byproducts are often underutilized. The types of solid waste generated by the food processing industry can be generally classified under different categories, with significant quantities of such wastes providing entrepreneurial opportunities for biogas or electrical generation and even new material developments.

Table 1 : List of Byproducts from different Sources/Industries and their Possible Applications (Source: Forschungsforum 3/97)

Industry	Petrochem. Raw Materials	Biogenic. Raw Materials	Examples of Raw Materials	Possible Applications
Reinforced Fibrous materials	Carbon fiber, Glass fiber, Polyamide	Plant fibers, Plant resins	Jute, Sisal, flax fibers	Toys, Housing, Furniture & Accessories
Floor Coverings	PVC	Barks, Vegetable oils, Plant fibers, Plant resins	Cork, linseed oil, Jute, Naval stores	Linoleum
Wood Glaze	Polyacrylate, glycol	Plant Resins, Essential oils	Dammar Resin, Lemon Peel oil	Natural resin oil, Glazes
Colors	Azo Paints	Coloring plants	Dyer's wood, Madder, etc.	Plant colors
Detergents	Linear Alkyl benzene Sulphonate (LAS)	Vegetable oils, Carbohydrates	Coconut fat, Sugary tensides	Household Detergents
Hydraulic Agents	Mineral Oils	Vegetable oils	Rapseed oil, Jojoba oil	Saw chain lubricants, top performance lubricants
Adhesives	Polyurethane	Natural Gums, Plant Resins	Natural Latex, Mastiche	Carpet Glue
Packaging Materials	Polyethylene	Polysaccharides	Biopol	Shampoo Bottles
Textiles	Polyester	Plant/Natural Fibers	Linen, Hemp, Cotton, silk, wool	Garments
Thermal Insulation	Styrofoam, Polystyrene	Plant Pulp, Straw, Lignin	Sunflower Pulp & Straw	New Absorbing Materials

Few examples of value-added useful products and applications of agro wastes include "green chemicals" such as solvents, additives, surfactants, adhesives, etc. and renewable materials such as cellulose, lignin, and biodegradable polymers.

Others include: handmade paper; geotextiles for soil erosion control; paneling composite materials using natural fibers; activated carbon from agricultural byproducts like rice husk as catalysts for bio remediation such as enzymes and adsorbents in water for removing organic chemical and metals of environmental or economic value (Daifullah *et al.*, 2003).

Similarly, during the last two decades or so there has been considerable interest in not only producing biodegradable plastics but also in the potential uses of those with thermoplastic properties to overcome some of the problems associated with waste disposal. In the same way, it is also said that use of industrial wastes in the manufacture of building materials can not only save energy but also take care of environmental concerns (Gupta 2000).

Of late, developed countries have accepted and followed the concept of "no waste", terming the wastes as "new resources" for new material development (Andre de Vries 1999). The strategy adopted here is their bulk utilization through large outputs and large production capacities by industrially feasible methodologies. These "new resources" can be utilized with appropriate processes to develop new materials useful for different sectors such as construction industry as building materials. However, to achieve this one has to take appropriate care of the local specific characteristics of the waste materials. It appears that the best chances for their utilization is as energy producers and/or in the new material development while addressing the environmental concerns. These concepts may also be followed by the developing countries since the utilization of waste materials is a global solution. In addition, it is interesting to note the expenses to be incurred by the developing countries to achieve industrial compliance with environmental norms with the traditional approach to reduce pollution and with one that stresses heavily on cleaner industrial production are similar. This is important since it is reported (Ralph Luken & Ann-Christin Freij 1994) that the developed countries presently spend about 80 per cent investment in "end-of-pipe" technology [end-of-pipe technologies are those which use wastes and emissions that have already been generated with a view to reduce them considerably] and 20 per cent investment in process controls while the pattern of investment in these two will be the same in developing countries as it has been in developed countries over the past two decades or so.

However, three factors seemed to be contributing for the non-realization of the value addition to these materials in the developing countries (Huisenga 2003), namely, nonexistence of a small number of capable ancillary industries to make use of specific byproducts; their level of market development and the lack of information about market opportunities for byproduct uses. Despite this, in recent years attempts have been made by some developing countries on the lines followed by the developed countries as mentioned above. For example, development of natural fibers incorporated composites as wood substitutes and highly efficient construction materials based on industrial wastes without the use of binders such as burnt refuses of coal mining and processing, mineral and metal slags, sludges from wastes of chemical, petrochemical, thermal power and other industries. These developments are significant as can be seen from examples such as construction industry. It is pointed out that the building materials contribute about 60-70% of the cost of construction with an increasing trend for shortage of conventional and traditional building materials. Further, the cost of construction using such materials is continuously increasing (15-18% per annum)(Suresh 1999). It is also known that many of these materials are energy intensive and hence the cost may further go up in addition to these materials causing environmental problems. With increasing shortage of building materials and increasing generation of large quantities of wastes as mentioned above, any attempt for developing new materials using wastes will help in value addition to these wastes along with promoting the employment generation. Similar concept may hold good in other application sectors as mentioned below: a number of natural fiber reinforced polymer composites are being utilized in automotives not only to make the vehicles lighter and thus working towards fuel saving but also to address environmental friendliness (Kandachar 2002, Netravali & Chabba 2003); fly ash, a byproduct of thermal power plants, is used in concrete and other cement base materials; recovery and recycling by thermal or chemical treatments of TiO_2, one of the solid wastes of paper industry (~30%), for its use in the manufacture of paints as a low cost material. It is reported that this material exhibits same hiding power and accounts for about 50% reduction in cost of the total raw material (Francisco - http://rko.zcu.cz/sme.php3). Use of recovered natural substances from these wastes in varnishes to protect the wood from UV light degradation is another example (Nicholas 1999).

The above are some of the solutions for socio-economic developments in general while attempting to use the agro and industrial wastes. Thus, the overall strategy adopted for using locally available agro products, industrial by-products and other wastes throughout the world has been to develop new materials/products through innovative processes. This is due to the fact that both a cost efficient processing system and inexpensive raw materials are needed for the low-cost nature of disposable products.

A number of International Conferences/Workshops/Training Programs have been held dealing with such developments. While a number of R&D organizations and academic institutions across the world are involved in such developments, promoting agencies such as UNIDO, UNDP and some of the nongovernmental agencies in several countries are in the forefront to bring awareness and utilization of these new materials. To work towards the above mentioned strategy, the first exercise to be undertaken is to make survey of the available wastes to assess the quantity available for their new uses, tabulate their available chemical, physical and useful properties and also current conventional uses, if any. This is to be followed by the characterization of these with respect to any other properties that have not been evaluated, which may lead to find possible new uses. Based on these, one can derive the research imperatives using Technology Forecasting and Technology Assessment methodologies to list out techno-economical products for development and then develop appropriate processing methods for these. Such exercises have been done successfully in many countries. In the following Sections, some of these are discussed with special attention to development of composites, construction/building materials and other societal useful materials. The first is one of the advanced materials with wide range of fields of application. On the other hand, development of new materials for construction sector will be significant. This is because this sector is one of the largest consumers of energy in many parts of the world although it helps directly the development, expansion, improvement and maintenance of housings in particular and economic growth in general. This can be understood from the energy content in the production of conventional materials such as bricks, mortar, cement, glass and wood, which range between 1-60GJ/ton (Gupta 2000). Hence, reduction of the total energy use in this sector through the development and use of environmentally sustainable and cost effective technologies/processes and materials will augur good for the future of mankind. Similarly, water-soluble biodegradable polymeric materials are developed based on feedstock derived from fossil fuel and renewable resources, the latter gaining increased attention for industrial development due to the reasons mentioned above. These materials have a wide range of applications in different industrial and other segments such as food, textiles, leather, coatings, paper, health-care, bio-fuels, wastewater treatment biomedical, and pharmaceutical fields. Accordingly, this Chapter gives details some of the developments made using the agro and industrial wastes and some perspectives for future towards working for environment management.

Agro by-products/Wastes

In addition to providing food and animal feed, other important function of agriculture and forestry has been to provide energy and raw materials. This may continue in future also with supply of sustainable raw materials to a very considerable extent. It is also said that any cycle goes round and use of renewable resources such as bio-resources is not outside this ambit. Thus, bio-resources on which extensive dependence was on at the turn of the last century were replaced in 70's by petroleum-based materials. Thanks to the ecological concerns followed by some of the recent developments, the bio-resources have been again drawing serious attention in recent times as alternate feed stock to produce some of the industrial products as mentioned earlier. However, sustained growth for this depends on the development of new materials of competitive performance, their costs and markets. Accordingly, production of such materials can be proved only by the efficient utilization of renewable agro resources along with their byproducts.

It is reported that nature produces about 170 billion tons of plant biomass annually (Forschungsforum 1997). Of the 3.15 billion (about 55%) of the world population who use biomass energy for cooking, more than 70% are in developing countries with this energy being about 80-90% of their total energy consumption (Nicholas 1999). Further, it is estimated that these sustainable raw materials can increase the production of heat, electricity and fuels from about 40 million tonnes oil-equivalent in 1995 to more than 130 million tonnes oil-equivalent by 2010. This expansion by 90 million tonnes oil-equivalent corresponds to an area of 14 million hectares (Gmunden 1998). Since these agro wastes are part of the closed cycle of biosphere, using them will provide an opportunity to supply all needed substances without polluting the biosphere with any hazardous substance. In addition, the plants possess "biologically active" agents of the secondary metabolism such as coloring substances, resins, tanning agents, waxing agents, essential oils, repellents, plant hormones, etc., which can be used for special purposes and according to their "natural function" (Forschungsforum 1997). For example, it is well known that plant colors look attractive and they are fast to light when applied on textile/natural fibers. The colors stick to these fibers with more harmless chemical auxiliary substances. Further, it is well known that agro processing yields many byproducts/wastes that have significant value, but may not be appropriately used. In fact, these biomass residues are one of the largest sources of renewable energy in view of their storage capability of solar energy. These can be easily converted into various convenient forms of energy and also can be used as adsorbents for the removal of traces of valuable heavy metals and toxics from industrial and municipal water effluents (Abia *et al.*, 2003, Quek *et al.*, 1998, Gardea-Torresdey *et al.*, 1999, Gang & Weixing 1998, Yaman 2003, Vamvuka *et al.*, 2003). Being organic in nature, they are also source for carbon and host of other useful materials and chemicals, particularly for the production of a new generation of "green" materials. For example, rice husk, which constitutes more than 10% of a world rice production of about 400 million tons (Sathy *et al.*, 2003), has been one of the most promising agro-wastes to produce a variety of inorganic materials such as silica, sodium and potassium silicates, zeolites, solar grade silicon metal, building and thermal insulating materials, polymer composites, etc.

The ash produced by the controlled burning of this is reported to be highly pozzolanic and hence suitable as one of the building material components (Ramachandran1995, Mehta & Kumar 1983, Aitcin & Pigeon 1986, Kihara & Bucher 1986, Sathy *et al.*, 2003). Similarly, straws of rice and sugarcane (bagasse), which are normally used as fuels in households or sometimes *in situ* burnt in the fields as means of disposal, can be utilized as fillers/reinforcements in polymer or cement to develop new materials for construction industry. For example, in Brazil, sugar cane is used for the production of polyhydroxyalkanates, which are being explored for medical applications (Ortega & Innocentini Mei 2003). Other by-products include sawdust, coir dust, shells of coconut, oil palm, etc.

Table 2 (a) lists some agro wastes of India, their availability and potential uses while Table 2 (b) shows list of options in the case of by-products from food processing (Huisenga 2003). Many of these materials have been characterized in respect of their chemical, physical and other properties. These are given elsewhere (Grover 1989, Sukumaran *et al.*, 2001).

Table 2 (a) List of Some Agro Wastes, their availability and potential uses (India) [Courtesy: BMTPC, India]

Waste	Industry	Annual production (mt)	Potential uses
Bagasse	Sugar Mills	5.0	Insulation boards
Corn Cobs and Stalks	Agricultural Farms	14.0	Building Boards
Coconut Husk & Byproducts	Coconut Based	3.0	Roofings, Composites& Brick
Jute Sticks and Bark	Jute Mills	5.0	Roofings & Composites
Rice Husk	Rice Mills	25.0	Particle Boards, Cement, Pozzalona, Chemicals and Condensed Silica Fume
Peanut Husk	Agricultural Farms & Peanut oil Mills	2.5	Composites / Insulation Boards
Straws of Rice&Wheat	Agricultural Farms	90.0 & 33.0	Composites/Fiber Boards

Table 2 (b) Options for Byproducts of Food Processing [Courtesy: Huisenga 2003]

Options	Examples
Food	Non-nutritive fiber from fruits, dried citrus peels, lecithin from soyabeans, etc
Feed	Desolvenized Cotton seed meal, sugar beef pulp, ,spent brewer's grains, etc.
Fuel	Sugarcane bagasses, peach and olive pits, rice,wheat and coffee hulls and straws, ethanol from cassava, etc.
Others	Essential oils from citrus, Tartrates from grape lees, anthocyanin dyes from grapes, cherries, starch derivaties.
Land Application	Farming land enriched by wastewaters containing substantial solid matters, land application pomace, stem waste and leaves, etc.
Disposal	Land filling of pit wastes, land fill of washing mud from vegetable processing.

Table 3 (a) lists chemical composition of various lignocellulosic fibers and Table 3 (b) shows a comparison of some characteristics with those of synthetic materials (BMTPC 1999). With a view to understand these properties and also to find new uses, attempts have been made to study their

Table 3 (a) Chemical Composition and Moisture Absorption of Some Lignocellulosic Fibers (Source: Sukumaran *et al.* 2001, Courtesy: Meshap Scientific Publishers, India)

Fibre	Cellulose (%)	Hemi-cellulose (%)	Lignin (%)	Moisture regain at 85% R.H	Transverse Swelling in water (%)
Banana	60-65	6-8	5-10	10-15	16-20
Coir	43	<1	45	10-12	6-15
Cotton Lint	90	6	-	7	20-22
Flex	70-72	14	4-5	7	20-25
Jute	61-63	13	5-13	12.5	20-22
Mesta	60	15	10	13	20-22
Palmyrah	40-50	15	42-45	10-12	--
Pine Apple Leaf	80	-	12	10-13	18-20
Ramie	80-85	3-4	0.5	5-6	12-15
Sisal	60-67	10-15	8-12	10-12	18-20
Straw	40	28	18		
Sun Hemp	70-78	18-19	4-5	10-11	18-20
Wood	45-50	23	27	-	-

microstructures. Some of these are shown in Fig. 1.

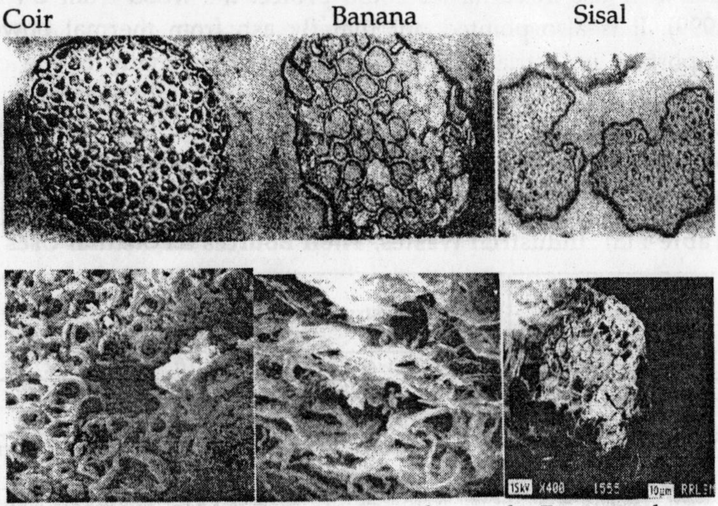

Fig. 1 Cross Sectional Micrographs and Fractographs of lignocellulosic Fibers (Source: Sukumaran *et al.*, 2001, Courtesy: Meshap Science Publishers, India)

Attempts have also been made (Gmunden 1998) to form networks and centers of know-how within such net works such as Bio fiber Net- a forum for industry and academy and IBAW - International Biodegradable Polymers Association & Working Groups in Europe, an organization networking in the fields of bio-based materials, especially biodegradable polymers. In such forums, the methodology adopted is to document the ecological, social and economic advantages of sustainability of these materials and then work towards the use and application of products based on these resources within the context of a broad program of research to be undertaken by them for the promotion of these raw materials, establishment of standards, waste legislation, etc. Here, concerned persons can exchange ideas about product development and find partners for research, product development and business activities. Through such net works it would be possible to enhance the utilization of lignocellulosic fibers for sustainable manufacture of industrial products as mentioned earlier.

Table 3 (b) Specific Strength, Cost and Energy Contents of Synthetic and Lignocellulosic Fibers (Courtesy: BMTPC, India)

Fiber	Specific Gravity	Specific Tensile Strength (GPa)	Specific Tensile Modulus (GPa)	Cost (US$/tonne)	Energy Content GJ/tonne
Plant Fiber	0.6-1.2	1.60-2.95	10-130	200-1000	4
Glass	2.6	1.35	30	1200-1800	30
Kevlar	1.4	2.71	90	7500	25
Carbon	1.8	1.71	130	12500	130

Industrial wastes

The rapid industrialization in many parts of the world has contributed to the piling up of large quantities of wastes, which pose a major problem for their disposal. These include both agro and industrial wastes such as bagasse from sugar industries, saw mill wastes cinder and fly ash from thermal power stations, coal washery wastes, mineral mining tailings red mud from aluminium plant, blast furnace slag, phosho gypsum from a by-product of the phosphoric acid based fertilizer industry [The interaction of ground phosphate rock with sulphuric acid produces 10-40% free moisture along with phosphogypsum] (BMTPC 1999), tannery sludge from leather industry, water glass from glass plant, TiO_2 [~30% and one of the solid wastes of paper industry and their use in varnishes could protect the wood from UV light degradation], etc. (Nicholas 1999). It is also pointed out that fly ash from thermal power plants constitute some of the important industrial wastes and are used in many of the countries for the manufacture of building materials such as bricks, tiles, light weight aggregates, polymer composites, etc. (Gupta 2000, Andre de Vries 1999). Table 4(a) shows the availability of some of these. In addition, there are also wastes from chemical, paint, petrochemical and other engineering industries. Chemical assays of some of these are given in Table 4(b).

Table 4 (a) Industrial Wastes, Their Sources &Potential Uses [Courtesy: BMTPC, India]

Waste	Industry	Annual production (mt)	Potential uses
Blast furnace Slag	Thermal Power Plants	10.0	Cement
Byproduct Gypsum	Fertilizer, Phosphoric & Hydroflouric Acids	5.0	Cement, Plaster of Paris, Plaster Plastic Boards
Cinder	Thermal Power Stations & Railways (coal)	2.5	Brick & Aggregate
Coal Washery Rejects	Coal Washeries	3.0	Brick, Lightweight Aggregate
Fly Ash	Thermal Power Plants	100	Cement, Bricks, Aggregates Pozzolana, etc.
Lime Sludge	Paper, Sugar, Fertilizer, Acetylene and Tannery	4.5	Lime, Cement masonry Mortar
Lime-kiln Rejects	Lime-kilns	0.4	Masonry Mortar
Mining Tailings	Zinc, Copper, Gold, Iron, etc.	6.0	Calcium-Silicate, Concrete, Bricks, Masonry, Cement
Red Mud	Aluminium	2.0	Brick, Ceramics and Cement

It has been reported (Suresh 1999, Gupta 2000, Mymrine 1999, 2001, 2003) that many of the above industrial solid wastes could be used as secondary raw materials for the development of new materials particularly for construction industry. It is found that the strength of such aggregates has increased when 20-30% of glass could be replaced with fly ash, while lightweight aggregate could be produced using waste glass having more compressive strength. Also, lightweight and fire resistant concrete could be produced from wood chip wastes with cement (http://www.techno-preneur.net/timeis/technology/sctech17.html].

Table 4 (b) Chemical Assay of Some Industrial Wastes [Courtesy: BMTPC, India]

Constituents (wt.%)	Blast Furnace Slag	Zn Tailings	Copper Tailings	Fly ash	Coal Washery Tailings	China Clay Waste Tailings
CaO	40-48	12.68	2.11	1.3-10.80	0.07	0.36-0.66
SiO_2	31-35	60.63	59.42	37.15-66.75	57.60-60.40	50.30
Al_2O_3	10-18	0.54	12.06	18.3-28.9	31.3	32.7
MgO	3-8	7.27	1.86	0.8-5.25	0.44-0.92	0.35
MnO or FeO/Fe_2O_3	0.5	3.02	19.82	3.2-21.90	3.86-7.03	2.37
Na_2O+K_2O	-	-	-	0.04-1.30	2.35-2.73	0.59
Loss on Ignition	-	14.75	2.24	0.05-16.6	1.8-16.9	11.20

Further, it is also reported (Gupta 2000) that these methods could help not only in achieving energy saving in construction sector but also improves some of the properties of the resulting materials (Table 5). This has been observed, for example in the case of concrete mixes with fly ash, resulting in discontinuous pore structure, which clogs pore channels with the addition of calcium silicate formed by the chemical reaction with fly ash. Use of such concretes resulted in high durability with impermeability whereby chemical attacks and water penetration could be prevented.

Table 5 Energy Savings in the manufacture of Building Materials through the Use of Industrial Wastes

Building Material	Composition (%)	Material Compared (%)	Energy Savings (%)
Portland Pozzalana Cement	75 Ordinary Portland Cement (OPC)	100 Ordinary Portland Cement	20
Portland Blast Furnace Slag (BFS) Cement	60 OPC	100 OPC	30
Masonry Cement	50 OPC; 40BFS	100 Masonry Cement; 60 OPC+50 Limestone	20
Lime-Pozzalana Mixture	25 Acetylene Gás lime; 25 Fly Ash	Ash 25 Lime; 75 Calcined Brick	75
Calcium Silicate Brick 10% Lime(Waste Source)	90 Fly Ash Tailings	Burnt Clay Bricks	40
Burnt Bricks	75 Clay; 25 Fly Ash	Burnt Clay Bricks	15

[Source: Building materials in India: 50 years-A commemorative Volume, Ed: T.N.Gupta, BMTPC, New Delhi, India, 1999]

Similarly, the slags from both ferrous and nonferrous metal industries have been used as sources for minerals in the manufacture of cement, which can replace substantial amount of expensive cement in concrete while giving better properties to the Portland cement mixes. Use of such cement has been reported (Gupta 2000) not only in reducing the water requirement by varying its content but also resulting in good workability.

Also, the mine tailings, which are available in large quantities while mining and also from mineral dressing industries of copper, zinc, etc. [for example India alone produces about 100 million tons of this] can be used in the Portland cement mixes resulting in better properties of masonry cement, concrete and as admixture to soil burnt clay and tiles.

Methodologies for the possible utilization of Agro and Industrial wastes

Many attempts have been made for utilization of the agro byproducts and industrial wastes in addition to their existing conventional uses. Several methodologies have been attempted for utilizing these wastes in combination with other materials either in the natural form or converting them into other intermediary or to develop the final product itself. The examples for the first category are lignocellulosic fibers, bagasse, fly ash, tannery waste, etc. to produce composites with polymer or ceramic matrices, as building materials, etc. The second one is extraction of cellulose or lignin from the agro wastes by some techniques or fusion of by-products/wastes and use them to produce the final useful products, such as rice husk ash for producing different products, binders from different mineral wastes and industry slags and sludges, etc. The third category is use of fly ash for land-fills, in agriculture, ceramics bricks, laterite wastes [iron oxide bearing soil] as bricks and tiles, aluminium or glass waste to produce foams, rice husk ash and coir pith for waste water treatment, etc. Specific examples of some of these are discussed in detail in the following pages. However, it may be noted that many of these have been attempted at laboratory scale only while the others have also been taken to pilot plant scale and further commercialized. Thus there is need for scaling up of the successful laboratory scale processes for the better utilization of the wastes in general while taking care of ecological and energy concerns.

Development of composites

A "composite" material is one that consists of two or more physically and/or chemically distinct phases suitably arranged or distributed. The constituents retain their separate identity after processing but exhibit superior properties than either of the constituents. Thus it is different from "alloys"- where although the latter also consists of two or more materials, the individual materials do not exist as before. Hence, a composite material is an homogenous mixture of two or more materials prepared in an engineered way.

The bulk/continuous phase is called matrix, while the other/distributed phase is called reinforcement or filler. The property of composite is sum of properties of both matrix and reinforcement and in most cases having synergistic effect. Also, three factors determine the characteristics of a composite: the reinforcement, the matrix and the interface between them. They have unique mechanical properties that are dependent on the amount and the manner in which their constituents are arranged.

Natural fiber-polymer composites : The synthetic fiber-polymer composites are well established all over the world for a wide variety of applications in view of their high specific strength and modulus compared to conventional materials such as metals and alloys. These have long durability. Despite their high cost, there has been continuous increase in their utilization with high volume uses. This is posing disposal problems and hence there is serious lookout for materials based on natural resources in view of their eco-friendly attributes and cost. Of course, use of natural fibers/plant fibers with clay/polymer is an old concept from the early days of the civilization [eg: clay bricks reinforced with rice straw]. However, there has been intense research on the natural fiber incorporated polymer composites (NFPC) over the last two decades or so for the use of natural materials to substitute synthetic fibers. This is because of their (1) abundant availability, which are under utilized and or amenable for value addition and thus it is mostly driven by need-cum-resource base; (2) there has been persistent quest for new materials, which support global sustainability and, finally, underlining the diversification of uses to new areas, addressing to global environmental and stable economy through agro-industrial activities. Also, being hollow and lignocellulosic in nature, natural fibers are better thermal and acoustic insulators. Further, the need for their use includes exploitation of potential lightweight of fibers, biocompatibility, reduction in energy consumption, etc. Also, these fibers possess low density (See Table 3b) and are biodegradable, not posing health or environmental hazards unlike man made fibers. These ligno-cellulosic composites exhibit brittle fracture, which is an advantage in automotive interiors. They are less abrasive for tooling and also do not cause any airborne particles as in glass fiber composites to cause respiratory problems for workers. Besides, they can be processed by general polymer/composite processing techniques. For instance, new techniques such as thermo forming have been developed (Bhattacharya *et al.*, 1999) for short wood fiber composites, which depend on the type, dimensions and quality requirement of the end product. Even cost considerations suggest they are commercially viable. For example, 5 Gallon buckets made of high density polyethylene (HDPE)-40% sisal composite even with existing mold design produced by injection moulding process costs US $ 1.50 compared to US $ 1.65 for the same bucket made of HDPE (Satyanarayana 2002).

Many composite systems studied to-date are produced by the conventional polymer or polymer-synthetic fiber processing techniques. The systems include jute-polyester, jute-epoxy, flax-polypropylene (PP), sisal-polyester, sisal-epoxy, coir-polyester/phenol formaldehyde, banana-polyester, banana-cotton-polyester, etc. It may be noted that these fibers have been upgraded both technologically (improvement of properties) and biologically (resistance to fungi, insects, etc) by chemical modification using different organic and inorganic solvents (Beldzki & Gassen 1999). Such modifications have resulted in improved properties (Table 6) in the fiber composites that allow them to be used in new applications, where traditional wood fiber based products could not be used due to their undesirable technical properties.

Table 6 : Properties of Lignocellulosic fiber-Polymer Composites with and without Surface Modifications to the Fibers. (Source: Satyanaryana 2002)

Fiber	Treatment	FPS (MPa)	TS (MPa)	Toughnes* (MPa)
Polyster-0.3VfBanana (Unidirectional)	Untreated	840	121	816
	Furfural	1175	108	--
	Polyester	--	121	--
Polyster-0.3Vf Sisal (Unidirectional)	Untreated	580	129	1250
	ZrMA	640	139	--
LDPE-0.25Vf	Polyester	-	135	--
Sisal (Oriented-10mm)	Untreated		21.2	--
Polypropylene-0.5Vf Sisal	Untreated			3200
Polyester-Coir	Untreated		60	
Polyester-Coir	Alkali	--	16	--
LDPE -Coir	Untreated		20.4	
Polyester-Pineapple		--	23.3	970
Polyester-Banana-cotton Woven fabric	Untreated		--	
	Untreated		28 –36	--
	--			
Polypropylene-0.5Vf-Jute	Untreated	--	73	--

FPS: Fiber Pull-out Stress; TS: Tensile Strength

Applications of the natural fiber-polymer composites include starting from the first cotton-polymer composite used by the military for radar aircraft during the World War II, to the latest ones such as semi-structural applications and several automotive components (Beldzki & Gassen 1999, Kandachar 2002, Netravali & Chabba 2003, Satyanarayana 2002). In Germany in particular, flax- PP composites have been used for the interior parts of the automobiles. Even wood flour-polymers have been studied and are being used in various applications. One of the drawbacks of using these natural fibers is their inconsistent properties.

Hence, in recent years, attempts have also been made to overcome this by producing genetically engineered plant fibers. In view of these, greater interest for their various societal applications in general and automotive applications in particular has been projected. In UK and USA, attempts have also been made to use NFPC in aerospace applications. This has been possible despite non- availability of exhaustive enough data on their properties for design of any product for any of the applications.

Being inexpensive and with load bearing potential, other extensive uses of the NFPC have been in building and construction industry as panels, false ceiling, roofing, etc., in transport sector such as automotive and railway coaches as body panels, seat and backrest, interior linings, as devises for grain storage, crates and boxes for fruits and as foot wear components such as puffs, insoles as cheaper replacements of leather and lining materials (Netravali & Chabba 2003, Kandachar 2002, Satyanarayana 2002). Some of these products are shown in Fig. 2 (a).

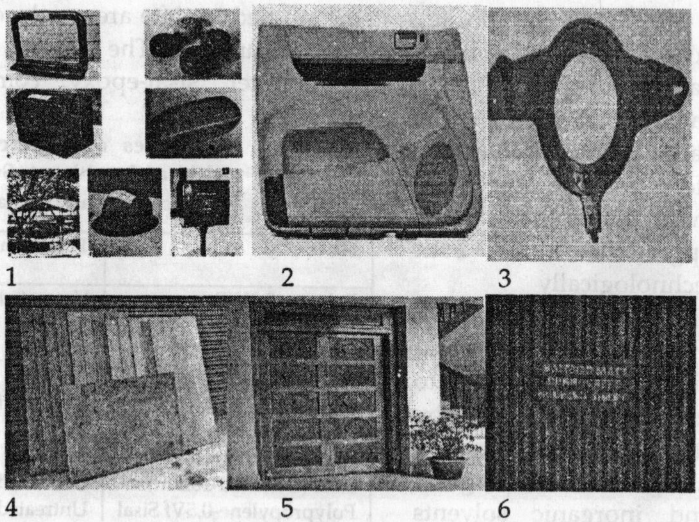

Fig.2 (a) Composite Products based on agro byproducts including their wastes. 1-covers for consumer products, roofing & post box;2-Automotive part; 3-Fan housing; 4-Banana Leaf sheath-polymer panels; 5-Coir-PF door;6-Bamboo mat Roofing (Sources: Pavithran et al 2001, Courtesy: Meshap Science Publishers, India and BMTPC India; Netravali & Chabba 2003 and Courtesy Elsevier Inc.Publishers)

Thus, NFPC with their increased applications as engineered products can contribute towards the sustainable product development through optimal combination of technical, economical and ecological feasibilities. However, even with the renewability and hence abundant availability of these fibers, limited data on their properties, fabrication processes and properties of their composites are limited to non-degradable polymeric matrices, such as polyester, PP, epoxy and PF. Further, in view of ever growing environmental consciousness all over the world, some countries such as those in Europe have already established the use of NFPC in automotive applications and presently a lot of NFPC products are being exported to USA and other countries with expected lead of 8 years over others (Kandachar 2002, Netravali & Chabba 2003, Marsh 2003, Peijis 2003).

While natural fibers are fully biodegradable, their polymeric matrices are not and hence this comes as a limitation for their extensive use (Kandachar 2002). Similarly, to overcome the problem of good bonding between these and the fibers, attempts are being made to develop natural resins and their composites based on renewable resources with a stress on ecological aspects as well as replacement of wood/GRP in various applications for wider acceptance (Wool & Shrikant 2002). This is particularly so in the western countries who are striving for the development of natural "green products" - completely bio-degradable composites based on natural sources for both matrices (biodegradable polymers such as starches) and reinforcements (natural fibers such as flax), which should be commercially viable for their industrial applications. While these fully green composites may find applications in mass-produced consumer products for short time uses, they may be useful in long term indoor applications. The

European Union has a directive for the "end of life" of vehicles requiring by 2015, with all new vehicles using 95% recyclable materials, which points to the promise "green" composites will hold for the future. It is believed (Netravali & Chabba 2003, Marsh 2003) that time is not far off for the use of NFPC as structural components of automotives, which are presently used only for the interior parts of cars. This does not seem to be exaggerated since unidirectional hemp fiber reinforced and yarns of ramie fiber starch based polymer composites have exhibited tensile strength in the range of 200-250MPa, comparable to that of steel, while attempts are being made to develop high strength "Bio Steel" fibers based on proteins such as spider silk (Netravali & Chabba 2003). It is interesting to note that many leading plastic industries in USA and other western countries are in the forefront in the development of biodegradable materials from natural materials including wastes to alleviate the environmental concerns. Figure 2(b) shows some of the biodegradable products made by various researchers.

Fig. 2 (b) *Some biodegradable products from natural materials including wastes. 1-NF-Polymer Trays; 2-Furniture; 3-various components; 4- Door Inner Trim Panel; 5-Helmet; 6-Plant pots. [Source: Netravali & Chabba 2003, Marsh 2003, Nickle & Riedel 2003 and Courtesy Elsevier Inc. Publishers; Kandachar 2002 and Courtesy NATPAC Biotech, UK;].*

Ricehusk/coconut wastes filled composites : Use of rice husk or its ash as filler in polymers to produce composites has been reported in a recently published review (Sathy *et al.*, 2003). While flexural modulus and thermal properties showed improvement, tensile strength and % elongation and oxidative stability of the composites were comparable with those filled with commercial silica, but they decreased with increasing filler content.

Coconut shell char (up to 30 vol.%), waste glass powder (up to 8wt%) and fly ash (12wt%) have been successfully used as dispersoids in the preparation of aluminium matrix composites by foundry technique, while coconut shell char-copper composites have been prepared by powder metallurgy technique (Gopakumar *et al.*, 1982, Satyanarayana *et al.*, 1984, Pai *et al.*, 2001). While both Al and Cu base shell char composites showed comparable adhesive wear properties, copper shell car composites showed good electrical properties and extruded Al-glass and Al-fly ash composites showed good mechanical and tribological properties for their use in structural applications (Pillai 2003).

The latter composite called "ash alloy", developed earlier in India but followed up more rigorously by the University of Wisconsin (USA) with the support of an industry is reported to have been taken up for implementation to manufacture products for automotive industry (Kumar & Jha 1999, Rohatgi *et al.*, 1995). This composite has shown very interesting mechanical properties, which suggests them to be one of suitable candidate for substituting expensive Al-SiCp composites in many applications. Interesting point in all these cases to be noted is that these wastes and the products made based on them possess appropriate physical properties (Table 7) making them viable alternatives for the expensive conventional metals and alloys [Al, Cu and their alloys] for such applications.

Table 7 : Properties of Aluminium Alloy and Copper–industrial waste incorporated Composites and their possible applications

Properties	Coconut Shell char	Mica	Fly Ash	Waste Glass	*Cu-Shell Char (PMMethod)
Dispersoid Wt.%	1-8	0-2.2	2-10	2.5-10	2-10 (vol%)
Density (Kgm^{-3})	2600-2400	2700-2600	--	--	7480-5800
UTS (MPa)	130-90	140	83.6-69.3	79.6-72	-
Elongation(%)	3-2	1.1	12.3-4.1	14.6-5.1	-
CS(MPa)	530-200	430	119-167	124-143	-
Hardness(BHN)	70-72	60	33-71	31-65	40-22.3
Fracture Toughness (KIc-MPa m$^{-1/2}$)	--	--	10.37-15.99	11.7-15.83	-
Possible Applications	Bearings, Seals and Electrical contacts	Bearings and Sliding Parts	Extruded Composites for Structural Applications	Extruded Composites for Structural Applications	Brushes and other Electrical Applications

UTS: Ultimate Tensile Strength; CS: Compressive Strength; * Annual Reports 1980-82, Regional Research Laboratory, Trivandrum (Kerala-India).

Development of Building Materials

Amongst various agro and industrial wastes used for the development of building materials, lignocellulosic fibers, fly ash, metal slags and rice husk ash are the prominent ones.

Agro Waste Based Building Materials : Composites made with lignocellulosic fibers with polymers or ceramic as matrices have been mostly used for roofing, partition walls and even as flooring tiles. In Philippines agro and forestry wastes such as wood chips, husks from rice, coffee and coconut have been used to develop several products starting from improved biomass cooking stove, a simple house hold gadget to more sophisticated agro industrial machines such as biomass pyrolyzer and using briquetting plants (Nicholas 1999). Basically all these use the thermo chemical conversion process (pyrolysis). They have also found a fungus activator [*Trichoderma harzianum*], which is known to hasten the decomposition of biomass. This along with briquetting using suitable binders has led to the better use of many of the above agro wastes rather than simple burning or disposing them as wastes.

Technologies for (i) cement based on rice husk ash, (ii) cement bonded boards (CBB) based on bagasse, bamboo shavings and other wastes and (iii) wood wool cement (WWCB) boards based on shredded wood, Portland cement, water and an accelerator have been commercialized by the Forest Product and Development Institute of Philippines. While, reduction of cost with disposal of the waste has been observed by the use of the rice husk ash cement in hollow blocks, CBB could be used as substitutes as roof tiles and WWCB could be raw material to produce modular and prefabricated structures. The latter exhibited excellent thermal insulating and acoustic properties, and also resistance to weather, fungus and termite (Nicholas 1999).

In the case of rice husk ash, knowing its chemical reactivity with lime imparts pozzolanic properties, a number of patents and processes have been reported (See Ref: Sathy *et al.*, 2003) for its use as a source for silica in cement manufacture and as possible substitute for expensive condensed silica fume in concrete to obtain improved strength properties. On the other hand, enhancement of strength has been reported (Cisse *et al.*, 2000) when the ash was used only as filler in concrete blocks. Similarly, 25% of silica waste from a geothermal plant used in Portland cement enhanced the cement hydration (Escalante *et al.*, 1999). It may be noted that condensed silica fume, which is produced by the hydrolysis of silicon tetrachloride vapors in a flame of oxygen, is used in concrete as a replacement for several admixtures in various countries (about 1 million ton per annum). Though expensive (2-3 times the cost of cement), this offers many advantages such as ultra strength properties, saving cement but increasing the age strength of concrete, controlled alkali silica reaction, reduced chloride associated corrosion and sulphate attack as well as better linear expansion (30-80%) in mortars.

Industrial Waste Based Building Materials : Fly ash, another the largest industrial byproducts available in many countries (China and India produce about 100 million tons per annum) has attracted the attention for the development of value added products – variety of construction materials such as bricks, activated slag cement, anhydrous self leveling floors and different kinds of concrete (Suresh 1999, Gupta 2000). Many of the developing countries in Asia, Africa and some of the Western countries have developed processes for the utilization of fly ash irrespective of local availability of this industrial waste. For example, Malaysia, Philippines, Indonesia and Singapore and the Netherlands (production - a few thousand tons per annum only) have developed appropriate technologies for its use in cement, concrete, advanced SIALON fibers, foam insulation products, etc. China and India have not only developed but also used the products based on fly ash in field (International Workshop 1999). Extraction and beneficiation of several high value minerals such as alumina, magnetite, gallium, germanium, cenospheres, and carbon have also been made by USA, UK, Hungary and Poland. Even ferrosilicon has been extracted in Canada by plasma furnace technology. Other high value products developed by many of these countries with fly ash as a raw material include high wear resistance ceramic tiles, insulation products, lightweight refractory and other products, heat insulation fire bricks, glazed floor and wall tiles, marble and mozzaic tiles, decorative glasses, mineral wool, adsorbent for toxic organics in waste waters and distempers for buildings. Some of these technologies are available for transfer from various countries such as India, China, Singapore, etc. Details of some of these developments are given below.

China seems to be the first country to find better utilization for "waste" fly ash as early as 1950 by using it in construction industry for dry concrete and as partial cement substitute. However, the resurgence in the research for its better utilization started again in late 70's following higher production of fly ash with its lower utilization. This resulted in requirement of large area for (about 40,000 hectares) of land for dumping. Hence regulations both at the regional and national levels were passed for its extended use (Wang 1999). It is reported that the increase of fly ash utilization has been on the raise with about 70% in 1970 to about 100 % in 1997.

Accordingly many uses have been developed and implemented which include more than 70% as building materials and the rest in agriculture, land filling, etc. Use in building sector include fly ash concrete in developing pumping/high strength concrete, which has been successfully used in large constructions such as big bridges and high rise buildings (up to 450 m altitude) with as many as 63 storeys. About 15 million m^2 area of residential buildings have been built with fly ash-lime-gypsum blocks (LFG), which showed compressive strength of 10-20MPa. Similarly, taking advantage of reduction of hydration heat in concrete by fly ash, a number of tall buildings have been constructed since 1980 using fly ash concrete mat (size: up to 4m thick) foundation particularly in Shanghai as well as nuclear shielding (1m thick mat) for nuclear power plants. Also, considering the improved impermeability of concrete by the addition of fly ash, big dams have been constructed with this in East China, which provided better resistance to chloride attack than the normal concrete. Also, with its pozzolanic property to provide hydraulic and retarding property, sub base of fly ash-lime and crushing stones [33%fly ash- RCC admixture] have been used in constructing all the highway pavements / embankments in Shanghai to take care of detrimental effects due to weathering, water infiltration and traffic. About 17.m tons of fly ash were expected to be utilized in a couple of years, if this were to be extended to other National Expressways in China. Another use as admixture with cement and/or clay has also been used for grouting the mine walls due to its high retainable strength. Chimney bricks developed (Xu Quing 1999) using about 55% fly ash with high plastic soil (30%), quartz (12-15%) and other admixtures exhibited density of 1500-1600Kg/cm^3, compressive strength of 20-28MPa, good acid (98%) and frost resistance, thermal conductivity of about 41% less than the common soil bricks and without any deformation. These bricks have been successfully used for the construction of power plants in several cities in China. Based on the high acidity (1.8Mk) of fly ash, and making use of its high melting point and viscosity at high temperature, its other interesting application is in the development of insulating fibers/tapes. These fibers (3-7μm in dia.) have 50-70% fly ash, apparent density of 30-100kgm^{-3}, more than 98% water repellent ratio and thermal conductivity of 0.038-0.04W/m.k with high fire retardancy (Class A1 - Chinese Standard). Hence, these fibers converted into tapes can be used for heat insulation for equipments, as composite plates for insulating building walls as well as fireproof panels and as ceiling material for acoustic, decoration and waterproof purposes. Similarly, artificial marble has been developed using fly ash and waste glass. The process consists of pouring the high alumina cement mortar on float glass [for improving the surface finish and to maintain proper level] with a base layer of common Portland cement. Then, fly ash is added to be the active mixer containing waste glass powder [this powder produces glassy materials in the surface, improves reflection, refraction, surface smoothness and wear resistance] between the surface and the base layer. Finally, superplasticizer is added to the above. Then, the whole system is subjected to vibration to avoid water of the base layer moving to the surface layer. The product is then stripped off (optimum time: 30-40hrs to get sooth products) and naturally cured for about 20 days with about 80% humidity. This resulted in artificial marble exhibiting very attractive and superior characteristics such as flexural strengths of 4.4-4.7MPa and 5.6-6.4MPa, compressive strengths of 20.2-24.8MPa and 47.6-53.8MPa (strength of 3 days and 28 days respectively), deformation of 0.05-0.13mm, surface water resistance of 0.014-0.121 gcm^{-2}, high surface finish, high fire resistance and moisture proof over other artificial marble materials.

In India, the fly ash has been used in various ways (Gupta 2000, Suresh 1999, Kumar & Jha 1999, Rao & Garg 1999). For example, as raw material for lightweight cellular blocks and slabs, precast concrete building elements, as ceramics, as admixture and filler such as cement concrete using ordinary Portland cement, fly ash-soil mixes for soil stabilization in road construction, bituminous concrete in surfacing of roads, as aggregate in lightweight aggregate concrete for blocks, walls, partitions and also as sintered light weight aggregate and, finally, as pozzolana in Portland cement and in lime-pozzolana binders as masonry cement. Some of these are shown in Fig. 3. These depend on the chemical constituents of the fly ash and their size. Some of the advantages of these fly ash bricks include savings in mortar, plastering, possibility of obtaining several load bearing grades and smart looking brickwork. Table 8 lists some of the properties of these products, which indicates comparable or even superior properties to the conventional products underlying the significance to the value addition for this industrial waste.

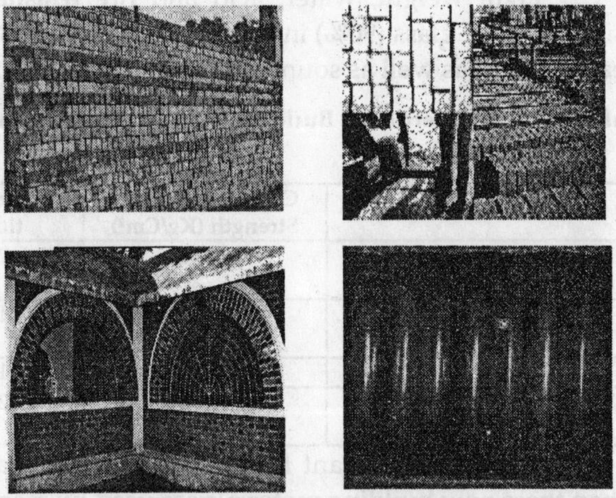

Fig. 3 Building Materials based on industrial wastes. 1- coir pith-clay light weight bricks; 2- Fly ash bricks; 3- Tannery sludge-clay bricks; 4- Redmud-Polymer Roofing [Source: Satyanarayana et al 1999; Courtesy: BMTPC, India].

Another interesting application of fly ash is the development of FaL-G (fly ash, lime and phosphogypsum, another industrial waste) similar to China's LFG (Suresh 1999), which can be used as cement, blocks, mortar, for roads as concrete, etc. This is a successful example of industrial entrepreneurial initiative supported by a federal government agency.

Incinerated fly ash has also been evaluated as blended cement and sintered sludge, as construction material and in recycled marine clay and industrial solid wastes as high value added engineering materials **http://www.ntu.edu.sg/cee/research/Research_groups/Env/env_group.htm#Waste**. Industrial wastes such as red mud (by product of aluminium extraction) with low curing agents like water glass and tannery sludge (waste from leather industry) have been successfully used in the development of ceramic building materials (International Workshop 1999). Tannery sludge contains organic matter between 50-60%, some amounts of sodium and calcium and Cr_2O_3 (VI) in the range of 1-2% as flocculated and 10-12% in filtered form, which is leachable. This poses both land and water pollution because it is usually dumped near the industrial sites. Hence, its utilization has to take care of converting of this leachable chromium (VI) to non-leachable chromium (III). Such processes have been developed in India, the Netherlands and Italy and used successfully for the development of bricks. In general, the patented process (International Workshop 1999, Gopakumar *et al.*, 2001) consists of subjecting the tannery sludge along with clay to a series of fabrication processes [mixing, extrusion and forming into shapes], followed by a heat treatment [oxidation (firing) and reduction (cooling)].

The organic matter present in the sludge helps in the fuel requirement for firing, while iron oxide present also gets reduced during the conversion of Cr (VI) to Cr (III). In view of this conversion of iron oxide, the resulting products will be black in color. Mechanical properties of these bricks are shown in Table 8.

Light weight, water, acid and fire resistant building materials have been developed using recycled glass (90%) in Germany. This is reported to be extremely stable, insulating to both heat and cold as well as sound absorbing (Science Express -India 2000).

Table 8 : Properties of Building Materials made using Industrial Wastes (Source: Satyanarayana et al 1999)

Product	Crushing/ flexural Strength (Kg/Cm²)	Water Absorption (%)	Density (Kg/m⁻³)	Firing Temp. (°C)	Fired Color
Light weight Bricks with coir pith	65-85	20-25	1200-1400	800-950	Brick red
Laterite tiles	60-90	15-20	1500	400	Dark Brick red
Flux Bonded Flyash Bricks	80-110	18-20	1400	900-950	Brick red
Tannery Sludge-Clay Blocks	70-90	20-25	1300-1500	850-900	Dark grey Black

Another important factor the energy management in buildings, can become operative, when these new building materials are used by resorting to appropriate building designs.

Development of Other Industrial Products or uses

Packaging Materials : Packaging of food has several important functions such as containment and protection of food, maintenance of its quality and safety, conferring convenience to food along with communicating information about food to consumers (Robertson 1993). Accordingly, the packaging materials should have some important characteristics, which include appropriate physical and other properties such as strength, gas and water permeability, sealing capability, thermoforming, machinability, resistance to water, temperature and light. Hence their design and manufacturing involve several steps. Presently, the packing materials are being largely made using non-renewable materials, such as aluminium, plastics (based on petroleum feed stock), glass and paper. It is reported that about 25% of the total 125 million tons of plastics are used annually for packaging purposes alone (Fang and Fowler 2003). Hence, packaging becomes one of the potential but competitive areas for the use of renewable bio resources in many attempts being made to find non-food uses for agro resources.

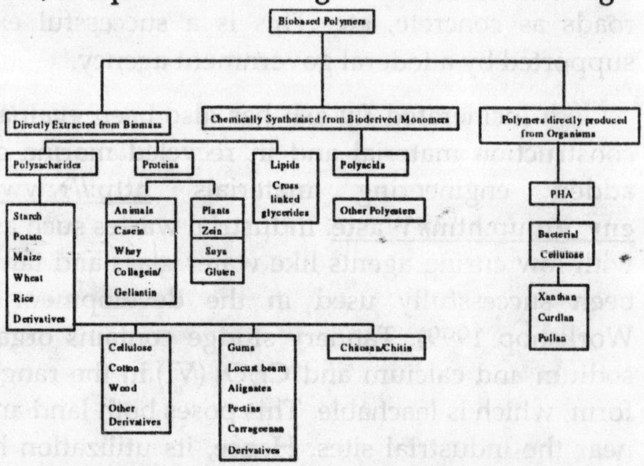

Fig. 4 Schematic presentation of three categories of Biobased polymers for packaging materials [Source: Robertson 1993; Courtesy: Marcel Dekker Inc.].

Accordingly, several countries have made serious efforts to evaluate their potential particularly for packaging (Coombs and Hall 2000, Mangan 1998). European countries have brought out recently a status report on bio based food packaging materials along with an outline of the future scenarios and developments, involving the producer and end user (industries of packaging materials) and concerned experts (Weber 2000). This report also underlines the resulting environmental effects along with the market for these materials.

In view of the important characteristics of the packaging materials mentioned above, the use of biomaterials for packaging gives very good opportunity for Materials' Scientists to incorporate these characteristics in to the bio-resources along with appropriate functionality such as durability, long product shelf life, compostability in view of the high degree of contamination by the accompanying food materials, leading to intensive recycling of bio-wastes, and their cost. Also, one has to take into account the entire dynamic interaction between materials of packaging, food and environment making the task of producing biomaterial based packaging material a challenging one for all concerned. Despite these, some of the bio-resource based packaging materials have already been commercialized.

Foam Materials Based on Industrial Waste Materials : It is well known that foam materials exhibit some special properties such as light weight (density of aluminium foams is 300-1000kg/m3, about 30-100% less than that of water) and superior compressive strength. Foam glass in particular is non- corrosive, noncombustible/nonabsorbent of combustible liquids and resistant to most of the industrial reagents. All the three types of foam materials, namely, metal, polymer and ceramics have been produced commercially and widely used (Gibson & Ashby 1997, Banhart 1999, 2001, Baumsteiner *et al.*, 1997), while attempts have also been made to produce and evaluate even the biodegradable polymer foam (Colney *et al.*, 1996, Ramsteiner *et al.*, 2001, Willet & Shogsen 2002). Industrial wastes such as metal scraps, recycled polymer wastes and waste glasses have been used to produce the respective foams with the above attributes using appropriate processing methods. Another type called "syntactic foams" is an artificially structured material offering high compressive strength-to-weight ratios (Erikson 2002, Pillai *et al.*, 2003). They are also consisting of manufactured hollow micro spheres (these are often resistant to crushing) dispersed in a matrix. In this case the bubble pores in the constructed foam (hence syntactic) are formed in advance separately unlike *in situ* as in the case of metal foam/rubber foam by gas bubbles. These hollow spheres resulting in pseudo-closed cell structure make the foam considerably stiffer and stronger than common open cell structure.

Table 9 lists some properties of such foams while Fig.5 shows some of the foam products including one of the bridges, which uses the Al metal foam. The polymer foams of polyurethane, styrene and melamine exhibit density of 1050-1410 kg/m^3, while polypropylene and styrene show Young's modulus of 1140-3500, fracture toughness of 1.35-3.7Mpam$^{1/2}$, (Ramsteiner *et al.*, 2001, Willet & Shogsen 2002).

Table 9 : Properties of Foam Materials

Properties	Glass Foam	Al-Metal Foam
Cell Size (mm)	0.1-2	Closed; Open
Bulk Density (Min) (Kgm^{-3})	220-260	200-1050; 0.16-0.25
Max Apparent Porosity (%)	85-90	--
Open Pores (%)	85-92	--
Compressive Strength (MPa)	1.69-2.0 (Kg.cm^{-2})	1.3-14; 0.9-3.0
Thermal Conductivity (Wm^{-1}C^{-1})	0.04-0.12	3.0-4.5; 6-11

[Source: Annual Reports 1980-82, Regional Research Laboratory, Trivandrum (Kerala-India), *Pillai et al. 2003; Courtesy: NPL, UK]*

Similarly, biodegradable poly lactic-co-glycolic acid foams of thickness 2-3 mm show porosity of 0.798-0.942, median pore dia of 71-154mm and shear modulus of 0.59-9.55 MPa (Colney *et al.*, 1996).

The application areas for different types of foams (Gibson & Ashby 1997, Banhart 1999, Sepulveda & Binner 1999, Binner *et al.*, 2004, Pillai *et al.*, 2003) include: hot gas filters and biomedical applications, thermal insulation, kiln furniture and catalyst supports; different types of pipelines including those for cryo and underground steam lines, hot oil/asphalt storage tanks, heat transfer fluid systems, hydrocarbon and sulfur processing systems, off-shore platforms, food processing/cold storage facilities such as breweries and cyclic or dual temperature applications.

In the case of metal foams, closed cell structures can be used in structural applications, while open celled structures for heat exchangers and heat sinks. Attempts are being made to use them as impact absorbers in automobiles, lifting and conveying systems and as stiff, light weight and fire resistance panels, as sound absorber in buildings, road and rail vehicles and in EMC shielding. It may be used as fillings in hollow sections in order to stiffen them against buckling (http://www.npl.co.uk /npl/cmmt/metal_foams). On the other hand, syntactic metal foams are used as energy absorbing parts for occupant safety in automobiles interiors functioning well even when the crash distance is limited. It is reported that syntactic Al foams are to be made of both solid and hollow spherical fly ash (called cenosphere) creating the voids or bubbles with significantly improved stiffness and crash resistance properties for lightweight truck frames. One such material, called ALFA foam is expected to reduce the weight of light trucks leading to fuel efficiency while providing crash worthiness (Pillai *et al.*, 2003). Other possible applications of this material include floor panel requiring sound dampening and firewalls exhibiting enhanced insulating values.

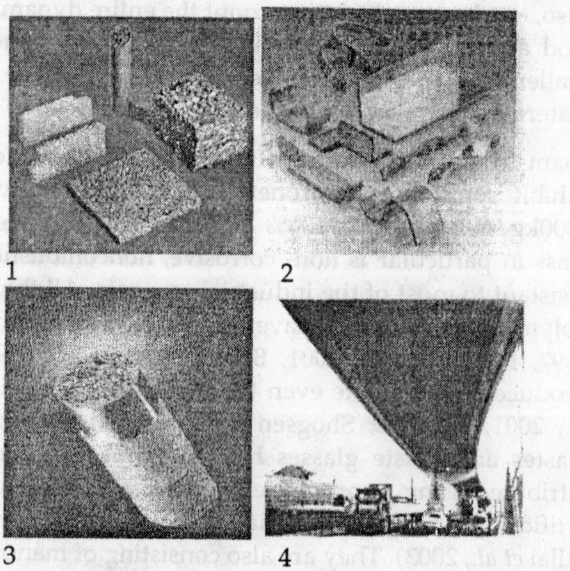

Fig. 5 Some Foam Products based on Waste Materials 1&2-Al-foams; 3-Products of foam; 4-Bridge with Al-Foam. (Source: Pillai et al 2003; Courtesy: NPL, UK).

Fuels and other applications in Metallurgy : Taking advantage of the chemical composition, physical properties and the energy content of some of the agro wastes, it is possible to find new application areas. For example, wastes of coconut industry such as coir pith and coconut shell possess low sulphur (0.2 and 0.02 wt% respectively), low phosphorous (0.11 and 0.35wt% respectively) and low ash content (7.10 and 1.05 wt%); but the calorific values of these two materials are in the range of 17-20MJ/kg (Mohandas, 2003). Based on these properties, successful attempts have been made to use them as reductants in metallurgy. Both iron ore and ilmenite [ore of titanium] have been reduced using these wastes in laboratory scale experiments. More than 85% of metallization in about 4 h at temperatures of about 1323K have been reported in the case of ilmenite and equally encouraging results have been obtained with iron ore in a rotary kiln (Annie George & Mohan Das 1985, Mohandas, 2003).

Similarly, making use of the calorific values of some of the biomass materials [16-21MJ/kg] (Grover 1989), many countries particularly the developing ones have developed briquetting technology with die and punch, screw or hydraulic presses, with or without the use of binders for the waste utilization (Grover & Mishra 1995). The agro wastes used to produce briquettes include rice husk, bagasse, peanut shell, stalks of various oil producing plants such as soya bean, oil palm, straws of various pulses and cereals and others, corn cob, cassava, residues of coconut, palm oil, rubber wood, coffee plantation, logging residue and wood processing. Various shapes of briquettes such as cylindrical, cubes and balls have been developed. Most of these briquettes possess attractive characteristics such as lower fuel cost, higher calorific value, efficiency of 60% and lower energy cost than those for other generally used fuels. Of course, the briquettes thus produced can be used in the country of origin (mostly developing and under developed countries) as well as for exporting particularly to Europe, Japan and other developed countries. However, it may be noted that many operational problems associated with this technology and the quality of raw materials are crucial in determining its success for commercialization. It has also been shown that briquettes can be gasified and this gas can be used to replace petroleum fuels like LPG, LDO and furnace oil. Further, it is reported that the cost of electricity generated from biomass gassifier-based diesel power plants is low compared to photovoltaic or wind generated electricity (Grover & Mishra 1995, http://www.rwedp.org/acrobat/rm23.pdf]. On the same lines, one successful demonstration in laboratory scale is the use of chars of coconut shell as foundry fuel. Use of both shell char and its briquettes for melting metals and alloys with considerable reduction in melting time has been reported (Pillai *et al.*, 1982). Such exercises with other agro wastes/byproducts if carried out at pilot plant level may go a long way in better utilization of wastes.

Other Uses : Attempts to develop other industrially important products include porous materials with molecule-sized pores based on transition metals in layer lattices, which can be used for environmental protection as absorbents and catalysts; extender of paints from fly ash, coatings based on cashew nut shell liquid and other similar waste materials. In the latter case, fly ash could be easily dispersed in the paints and they did not show any hard settling as is normally observed with commercial ones (http://www.ntu.edu.sg/cee/research/Research_groups/ Env/env_group.htm#Waste).

On other hand, 1nm wide and 5nm high porous pillared materials have been made using Si, Ti and other metal oxides through the development of organic/inorganic composites using layered polysilicic acid taking advantage of its ability to form an orderly configuration between the layers (www.nire.go.jp/ eco_tec_e/new-sigen_e.htm).

A number of patents and processes have been reported (See Ref in Sathy *et al.*, 2003) on the use of rice husk ash in various industrial products such as advanced materials (SiC, SiN, etc), synthesis of solar grade silicon, zeolites, Forsterite (2 MgO SiO_2)[material with low dielectric loss and used as high frequency insulator at elevated temperatures], as supporter for Ni catalyst for various organic reactions, as absorbent of metals and in ceramics for white wares substituting quartz. Use of unburnt carbon recovered from fly ash as substitute filler in tyres in place of carbon block has also been investigated in USA (Kumar & Jha 1999).

It may be noted that the above is not an exhaustive list of all the reported successful attempts for the utilization of industrial wastes.

Development of new Methodologies

With a view to find new uses for the wastes through new material development and/or other possibilities, several attempts have been made to develop new methodologies taking into account that conventional methods can not eventually yield the desired results such as higher yields and improved properties. Some examples are given below:

Chemical pathways : Two steps selective catalysis of hydrolysis followed by dehydration reactions has been used to increase the furfural production using agro wastes with minimum destruction of cellulose (Gravitis *et al.*, 1989). This new methodology uses small amounts of strong catalysts. Furfural yield is reported to have increased to 75% from 55%, while the cellulose destruction decreased by five fold. Also, preparation of valuable monomers with cellulose residues such as levoglucosan has been carried out using different biomass treatment methods including steam explosion. This approach has helped in utilizing the resultant lignin as a value added product to produce fiber boards and composites, and the carbonization of these to produce wood ceramics. Figure 6 shows the flow diagram for a possible integrated process to obtain value added products from agro wastes as explained above. Some of these have been already commercialized.

Other technique generally used in the case of agro wastes/renewable resources is the pretreatment of lignocellulosics. Saccharification of wood and agricultural residues has been always considered a promising way to produce fuels and chemicals from renewable resources (Parisi 1989, Lynd *et al.*, 1991, Ghosh & Singh 1993). In this regard, dilute acid hydrolysis has been investigated using a wide range of catalysts such as hydrogen fluoride (Hawley 1986), sulfuric acid, nitric acid, and hydrochloric acid (Parisi 1989, Wayman 1986) but several operating problems, such as acid corrosion and the need of extensive effluent treatment, had a negative impact on further developments of this technology (Wayman 1986).

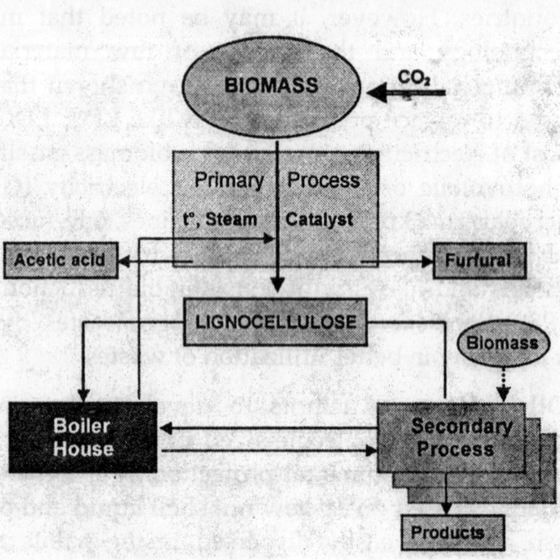

Fig 6 Schematic flow diagram for the integration of the processes to obtain value added products from agro wastes [Source &

Concentrated acid processes involving 72% (w/v) sulfuric or 41% (w/v) hydrochloric acids have also been used to solubilize plant polysaccharides at low temperatures. A second stage at 100-120°C for 30-360 min, after dilution to a 3-6% (w/v) acid concentration (Wayman 1986), provides a close to theoretical recovery yield of fermentable sugars that are considered suitable for chemical or biological conversion to fuels and chemicals.

Even though some of these processes have already been made commercially available, questions still remain about the high capital investment, acid consumption and acid recovery costs that are required to implement this technology (Parisi 1989).

Successful saccharification of cellulosic residues has also been accomplished using enzyme technologies (Fan 1987, Wood & Saddler 1988, Wood & Garcia-Campayo 1990). Lignocellulosics, however, require some form of pretreatment to increase accessibility because they have been primarily designed by nature to act as structural materials. In order to make pretreatment an economically competitive process, the method must also result in high recovery yields of hemicelluloses and lignin for further utilization as chemical feed stocks (Nguyen 1991).

A variety of biological, physical and chemical pretreatments have been assessed for their technical and economical effectiveness at pretreating lignocellulosic residues (Parisi 1989, Lynd *et al.*, 1991, Ghosh & Singh 1993, Saddler *et al.*, 1993). The relative success of each method usually depends on how much of the starting material is recovered after the pretreatment step and/or the extent to which the enzymatic hydrolysis of cellulose (and the fermentability of the sugars derived thereof) is improved.

Biological pretreatments result in partial delignification of lignocellulosics using lignin-degrading microorganisms such as fungi and bacteria (Ghosh & Singh 1993). However, lignin biodegradation is a very slow process that can be considered cost effective only if applied in conjunction to other physical and/or chemical methods such as thermo-mechanical pulping (Akhtar *et al.*, 1998) and steam explosion (Sawada1995).

Chemical pretreatments tend to solubilize and/or extract hemicellulose and lignin in order to expose the cellulose component to enzymatic hydrolysis (Fan, 1987, Wood & Saddler 1988). In this regard, a wide variety of chemicals have been suggested in the literature and these include sodium hydroxide, sulfur dioxide, aqueous ammonia, calcium hydroxide plus calcium carbonate, phosphoric acid, alkaline hydrogen peroxide, inorganic salts with acidic properties, ammonium salts, Lewis acids and organic acid anhydrides, acetic acid, formic acid, sulfuric acid, n-butylamine, n-propylamine and alcohols (methanol, ethanol or butanol) in the presence of either acid or alkaline catalysts (see review by Ramos 2003). Physical pretreatments, such as milling (Rivers & Emert 1987) and microwave irradiation (Nakayama & Okamura 1989), have also been utilised to enhance the hydrolyzability of lignocellulosic materials (Ghosh & Singh 1993). However, the major disadvantage of these methods is the requirement of high energy.

The best pretreatment options for bioconversion purposes are those, which combine elements of both physical and chemical methods (Ghosh & Singh 1993, Saddler *et al.*, 1993, Ramos & Saddler 1994a). In this regard, high-pressure steaming, with or without rapid decompression (explosion), has been claimed as one of the most successful option for fractionating wood into its three major components and enhancing the susceptibility of cellulose to enzymatic attack (Schwald *et al.*, 1989a 1989b, Ramos *et al.*, 1992a, 1992b, Saddler 1982, Brownell & Saddler 1984, Brownell 1989, Excoffier *et al.*, 1991, Ramos, 2003). Several patents (DeLong 1983, Foody 1984, DeLong *et al.*, 1990) have been granted to this process and many pilot plants of different capacities have been developed in Canada, USA, Spain, Sweden, France, Italy, Japan and Brazil for either commercial or research purposes.

Steam explosion (or autohydrolysis, if no exogenous acid catalyst is used in the process) has also been used to prepare value added products from lignocellulosic materials such as microcrystalline cellulose preparations that can be incorporated into composite materials. Table 10 shows different agro materials treated by this process and some of the process parameters. Steam pretreatment in a batch reactor involves heating agro materials at high temperatures and pressures, followed by mechanical disruption through a violent discharge into a collecting tank (explosion) (Schwald *et al.*, 1989a, 1989b, Ramos *et al.*, 1992a, 1992b, Saddler 1982, Brownell & Saddler 1984, Brownell, 1989, Excoffier *et al.*, 1991, DeLong 1983, Foody 1984, DeLong *et al.*, 1990, Ramos & Saddler 1994, Ramos 1992).

Table 10 : Different agro materials treated by steam explosion process with process parameters (Source: Saddler et al 1993; Ramos 1992).

Material	Pretreatment Conditions: Catalyst	Temperature(°C) (Pressure), time(min)	
Wheat straw	-	247	0.75
	-	200-230	0.5-10
	35mM H_2SO_4	200	2
Rice straw	-	225-251	0.5-10
Corn stover	2-3% SO_2	150-208	0.75-20
	-	215	2
Corncobs	-	0.7-2.5 MPa	3–5
	1% H_2SO_4; 1% H_3PO_4	0.7-2.5 MPa	3-5
Sugarcane bagasse	-	188-250	0.5-44
	1-2% H_2SO_4	170-250	2s-1.5
Onopordum nervosum, *Cynara cardunculus*	0.4-1% H_2SO_4	170-230	1-8
Bamboo grass culms	5-30 mmol.kg^{-1} of various catalysts [(1)]	197	10
Peanut hull	special catalysts [(2)]	137	1-2
Wheat straw, wheat chaff, alfafa steam	-	240	1.3
Woody hemp chènevotte	0.1% H_2SO_4	200-240	3
Potato	-	151-233	10

** For the references, please refer to the literature cited at the end of this chapter; (1) Lewis acids, NH_4Cl, $(NH_4)_2SO_4$, acetic acid and organic acid anhydrides; (2) Acetic acid, sulfuric acid, aluminium sulfate, water and sodium hydroxide at various loadings.*

The high-pressure steam radically modifies the plant cell wall structure (Ramos 1992). Partially hydrolysed hemicelluloses are easily recovered by water-washing, leaving a water-insoluble fraction composed of cellulose, residual hemicelluloses and a chemically modified lignin that can be further extracted by mild alkali (Schwald *et al.*, 1989a, 1989b, Ramos *et al.*, 1992a, 1992b, Brownell, 1989, Ramos 1992), dioxane, ethanol (Brownell, 1989), or oxidative agents such as alkaline hydrogen peroxide (Schwald *et al.*, 1989a 1989b, Ramos *et al.*, 1992a, 1992b) and sodium chlorite (Saddler *et al.*, 1982, Brownell & Saddler, 1984). However, it is important to realize that the recovery yield of alkali-soluble lignin strongly depends on the type of raw material used for pretreatment.

Compared to the batch process, the development of continuous reactors for steam explosion allowed for a better control of the pretreatment variables, critical to achieving optimal processing conditions at high temperatures, as well as higher purity of extracted components because heat transfer limitations are partially overcome, leading to lower accumulation of undesired degradation by-products (Ropars *et al.*, 1992, Zimbardi *et al.*, 2002, Heitz *et al.*, 1991). In general, continuous steam reactors are fed by an upstream screw conveyer, which moves the biomass through a compression tube that helps pressurizing the vessel. The densified biomass plug, upon entering the digester from the feeder compression tube, is forced onto a conical choke that breaks up the plug, forcing the undensified material to fall into the retention screw. Once fed, the digester retention screw conveys the material towards the discharge end in such a way that a precise retention time is achieved for a desired processing condition.

The discharge screw at the end of the digester conveys the processed material into the discharge valve, a rotary ball valve with a variable opening dwell time that is controlled by a timer according to the production rate. Depending on production conditions, this valve will typically open every 2 to 8 seconds, thus providing the desired explosion effect which empties the discharge tube and moves the pretreated biomass to the next processing step. Several demonstration plants based on this technology are already available in research institutes around the globe and the technology seems to have reached the status of full commercialization.

Steam explosion can be carried out with a great variety of plant biomass including forest (Saddler *et al.*, 1993, Ramos *et al.*, 1992b, Dekker 1987) and agricultural residues such as sugarcane (Dekker & Wallis 1983, Kling *et al.*, 1987, Jollez *et al.*, 1994, Kaar *et al.*, 1998) and cassava (Carta *et al.*, 1999) bagasses, wheat straw (Schell *et al.*, 1991, Chang 1998, Ternrud *et al.*, 1989, Beltrame 1992), potatoes (Kobayash *et al.*, 1998), corn residues (Ropars 1992, Torget, 1991, Weil 1997), hemp fibers (Vignon *et al.*, 1995), peanut hulls (Sewalt *et al.*, 1992), Onopordum nervosum and Cynara cardunculus (Martinez *et al.*, 1990), bamboo grass culms (Tsuda *et al.*, 1998), rice straw (Tanaka & Matsuno 1990, Moniruzzaman 1996), Brassica carinata (Ballesteros *et al.*, 2002), sunflower stalks (Sharma *et al.*, 2002), olive stones (Fernández-Bolaños *et al.*, 1999), and cotton gin waste (Jeoh & Agblevor 2001). In this regard, it is generally accepted that the wood residues derived from young trees are more easily and readily fractionated during pretreatment, yielding better substrates for bioconversion through enzymatic hydrolysis (Saddler *et al.*, 1993, Ramos *et al.*, 1992b). Therefore, the fast growing rates and short rotation times that can be obtained with several energy crops suggest a potential application of these species for bioconversion, particularly in the Southern Hemisphere.

Lignin is one of the most valuable and abundant renewable resource found on Earth and steam explosion followed by mild alkali extraction is one of the methods most commonly used to recover bulk amounts of lignin from wood and other lignocellulosic materials. It has been claimed that the amount of lignin extracted by alkali increases with pretreatment severity (Ramos *et al.*, 1992b, Brownell & Saddler 1984, Ramos 1992) and that the resulting alkali-soluble lignin still retains most of its reactivity and may be used as a raw material for many industrial applications (Tanahashi 1990) such as binding materials in fiber boards or as a source of phenolic compounds for the chemical industry. Alkali washing is also considered a valuable strategy to reduce the bulk volume of steam-treated materials, thus providing substrates with higher cellulose contents that are more amenable to hydrolysis and fermentation. Therefore, methods of enzyme recycling are facilitated (Ramos & Saddler 1994b) and the theoretical yield obtained through processes such as the simultaneous saccharification and fermentation (SSF process) of biomass hydrolysates are considerably increased (Holtzapple *et al.*, 1991).

Other agricultural residues such as sugarcane bagasse have also been steam-treated using aqueous H_3PO_4 as the acid catalyst of choice (Fontana *et al.*, 1995, Deschamps *et al.*, 1996). Being a weaker acid ($pKa_1 = 2.1$), H_3PO_4 generates less carbohydrate dehydration and does not have to be washed out prior to fermentation because phosphate can act as an important co-nutrient for microbial growth, particularly after partial neutralization with ammonia.

Another method used to convert lignocellulosic materials is the carbonization of fiberboard and slabs of charcoal fines with phenol formaldehyde binder, which yielded interesting carbonaceous building materials. Similarly, synthetic phenol formaldehyde has been substituted by the lignin as binder produced from agro wastes using steam explosion (Gravitis *et al.*, 1989).

An extensively researched chemical method (Wypych 2004) is the synthesis of "layered materials" to produce nanocomposites [these are combination of two or more components with at least one of the components have dimensions in the range of nanometers], which may have a bright future for the utilization of agro and industrial wastes. These materials belong to a special class of compounds in which the crystals are built by the stacking of "two-dimensional" units known as layers that are bond to each other through weak forces (Schöllhorn 1982 & 1994). Three methodologies are used, viz. intercalation, exfoliation and grafting.

Intercalation reactions occur by the topotactic insertion of mobile guest species (neutral molecules, anhydrous or solvated ions) into the accessible crystallographic defined vacant sites located between the layers (interlayer spacings) in the layered host structure. In the intercalation compounds, strong covalent bonds occur in the layers and weak interactions, between host lattice and guest species or co-intercalated solvents. Ionic and solvent exchange reactions are related to the replacement of solvated guest species (cations and anions) located into the interlayer spacings. In this case, only the solvent, the cations or the solvated cation can be replaced, depending on the reaction conditions.

Exfoliation reactions refer to the process of separation of individual layers in an appropriate solvent. Literally, it is a process of rupture of a layered crystal in such a way that stacked single layers are removed of the crystal and taken to suspension. The exfoliated single layers suspension can be used to surface adsorb polymers and help in the dispersion of the reinforcement agent into the polymeric matrix. Potentially, the process of separation of individual layers can be used for reactions of direct functionalization (grafting) since in those conditions the bonds do not exist among the layers that hinder the access of the reactants to the interlayer spacings.

Grafting reactions occur by establishing covalent bonds between the reactive groups of the layer and an adequate reactant molecule, which ensures higher chemical, structural and thermal stability for the compound. These reactions can be restricted to the crystal surface (the basal spacing remains unchanged) or layer surface (in this case an interlayer expansion occurs). These compounds can be collectively defined as hybrid materials, or more specifically, surface-modified inorganic layered materials.

There have been several attempts (Wypych 2004) to use the above methods particularly to develop a bio-resource-starch nanocomposites reinforced with natural materials. For example, a multi-agency project coordinated by the TNO Institute of Industrial Technology, Eidenhoven, aimed at the development of biodegradable food packaging material based on starch and clay, has already produced the nanocomposite flat films of thickness 500μm with desired properties of packaging (Progress Report 2001). These films have been evaluated for their performance as packaging materials.

In Japan, next generation innovative waste treatment technology has been developed by Mitsubishi through the integration of the waste from the plasma induction melting with other technologies developed by them. This has helped in tackling their environmental problems by increasing recycling yields in process operations (http://www.mmc.co.jp/english/business/rd.html#01).

Transformations : Of the several techniques used for the development of traditional building materials using agro and industrial wastes, the novel method of "liquid phase assisted processing" developed by the Regional Research Laboratory, Trivandrum (CSIR-India), in collaboration with the TNO Institute of Applied Physics, the Netherlands, seems to offer greater opportunity with several advantages. The process consists of modifying the composition of raw materials such as fly ash, clay and small amounts of oxide additives (TiO_2, Fe_2O_3) in such a way that the liquid phase is developed during sintering at reasonably low temperature [1273K] to produce tailored properties. This process uses about 85wt% waste material (fly ash) unlike other processes developed in India and abroad (Gopakumar 2001). The mechanism called "flux bonding" involved in the process is reported to be consolidation of fly ash holding together in the presence of binder, formation of low melting glass phase by the additives, which would activate the surface of the fly ash (coating by this) and generate more liquid phase to hold the fly ash together. In the early stage of firing of this fly ash-binder-additive mixture, the fine carbon coating present would get oxidized resulting in availability of fresh silica surface for flux bonding while becoming source for internal energy for the process. This is reported to be responsible for the higher fly ash content in the resulting ceramic products and the higher strength properties [9.8MPa] due to uniform fluxing, which could further be modified depending on the extent of liquid phase formed. These sintered products were "brick red" in color [See Fig. 3] unlike with other processes reported elsewhere. More than 1million bricks [size: $22x10x7cm^3$] and several tiles [size: $50x50x10mm^3$] have been made successfully in an industry. The process is ready for technology transfer.

Others : Other attractive methods generally adopted for the utilization have been the pyrolysis of biomass and the chemical methods to modify the biomass. The former method not only produces fuels (solid bio-fuels) but also economically and environmentally viable chemical feed stocks (liquid fuels). For more details, on the reader is recommended to see the recent review by Yaman (2003) and other publications (Vamvuka *et al.*, 2003, Hoogwijk *et al.*, 2003, Kim and Dale 2003). On the other hand, the latter method enhances the absorption capability of biomass for the removal of some of the metal ions such as Cd from aqueous solutions as well as from wastewater effluents (Abia *et al.*, 2003, Daifullah *et al.*, 2003).

Perspectives and Conclusions

While talking of waste utilization and recycling to help environment management, the first thing that comes to one's mind is why should so much wastes be generated? Firstly, is it because we do not bother for its generation and secondly is it not possible to reduce it? The answer for these could be positive leading to develop appropriate ways to tackle these problems. In doing so, it is relevant to think and analyze whether recovery and recycling of agro and industrial wastes would lead to save energy, and /or natural resources and /or environmental pollution by reducing the emissions. The systematic way of doing this is called "Life Cycle Assessment" (LCA), which has been practiced in an appropriate form applicable to waste management for at least a decade or so (Weber 2000, Roland 2004). Therefore, such an assessment for all the by-product utilization processes and their products should be made to compare the alternate products and processes developed with those of conventional ones. This should enable to further identify further improvement options and to take appropriate corrective actions to meet the objective of environment management. But, while the concept of LCA is increasingly used, the State legislation for using this is still awaited.

Thus, from the foregoing it becomes evident that the approach to utilize the wastes from agro and industrial sectors with a view to tackle the environmental problem through non-invasive technologies and sustainable development will certainly hold the key for a better future. It certainly offers greater opportunities for development of a variety of new materials and new methodologies with an appropriate coordination of supply and disposal of raw materials and their recycling to fully take care of ecological problems. One of the options to meet this, particularly in respect of agro and forestry wastes, is resorting to "soft chemistry"- defined in ROMMPS Dictionary of Chemistry as "processing of exclusively hazardous materials without by products or waste products except those that can be recycled" (Forschungsforum 3/97). The basic principle of this approach is "searching before constructing". However, this should fulfill certain prerequisites such as the continuous supply of such wastes through favorable logistics and a long term support by the people and governments. For this purpose, it is necessary to have technology clusters or groups consisting of experts and appropriate representatives from society such as agriculture, industry etc. The tasks of these bodies should be the collection of data on the availability of wastes, assessment of their quality, coordination of R&D efforts, conducting market analysis for requirements and specifications as well as of the competitiveness of the new products developed and even look into the certification methods to achieve the desired result of waste utilization and ecological balance. Some countries have already followed this approach. Examples include the continued support for 6 projects by the Danish Government for the improved applications of renewable resources for industrial non-food purposes, such as development of packaging material, improvement of potato starches for better uses, etc. (Forschungsforum 3/1997); the Fly Ash Mission of the Department of Science & Technology, Government of India is also a good example.

While talking of soft chemistry, it is relevant to note that in the case of lignocellulosic materials, some applications are limited by the close association that exist among the three main components of the plant cell wall, namely, cellulose, hemicellulose and lignin. Therefore, it is only through a clear understanding of this chemistry that one can identify the reasons as to why lignocellulosics are so resilient to biological and/or chemical processes.

Since many of the available characteristics of the waste materials have been evaluated using conventional techniques and instruments, it is highly beneficial to characterize them using some of the modern analytical tools such as NMR, XPS, LEED, ESCA, AUGER, AFS, STEM, etc. so that more information on their internal structure and the other details could be better understood leading to widening of their area of better utilization.

Following the development of certain products and process for the utilization of wastes some specific perspectives are given in the following:

One of the exciting possibilities of chemical pathways for the utilization of agro and industrial wastes could be the development of composites / nanocomposites using the concept of "layered materials" mentioned earlier. Development of starch based composite materials using lignocellulosic fibers with or without chemical modification is example (Chen *et al.*, 2003, Cinelli *et al.*, 2003, Lazaridou & Biliaderis 2002, Carvalho *et al.*, 2001, 2003, Angles & Dufresne 2000, Dufresne 2000, Warth 1997, Alvarez & Vázquez 2003, Curvelo *et al.*, 2001). While synthesizing layered composites, the following possibilities as shown in Fig. 7 exist:

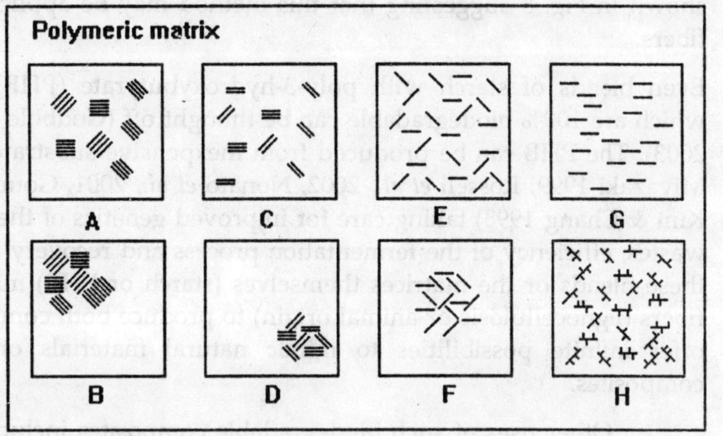

Fig. 7 Possible distribution of the reinforcement agent in a polymeric matrix.

(A) – the dispersion is appropriate but the crystals have the layers stacked as in the original structure. This represents intermediary condition because the layered crystals have a non-optimized interaction with the matrix; (B) – the crystals retain their structure and are segregated in only part of the matrix and hence it is not acceptable; (C) – the crystals are partially delaminated and the dispersion is good (an intermediary condition); (D) – although the crystals are partially delaminated this is not acceptable due to dispersion shown; (E) – the crystals are exfoliated and the dispersion is maximized and hence it is acceptable; (F) – although the exfoliation has occurred, the single layers are segregated and hence a non-ideal condition; (G) – The crystals are exfoliated and there is an ordered arrangement of the single layers in the matrix. This state may be

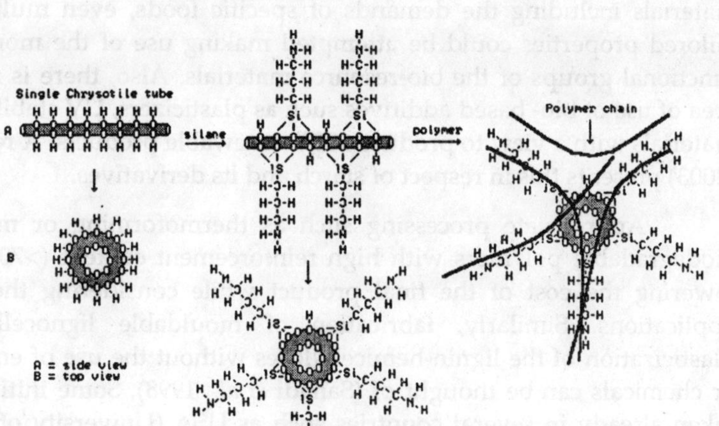

Fig. 8 Chrysotile fiber functionalization with 3-aminopropil-trietoxisilano and obtaining of the nanocomposite

acceptable when anisotropic properties of the nanocomposites are needed and, finally, (H) – single layers are totally dispersed and chemically bonded to the matrix due to the organofunctional and hence this is a highly acceptable condition.

However, the best condition is as in "H" since this has been successfully carried out even with nanotubes in place of layered crystals. Further, with inexpensive chrysolite fibers (nanotube) or other fibrous minerals, this has been attempted with organic functionalization as shown in Fig. 8 suggesting that this method may be applicable with lignocellulosic or animal fibers.

Even blends of starch with poly-3-hydroxybutyrate (PHB), polyester produced by bacteria, which are 100% biodegradable can be thought off (Godbole *et al.*, 2003, Imam 1998, Innocentini 2003). The PHB can be produced from inexpensive substrates such as sugarcane (Quagliano & Miyazaki 1999, Rossell *et al.*, 2002, Nonato *et al.*, 2001, Gouda *et al.*, 2001) or starch (Kim 2000, Kim & Chang 1998) taking care for improved genetics of the bacterial strains, utilization of bio-wastes, efficiency of the fermentation process and recovery and purification of polymer. Either these blends or the matrices themselves (starch or PHB) may also be used with other natural fibers (lignocellulosic or animal origin) to produce both composites and nanocomposites, which offer infinite possibilities to mimic natural materials or create new "biomimetic green" composites.

Other uses of such biodegradable composites include eco-friendly packaging materials for the food industry, super absorbent materials, medical applications, polymeric films with selective permeability for gases for coatings, for various substrates materials for encapsulation and controlled release of pesticides and herbicides, green containers to replace petroleum based materials and composite materials for automotive applications.

With a view to produce high performance bio-resource based polymer packaging materials including the demands of specific foods, even multi-layered or composites having tailored properties could be attempted making use of the more diverse side chains and more functional groups of the bio-resource materials. Also, there is need to carryout research in the area of use of bio- based additives such as plasticizers, UV stabilizers and colorants to packaging materials with a view to produce 100% renewable products. A recent review by Fang and Fowler (2003) presents this in respect of starch and its derivatives.

Appropriate processing such as thermoforming or melt processing of thermoplastic biodegradable polymers with high reinforcement content (>70wt%) should be looked into for lowering the cost of the final product while considering them useful in high performance applications. Similarly, fabrication of mouldable lignocellulosic composite by thermo-plasticization of the lignin-hemicelluloses without the use of environmental hazardous solvents or chemicals can be thought of (Sanadi *et al.*, 1998). Some initiation on all these lines has been taken already in several countries such as USA (University of Wisconsin and Forest Products Laboratory), New Zealand (University of Auckland), etc.

Attempts should be made to produce flexible, low density mats and other forms of the chemically modified fibers to be suitable for fabricating composites with appropriate matrix material. This may call for different type of machinery to produce such raw materials.

Recycling of used rubber tyres should be taken up with a view to develop high impact and noise insulate sheets or floor tiles by looking at the existing variables in rubber agglomeration, grain size, density of the floor tiles, etc., which are linked to dimensional stability of the tiles. (Francisco, RKO-ZC - http://rko.zcu.cz/sme.php3).

Similarly, on-going attempts for the development of new and improved processing technologies with chemical processing, bio-processing, thermo-mechanical and mechanical processing methods for the utilization of wastes should be continued; integrated or hybrid processes may be attempted, which may offer better opportunities for commercial success. However, to achieve this task, appropriate separation technologies should also be developed simultaneously. Some countries are practicing this already.

While talking of wastes, other interesting and important areas of waste utilization are those of nuclear and hospital wastes. It is high time that some serious thinking on this goes on and a serious effort towards this is taken for the sake of humanity in particular and environment in general. It is gratifying to note that there have been some serious attempts in the case of treatment of nuclear wastes. A recent publication by Forano (2004) reviews these. Basically, the processes for the disposal of hazardous nuclear wastes involve the immobilisation of these in cementitious materials (as barriers or absorbers of actinides of U, Th, etc.) such as hydrotalcite-like compounds and hydrocalumite or ettringite, the calcium form of hydrotalcite (Rousselot *et al.*, 2002, Abdoulas *et al.*, 1994, Altenheim-Haese *et al.*, 1994, Marx *et al.*, 1995, Serrano *et al.*, 2002). Attempts have also been made to improve the reliability of underground disposal systems used in the retention of nuclear waste for long periods of time by adopting higher adsorption capacity for anionic species than that of oxide and hydroxides materials as well as use of layered double hydroxides [LDH] (Kang *et al.*, 1996, Balsley *et al.*, 1998, Serrano *et al.*, 2000).

Last but not the least is recognition that a gap exists between the developers of the technologies for the utilization of wastes and the intended users. It is therefore essential that the following steps are taken: (1) documentation of all the technologies developed on the recovery, utilization and recycling of the wastes and their upgradaing; (2) maintenance of close linkages between various technology developers, manufacturers and users to enhance appropriate technology transfer and their continued utilization through some mechanism such as publications (Nicholas 1999); (3) while environmentally sound products can affect the market for products from developing countries, there should be more life-cycle analysis and research on environmentally compatible products. With the above developments and suggested lines for future action, it is hoped both exciting and promising future lies ahead for the ecology and energy management in general and for the utilization of wastes in particular.

Acknowledgements

The authors express sincere thanks to Dr. T.K.Ghosh, one of the Editors for inviting to contribute this Chapter thereby giving an opportunity to share their experience and views in the area of utilization of agro and industrial wastes through fermentation and other techniques. They acknowledge the permission received for the use of tables and figures from M/s. Elsevier, M/s Marcel Dekker, M/s NPL, UK, Dr.T. N. Gupta, Executive Director, BMTPC, India, and other authors from whose published work the figures/tables are included in the Chapter. They are thankful to Mr. Gregorio Guadalupe Carbajal Arizaga of Departmento de Quimica, UFPR,

Curitiba and Dr. T.P.D. Rajan of Regional Research Laboratory, Thiruvananthapuram (Kerala-India) for their help in the preparation of this Chapter. One of the authors (KGS) acknowledges some of his erstwhile colleagues of Regional Research Laboratory, Thiruvananthapuram and the Laboratory authorities to enable him to work in the diversified areas and whose work has been cited in the Chapter.

References

Abdelouas, A., Crovisier, J. L., Lutze, W., Fritz, B., Mosser, A. and Mueller, R. (1994) Formation of hydrotalcite-like compounds during R7T7 nuclear waste glass and basaltic glass alteration, Clays and Clay Minerals 42(5): 526-33.

Abia A.A., Horsfall, M. Jr and Didi, O. (2003) The use of chemically modified and unmodified cassava waste for the removal of Cd, Cu and Zn ions from aqueous solution, Bioresource Techn., 90(3): 345-348.

Aiticn, A. and Pigeon, M. (1986) Performance of Condensed Silica Fume Concrete used In Pavements and Sidewalks. Durability Build. Mat. 3 (4): 353-368.

Akhtar, M., Blanchette, R.A., Myers, G. and Kirk, K. (1998) p. 309. In Environmentally Friendly Technologies for the Pulp and Paper Industry; eds. Young, R.; Akhtar, M., John Wiley: New York.

Altenhein-Haese, C., Bischoff, H., Fu, L., Mao, J. and Marx, G. (1994) Adsorption of actinides on cement compounds. J. of Alloys and Compounds 213-214: 554-556.

Alvarez, V.A. and Vázquez, A. (2003) Thermal degradation of cellulose derivatives/starch blends and sisal fibre biocomposites, Polym. Degrad. and Stability (In Press).

Andre de Vries, A.H. (1999) Utilization Of Waste Materials in the Proceedings of the International Workshop on Building Ceramics from Industrial Wastes, Organized by International Center for Science and High Technology [UNIDO], Trieste-Italy and Regional Research Laboratory (CSIR), Thiruvananthapuram (India), March 22-26,.

Angles, M.N. and Dufresne, A. (2000) Plasticized starch/tunicin whiskers nanocomposites 1. Structural analysis, Macromol., 33(22): 8344-8353.

Annie, G. and Mohan Das, P.N. (1985) Relative Merits of various processes available for the production of rutile. J. Sci. Ind. Res., 44: 443-450.

Annual Reports 1980-82, Regional Research Laboratory, Trivandrum (Kerala-India).

Ballesteros, I., Oliva, J.M., Negro, M.J., Manzanares P. and Ballesteros M. (2002) Enzymic hydrolysis of steam exploded herbaceous agricultural waste (Brassica carinata) at different particule sizes. Process Biochem. 38: 87-192.

Balsley, S.D., Brady, P.V., Krumhansl, J. L. Anderson and Howard L. (1998) Anion scavengers for low-level radioactive waste repository backfills, J. Soil Contamination 7(2): 125-141.

Banhart, J. (1999) Metallic Foams, Euro Physics News, 1: 1-4.

Banhart, J. (2001) Manufacture, characterization and application of cellular metals and metal Foams. Progress in Mater. Sci. 46 (6): 559-U3.

Baumeister, J., Banhart J. and Weber, W. (1997) Materials & Design, 18 4/6: 217

Beldzki, A.K. and Gassen, J. (1999) Composites reinforced with cellulose based fiber, Prog. Polym. Sci., 24: 221-274.

Beltrame, P. L., Cartini, P. Visciglio, A. Focher, B. and Marzetti, A. (1992) Fractionation and bioconversion of steam-exploded wheat straw. Biores. Technol. 39: 165-171.

Bhattacharya D., Jayaraman, K Bowis, M and Boyle, S. (1999) Waste to product manufacturing: a scenario with waste plastics/natural fiber composites p. 304-310 in Proceed. International exhibition and Conference, 13-16 April, New Delhi.

Binner J. and Sambrook, R. (2004) http://www.azom.com/details.asp?Article ID=1869.

BMTPC [Building Materials & Technology Promotion Council] (1999), Government of India, News Letters, Brochures and website: www.bmtpc.org.

Brownell, H.H. and Saddler, J.N. (1984) Steam-explosion pretreatment for enzymatic hydrolysis. Biotechnol. Bioeng. Symp. 14: 55-68.

Brownell, H.H. (1989) Progress report for the Project No. 04-53-12-402, Forintek Canada Co., Ottawa, Canada,.

Bujan, F. RKO-ZC (http://rko.zcu.cz/sme.php3).

Carta, F.S., Soccol, C.R., Ramos, L.P. and Fontana, J.D. (1999) Production of fumaric acid by fermentation of enzymatic hydrolysates derived from cassava bagasse. Biores. Technol. 68: 23-28.

Carvalho A.J.F., Curvelo A.A.S. and Agnelli J.A.M. (2001) A first insight on composites of thermoplastic starch and kaolin, Carbohydr. Polym., 45 (2): 189-194.

Carvalho, A.J.F., Zambon, M.D., Curvelo, A.A.S. and Gandini, A. (2003) Size exclusion chromatography characterization of thermoplastic starch composites: 1-Influence of plasticizer and fiber content, Polymer Degradation and Stability, 79: 133-138.

Chang, V.S., Nagwani, M. and Holtzapple, M. (1998) Lime pretreatment of crop residues bagasse and wheat straw. Appl. Microbiol. Technol. 74: 135-158.

Chen, Y., Ishikawa, Y., Zhang, Z.Y. and Maekawa, T. (2003) Properties of extruded acetylated starch plastic filled with cellulose Acetate, Transactions of the ASAW, 46 (4): 1167-1173.

Cinelli, P., Chiellini, E., Gordon S.H. and Imam, S.H. (2003) Characteristics and degradation of hybrid composite films prepared from PVA, starch and lignocellulosics, Macromol. Symp., 197: 143-155.

Cisse, I.K. and Laquerbe, M. (2000) Mechanical characterization of filler sandcretes with rice husk ash additions - Study applied to Senegal. Cem. Concrete Res. 30 (1): 13-18.

Colney, M.W., Gupta, P.K. and Mikos, A.G. (1996) Fabrication of Pliable biodegradable polymer foams to engineer soft tissues. Cell Transplantation, 5(4): 465-471.

Coombs, J. and Hall, K. (2000) Non-Food Agro-Industral Research information, CD-rom version 1.2, issue 3, 2000 (ISSN 1368-6755, ISBN 1-872691-27-7), CPL Publishing Services, Newbury, UK.

Curvelo, A.A.S., Carvalho A.J.F. and Agnelli, J.A.M. (2001) Thermoplastic starch-cellulosic fibers composites: Preliminary results, Carboh. Polym., 45 (2): 183-188.

Daifullah, A.A.M., Girgis B.S. and Gad, H.M.H. (2003) Utilization of agro-residues (rice husk) in small wastewater treatment plans. Materials Letters 57: 1723–1731

Dekker, R.F.H., Karageorge, H. and Wallis, A.F.A. (1987) Pretreatment of hardwood (Eucalyptus regnans) sawdust by autohydrolysis explosion and its saccharification by trichodermal cellulases. Biocatalysis, 1: 45-59.

Dekker, R.F.H. and Wallis, A.F.A. (1983) Enzymic saccharification of sugarcane bagasse pretrated by autohydrolysis-steam explosion. Biotechnol. Biotechnol. Bioeng., 25: 3027-3048.

DeLong, E.A. (1983) Method of rendering lignin separable from cellulose and hemicellulose and the product so produced. Canadian Patent No. 1, 141, 376.

DeLong, E.A., DeLong, E.P., Ritchie, G.S. and Rendall, W.A. (1990) Method for extracting the chemical components from dissociated lignocellulosic material. United States Patent No. 4,908,098.

Deschamps, F.C., Ramos, L.P. and Fontana, J.D. (1996) Pretreatment of sugarcane bagasse for enhanced ruminal digestion. Appl. Biochem. Biotechnol. 57/58, 171-182.

Dowall, D.E. (1997) Future Cities, Building Materials News (India), Special Issue, 6/10/1997.

Dufresne, A., Dupeyre, D. and Vignon, M.R. (2000) Cellulose microfibrils from potato tuber cells: Processing and characterization of starch-cellulose microfibril composites J. Appl. Polym. Sci. 76 (14): 2080-2092.

Erikson, R. (2002) Syntactic Metals: Flight Weight Materials, Adv. Mater. Process. 160 (12): 44-47.

Escalante J.I. Mendoza, G. Mancha, H, Lopez, J and Vargas, G. (1999) Pozzolanic properties of a geothermal silica waste material. Cem. Concrete Res. 29 (4): 623-625

Excoffier, G.,Toussaint, B. and Vignon, M.R. (1991) Saccharification of steam-exploded poplar wood. Biotechnol. Bioeng. 38: 1308-1317.

Fan, L.T., Gharpuray, M.M. and Lee, Y.H. (1987) Cellulose hydrolysis; Springer Verlag: New York.

Fang, J. and Fowler, P. (2003) The use of starch and its derivatives as biopolymer sources of packaging materials Food agriculture & Environment, 1 (3&4): 82-84.

Fernández-Bolaños, J., Felizón, B., Heredia, A., Guillén R. and Jiménez, A. (1999) Characterization of the lignin obtained by alkaline delignification and of the cellulose residue from steam-exploded olive stones. Biores. Technol. 68: 121-132.

Fontana, J.D., Ramos, L.P. and Deschamps, F.C. (1995) Pretreated sugarcane bagasse as a model for cattle feeding. Appl. Biochem. Biotechnol. 51/52: 105-116.

Foody, P. (1984) Method for increasing the accessibility of cellulose in lignocellulosic. United States Patent No. 4, 461, 648.

Forano, C. (2004), Environmental Remediation Involving Layered Double Hydroxides in Clay Surfaces-Fundamentals and applications, eds. F. Wypych and K.G.Satyanrayana), Pub. Elsevier, USA (To be published in 2004).

Forschungsforum 3/97, http;//www.nf-2000.org.secure/contacts/C1103320.htm

Gang and Weixing, S. (1998) Sunflower as adsorbents for the removal of metal ions from wastewater. Ind. Eng. Chem. Res. 37: 1324-1328.

Gardea-Torresdey, J.L., Tang L. and Salvador, J.M. (1996) Copper adsorption by esterified and unesterified fractions of Sphagnum peat moss and its different humic substances. J. Hazard. Mater. 48: 191-206.

Gardea-Torresdey, J.L., Tiemann, K.J., Gamez, G. and Dokken. (1999) Effect of chemical composition for multi-metal binding by Medicago sativa (Alfalfa). J. Hazard. Mater. B 69: 41-51.

Ghosh, P. and Singh, A. (1993) Method for increasing the accessibility of cellulose in lignocellulosic. Adv. Appl. Microbiol., 39: 295-333.

Gibson L.I. and Ashby, M.F. (1997) Cellular Solids, Oxford.

Gmunden Address on Declaration of Gmunden for the encouragement of sustainable resources at the Conference "Crops for a Green Industry" Gmunden, 6-8 October 1998.

Godbole, S., Gote, S., Latkar, M. and Chakrabarti, T. (2003) Preparation and characterization of biodegradable poly-3-hydroxybutyrate-starch blend films. Biores. Technol., 86 (1): 33-37.

Gopakumar, K., Murali, T.P., and Rohatgi, P.K. (1982) Metal-shell char particulate composites using copper coated particles. J. Mater. Sci. 17: 1041-1048.

Gopakumar Warrier, K, Muralidharan Nair, K, Krishna Pillai, K and Perumal, P. (2001) Development of traditional ceramics from Industrial solid wastes through liquid phase assisted processing. Metal Materials and Processes 13 (2-4): 153-160.

Gouda M.K., Swellam, A.E. and Omar, S.H. (2001) Production of PHB by a Bacillus megaterium strain using sugarcane molasses and corn steep liquor as sole carbon and nitrogen sources, Microbiol. Res. 156 (3): 201-207.

Gravitis, J., Vedernikov, N., Zandersons, J. and Kokorevics A. (1989) http://cwch.uoregon.edu/ZeriFolder/10_ZERI_Projects_Package.pdf.

Grover, P. D. ed. (1989) Thermo-chemical characterization of Biomass residues for gasification, Indian Institute of Technology, New Delhi.

Grover, P.D. and Mishra, S.K. ed. (1995) Proceedings of the Workshop on Biomass Briquetting, New Delhi, India, 3-6 April, [http://www.rwedp.org/acrobat/rm23.pdf].

Gupta, T.N. (2000) Materials for the human habitat. MRS Bulletin, 25 (4) 60-63.

Hawley, M.C., Downey, K.W., Selke, S.M. and Lamport, D.T.A. (1986) p. 297 - In Cellulose - structure, modification and hydrolysis; ed. Young, R. A. and Rowell, R. M.; John Wiley, New York.

Heitz, M., Capek-Ménard, E., Koeberle, P.G., Gagné, J., Chornet, R.P., Taylor, J.D. and Yu, E. (1991) Fractionation of Populus tremuloides at the pilot plant scale: optimization of steam pretreatment conditions using the STAKE II technology. Biores. Technol. 35: 23-32.

Holtzapple, M.T., Jun, J., Ashok, G., Patibandla, S.L and Dale, B.E. (1991) The ammonia freeze explosion (AFEX) process. A practical lignocellulose pretreatment. Appl. Biochem. Biotechnol. 28/29: 59-74.

Hoogwijk, M., Faaij, A., van den Broek, R., Berndes, G., Gielen, D. and Turkenburg, W. (2003), Exploration of the ranges of the global potential of biomass for energy, Biomass & Bioenergy, 25:119-133.

http://www.techno-preneur.net/timeis/technology/sctech17.html.

http://www.nf-2000.org/secure/Air/S253.html.

http://www.techno-preneur.net/timeis/technology/sctechFeb-March/ scitech17.html

http://www.nire.go.jp/eco_tec_e/new-sigen_e.htm.

http://www.mmc.co.jp/english/business/rd.html#01.

http://www.npl.co.uk/npl/cmmt/metal_foams).

Huisenga M. (2003) Smart Business: The effective use of agricultural byproducts, in A Guide to Developing Agricultural Markets and Agro-enterprises edited by Daniele Giovannucci, ESSD/ardext.nsf/26 ByDocName/Smart Business.

Imam, S.H., Chen, L., Gordon, S.H., Shogren, R.L., Weisleder D. and Greene, R.V. (1998) Biodegradation of injection molded starch-poly (3-hydroxybutyrate-co-3-hydroxyvalerate) blends in a natural compost environment. J. Enriron. Polym. Degrad., 6 (2): 91-98.

Innocentini-Mei, L.H., Bartoli J.R. and Ballieri, R.C. (2003) Mechanical and thermal properties of poly(3-hydroxybutyrate) blends with starch and starch derivatives, Macromol. Symp., 197: 77-87.

International Workshop (1999) on Building Ceramics from Industrial Wastes, Organized by International Centre for Science and High Technology [UNIDO], Trieste-Italy and Regional Research Laboratory (CSIR), Thiruvananthapuram (India), March 22-26.

Jeoh, T. and Agblevor, F.A. (2001) Characterization and fermentation of steam exploded cotton gin waste. Biomass Bioenerg. 21: 109-120.

Jollez, P., Chornet, E., Overend, R.P. and Bridgwater, A.V. (1994) Steam-aqueous fractionation of sugar cane bagasse: an optimization study of process conditions at the pilot plant level. Advances in Thermochemical Biomass Conversion 2: 1659-1669.

Kaar, W.E., Gutierrez, C.V. and Kinoshita, C.M. (1998) Steam explosion of sugarcane bagasse as a pretreatment for conversion to ethanol. Biomass Bioenerg. 14: 277-287.

Kandachar, P.V. (2002) Overview and Advances on Science, Technology and Application of "Composite Materials based on Natural Sources", Proc. Materials Design and Production Processes for Low-cost Housing, 25-29 Nov., Bangalore (India) organized by UNIDO, ICS (Italy),ICAMT and BMTPC (India).

Kang, M.J., Rhee, S.W., Moon, H. Neck, V. and Fanghaenel, Th. (1996) Sorption of MO_4^- (M = Tc, Re) on Mg/Al layered double hydroxide by anion exchange, Radiochim. Acta, 75 (3): 169-173.

Kihara, Y. and Bucher, H.R.E. (1986) Microssílica uma nova Pozolana Artificial. Cerâmica, 32 (200): 211 -214.

Kim, B.S. (2000) Production of poly(3-hydroxybutyrate) from inexpensive substrates, Enzyme Microb Tech., 27 (10): 774-777.

Kim, B.S. and Chang, H.N. (1998) Production of poly (3-hydroxybutyrate) from starch by Azotobacter chroococcum, Biotechnol. Lett., 20(2), 109.

Kim, B. S and Dale, B.E. (2003) Global potential bioethanol production from wasted crops and crop residues, Biomass & Bioenergy, 26 (4):361-375.

Kling, S.H., Carvalho Neto, C., Ferrara, M.A., Torres, J.C.R., Magalhães, D. B. and Ryu, D.D.Y. (1987) Enhancement of enzymatic hydrolysis of sugar cane bagasse by steam explosion pretreatment. Biotechnol. Bioeng. 29: 1035-1039.

Kobayashi, F., Sawada, T., Nakamura, Y., Ohnaga, M., Godliving, M. and Ushiyama, T. (1998) Saccharification and alcohol fermentation in starch solution of steam-exploded potato. Appl. Biochem. Biotechnol. 69: 177-188.

Kumar, V. and Jha, C.N. (1999) High value added utilization of fly ash in the Procedings of the International Workshop on Building Ceramics from Industrial Wastes, Organized by International Center for Science and High Technology [UNIDO], Trieste-Italy and Regional Research Laboratory (CSIR), Thiruvananthapuram (India), March 22-26.

Lazaridou, A. and Biliaderis, C.G. (2002) Thermophysical properties of chitosan, chitosan-starch and chitosan-pullulan films near the glass transition. Carboh. Polym., 48 (2): 179-190.

Luken, R. and Freij, A.C. Paper presented in the OECD Workshop on Development Assistance and Technology Cooperation for Cleaner Industrial Production in Developing Countries, Hannover, Germany, 28 to 30 September 1994, http://www.unido.org/userfiles/PuffK/clean.pdf.

Lynd, L. R.,Cushman, J. H., Nichols, R. J. and Wyman, C. E. (1991) Fuel ethanol from cellulosic biomass. Science 251: 1318-1323.

Mangan, C. (Ed) (1998) The green chemical and polymers chain. European Commission, DG12, DG6, Luxenbourg. Office for Official Publications of the European Union, (ISBN 92-828-6116-3), Belgium.

Marsh, G. (2003) Next step for Automotive Materials. Materials Today, 6 (4) 36-43.

Marstrander, R. Souvenir of INCAL'98, New Delhi, February 11-13, 1998.

Martinez, J., Negro, M. J., Saez, F. Manero, J., Saez, R. and Martin, C. (1990) Effect of acid steam explosion on enzymatic hydrolysis of O. nervosum and C. cardunculus. Appl. Biochem. Biotechnol. 24/25: 127-134.

Marx, G., Altenhein-Haese, C., Bischoff, H., Engelhardt, J. and Feldmaier, F. (1995) Investigations into speciation; sorption and corrosion of actinides in systems of high ionic strength, J. Radioanal. Nucl. Chem. 194 (1): 95-105.

Mehta, P. and Kumar, (1983) Fly ash, silica fume, slag and other mineral by products, Sp-Am. Cocr. Inst. 79, Cocr.1: 1.

Mohandas, P.N. (2003) Beach Sand Mineralsp. 104-114 in RRL-T 25 Years [see Web site: http//:w3rrlt.csir.res.in].

Moniruzzaman, M. (1996) Effect of steam explosion on the physicochemical properties and enzymatic saccharification of rice straw. Appl. Biochem. Biotechnol. 59: 283-297.

Mymrine, V. (2003) Environment friendly method of high alkaline Bauxite's Red Mud and Ferrous Slag utilization as example of green chemistry. Green Chemistry (USA) 5: 357-360.

Mymrine, V. (2001) Red mud of aluminium production waste as basic component of new construction materials. Intern J. Waste Management & Research, Copenhagen (19): 465-469.

Mymrine, V. and Vázquez Vaamonde, A.J. (1999) New construction materials from Spanish Jarosite processing waste. Journal of Mineral Engineering. Cornwall. 12 (11): 1399-1402.

Nakayama, E. and Okamura, K. (1989) Influence of a steam explosion and microwave irradiation on the enzymatic hydrolysis of a coniferous wood. Mokuzai Gakkaishi 35: 251-260.

Netravali A.N. and Chabba S. (2003) Composites get Greener. Materials Today, 6 (4): 22-29.

Nickel J. and Riedel, U. (2003) Composites for Recyclability, Materials Today, 6 (4): 44-48

Nicholas, J.S. (1999) Biomass Based Recycling and Recovery Technologies: A Review of GO'S and NGO'S Works in the Philippines' Rural Enterprise Development Foundation,Inc., paper presented at the Biomass based Recycling and Recovery Technologies, Los Baños, Laguna, Philippines, Paper presented at the R'99 Congress (Recovery, Recycling, Re-integration), February.

Nguyen, Q.A. and Saddler, J.N. (1991) An integrated model for the technical and economic evaluation of an enzymatic biomass conversion process. Biores. Technol. 35: 275-282.

Nonato R.V., Mantelatto, P.E., Rossell, C.E.V. (2001) Integrated production of biodegradable plastic, sugar and ethanol Appl. Microbiol. Biotech. 57 (1-2): 1-5.

NPL News Letter, http://www.npl.co.uk/npl/cmmt/metal_foams/110. Nonato R.V., Mantelatto, P.E., Rossell, C.E.V. (2001) Integrated production of biodegradable plastic, sugar and ethanol Appl. Microbiol. Biotech. 57 (1-2): 1-5.

Ortega, S. and Innocentini Mei, L. (2003) in REPORT on Expert Group Meeting On Environmentally Degradable Plastics and promotion of related projects, Trieste, Italy, July 10-11, organized by UNIDO.

Pai, B.C., Pillai, R.M., Satyanarayana, K.G. Sukumaran, K. Pillai, U.T.S., Pillai S.G.K and Ravikumar, K.K. (2001) Dicontinuously Reinforced Aluminium Alloy Matrix Composites. Metal Materials and Processes, 13 (2-4): 255-278.

Parisi, F. (1989) Advances in lignocellulosics hydrolysis and in the utilization of the hydrolysates. Adv. Biochem. Eng. Biotechnol. 38: 53-87.

Pavithran, C Bramhakumar, M. Satyanarayana, K.G., Sukumaran, K., Pillai, S.G.K. and Ravikumar, K.K. (2001) Plant fiber reinforced Polymer Composites. Metals, Materials and Processes, 13 (2-4): 137-152.

Peijs, T. (2003) Composites for recyclability. Materials Today, 6 (4): 30-35.

Pillai, R.M., Seshadri, S.K. and Rohatgi, P.K. (1982) Coconut based materials for melting metals and alloys. Appr. Techn. (UK) 8(4): 28-30.

Pillai, R.M., Pai, B.C. and Satyanarayana, K.G. (2003) Lighter, Foamy, Tailorable, Smarter and Functional Aluminium Proceedings of International Conference on Auminium-INCAL'03, 23-25 April, New Delhi.

Pillai, R.M. (2003) Metal Processing p.159-174 in RRL-T 25 Years [see Web site: http//:w3rrlt.csir.res.in].

Progress Report, January 2001 - http://www.nf-2000.org/secure/Fair/F908.htm.

Quagliano J.C. and Miyazaki S.S. (1999) Biosynthesis of poly-beta-hydroxybutyrate and exopolysaccharides on Azotobacter chroococcum strain 6B utilizing simple and complex carbon sources, Appl. Biochem. Biotech. 82 (3): 199-208.

Quek, Y., Wase D.A.J. and Forster, C.F. (1998) The use of sago waste for the sorption of lead and copper. Water SA 243: 251-256.

Ramachandran, V.S. ed. (1995) Concrete Admixtures Handbook, 2nd edition, Noyes Publications, USA.

Ramos, L.P. (1992) Steam pretreatment and enzymatic hydrolysis of Eucalyptus viminalis Labill, Ph.D. Thesis, University of Ottawa, Canada.

Ramos, L.P. (2003) The chemistry involved in the steam treatment of lignocellulosic materials, Química Nova 26: 863-871.

Ramos, L.P., Breuil, C., Kushner, D.N. and Saddler, J.N. (1992a) Steam pretreatment conditions for effective enzymatic hydrolysis and recovery yields of Eucalyptus viminalis wood chips. Holzforschung, 46: 149-154.

Ramos, L.P., Breuil, C. and Saddler, J.N (1992b) Comparison of steam pretreatment of eucalyptus, aspen and spruce wood chips and their enzymatic hydrolysis. Appl. Biochem. Biotechnol. 34/35: 37-48.

Ramos, L.P. and Saddler, J.N. (1994a) Bioconversion of wood residues: Mechanisms involved in pretreating and hydrolyzing lignocellulosic materials. p. 325-341. In: Enzymatic Conversion of Biomass for Fuels Production,eds. Himmel, M.E. Baker, J.D. and Overend, R.P. American Chemical Society Symposium Series 566: Washington.

Ramos, L.P. and Saddler, J.N. (1994b) Enzyme recycling during fed-batch hydrolysis of cellulose derived from steam-exploded Eucalyptus viminalis. Appl. Biochem. Biotechnol. 45/46: 193-207.

Ramsteiner, F. Fell, N and Forster, S. (2001) Testing the behaviour of polymer foams, Polymer Testing, 20: 661-670.

Rao, O.P. and Garg, S.D. (1999) Fly ash & alternative building materials: Programmes in Mission mode in the Proceedings of the International Workshop on Building Ceramics from

Industrial Wastes, Organized by International Centre for Science and High Technology [UNIDO], Trieste-Italy and Regional Research Laboratory (CSIR), Thtiruvananthapuram (India), March 22-26, 1999.

Rivers, D.B. and Emert, G.H. (1987) Lignocellulose pretreatment: a comparison of wet and dry ball attrition. Biotechnol. Lett. 9: 365-368.

Robertson, G.L. ed. (1993). Food Packaging: Principles and Practice; Marcel Dekker, New York.

Rohatgi, P.K., Guo, R.Q. and Keshavaram, B.N. (1995) Cast Al alloy-fly ash composites p.283-292 in Key Engg. Materials Vol. 104-107, ISBN=0-87849-697-1.

Roland, C. (2004) What a waste. Materials Today, 7(2): 64.

Ropars, M., Marchal, R., Pourquié, J. and Vandecasteele, J.P. (1992) Large-scale enzymatic hydrolysis of agricultural lignocellulosic biomass. Part 1: Pretreatment procedures. Biores. Technol. 42: 197-204.

Rossell C.E.V., Nonato, R.V., Mantelatto, P.E. and Leal, M.R.L.V. (2002) Production of biodegradable plastic (PHB), sugar and ethanol in a sugar mill, Int. Sugar J., 104 (1243): 321-323.

Rousselot, I., Taviot-Gueho, C., Leroux, F., Leone, P., Palvadeau, P., Besse, J-P. (2002) Insights on the structural chemistry of hydrocalumite and hydrotalcite-like materials: investigation of the series $Ca_2M_3+(OH)_6Cl.{}_2H_2O$ (M^{3+} : Al^{3+}, Ga^{3+}, Fe^{3+} and Sc^{3+}) by X-ray Powder Diffraction, J. Solid State Chem. 167, 137-144.

Saddler, J.N., Brownell, H.H., Clermont, L.P. and Levitin, N. (1982) Enzymatic hydrolysis of cellulose and various pretreated wood fractions. Biotechnol. Bioeng. 24: 1389-1402.

Saddler, J.N., Ramos, L.P. and Breuil, C. (1993) Steam pretreatment of lignocellulosic residues. p. 73-92 In Bioconversion of Forest and Agricultural Plant Wastes, ed. Saddler, J.N., C. A. B. International, London.

Sanadi, A.R, Caulfield, D.F. and Rowell, R.M. (1998) Lignocellulosic/Plastic Composites http://www.fpl.fs.fed.us/documnts/pdf1998/sanad98a.pdf.

Sathy, C., Satyanarayana, K.G., Pramada, P.N. and Raghavan, P. (2003) Processing, Properties and applications of reactive silica from rice husk-An overview. J. Mater. Sci. 38: 3159-3168.

Satyanrayana, K.G., Mohan Das, P.N., Pillai, C.K.S., Pavithran, C., Gopakumar K. and Rohatgi, P.K. (1984) Characterization and utilization of coconut tree based materials. Indian Coconut J., xiv (9): 18-29.

Satyanarayana, K.G., Warrier K.G.K. and Pavithran, C. (1999) Development of building materials based on Agro and Industrial wastes and innovative processes, Proc. International Conference on Building Materials based on Agro and Industrial Wastes, eds: T.N. Gupta and Mohan Rai, 13-16 April 1999, New Delhi.

Satyanarayana, K.G. (2002) Perspectives in Natural Fiber Incorporated Polymer Composites, (Keynote Address), 3-15 in Proc. International Conf. And Exhibition on Reinforced Plastics (ICERP- 2002), ed: P. Alagusundaramooorthy and Rajaram Nagappa, India, 7-9 Feb. Allied Publishers Pvt. Ltd., Chennai.

Sawada, T., Nakamura, Y., Kobayashi, F., Kuwahara, M. and Watanabe, T. (1995) Effects of fungal pretreatment and steam explosion pretreatment on enzymatic hydrolysis of plant biomass. Biotechnol. Bioeng. 48: 719-724.

Schell, D.J., Torget, R., Power, A., Walter, P.J., Grohmann, K. and Hinman, N. D. (1991) A technical and economic analysis of acid-catalysed steam explosion and dilute sulfuric acid pretreatments using wheat straw or aspen wood chips. Appl. Biochem. Biotechnol. 28-29: 87-97.

Schöllhorn, R. (1982) Solvated intercalation compounds of layered chalcogenides and oxide bronzes. p.315-357 in Intercalation Chemistry, ed. M.S. Whittingham and A.J. Jacobson, Academic Press, New York.

Schöllhorn, R. (1994) Materials and Models: Faces of intercalation Chemistry, p.1-81 in Progress in Intercalation Research, ed. by W. Müller- Warmuth. and R. Schöllhorn, Kluver Academic Publishers, Dordrecht.

Schwald, W., Breuil, C., Brownell, H.H., Chan, M. and Saddler, J.N. (1989a) Asessment of pretreatment conditions to obtain fast complete hydrolysis on high substrate concentrations. Appl. Biochem. Biotechnol. 20/21: 29-44.

Schwald, W., Smaridge, T., Chan, M., Breuil, C. and Saddler, J.N. (1989b) The influence of SO_2 impregnation and fractionation on product recovery and enzymic hydrolysis of steam-treated sprucewood. p. 231-242. In Enzyme Systems for Lignocellulose Degradation; Coughlan, M. P., ed.; Elsevier: New York.

Sepulveda, P. and Binner, J.G.P. (1999) Processing of Cellular Ceramics by Foaming and in situ Polymerisation of Organic Monomers, J. Eur. Ceram. Soc. 19: 2059-2066.

Serrano, J., Bertin, V. and Bulbulian, S. (2000) ^{99}Mo Sorption by Thermally Treated Hydrotalcites. Langmuir 16 (7): 3355-3360.

Serrano, J.; Bertin, V.; Bulbulian, S. (2002) Elution of 235U fission products through a calcined hydrotalcite-packed column, J. Radioanal. Nucl. Chem. 254 (1), 91-94.

Sewalt, V.J.H., Glasser, W.G. and Fontenot, J.P. (1992) Fiber composition and in vitro digestibility of steam-exploded peanut hulls. Animal Science Research Report, Virginia Agricultural Experiment Station 10: 111-112.

Sharma, S.K., Kalra, K.L. and Grewal, H.S. (2002) Enzymatic saccharification of pretreated sunflower stalks, Biomass Bioenerg. 23: 237-243.

Sukumaran, K., Satyanarayana, K.G., Pillai, S.G.K. and Ravikumar, K.K. (2001) Structure, Physical and Mechanical Properties of Plant fibers of Kerala. Metals, Materials and Processes 13 (2-4): 121-136.

Suresh, V. (1999) Building Technology for Sustainable Development, State of the art Lecture, Second All India People's Technology Congress, 14-17 Feb., Organized by Forum of Scientists, Engineers & Technologists (FOSET), Kolkata (India).

Tanahashi, M. (1990) Characterization and degradation mechanisms of wood components by steam explosion and utilization of exploded wood. Wood Res. 77: 49-117.

Tanaka, M. and Matsuno, R. (1990) n-Butylamine and acid-steam explosion pretreatments of rice straw and hardwood: Effects on substrate structure and enzymatic hydrolysis. Enzyme Microb. Technol.12: 190-195.

Ternrud, I. E., Theander, O., Torneport, L. and Vallander, L. (1989) Changes in chemical composition of steam-exploded wheat straw during enzymic hydrolysis. Enzyme Microb. Technol. 11: 500-506.

Torget, R., Walter, P., Himmel, M. and Grohmann, K. (1991) Dilute acid pretreatment of corn residues and short rotation woody crops. Appl. Biochem. Biotechnol. 28/29: 75-86.

Tsuda, M., Aoyama, M., Cho, N. and Cho, N.S. (1998) Catalyzed steaming as pre-treatment for the enzymatic hydrolysis of bamboo grass culms. Biores. Technol. 64: 241-243.

Vamvuka D., Kakaras, E., Kastanaki, E. and Grammelis, P. (2003) Pyrolysis characteristics and kinetics of biomass residuals mixtures with lignite. Fuel, 82: 1949-1960.

Vignon, M.R., Garcia-Jaldon, C. and Dupeyre, D. (1995) Steam explosion of woody hemp chènevotte. Biotechnol. Bioeng. 12: 22-42.

Wang P. (1999), State of art of fly ash utilization in China in the Proceedings of the International Workshop on Building Ceramics from Industrial Wastes, Organized by International Center for Science and High Technology [UNIDO], Trieste-Italy and Regional Research Laboratory (CSIR), Thiruvananthapuram (India), March 22-26, 1999.

Warth, H., Mulhaupt R. and Schatzle, J. (1997) Thermoplastic cellulose acetate and cellulose acetate compounds prepared by reactive processing, J. Appl. Polym. Sci., 64 (2): 231-242.

Wayman, M. (1986) p. 265 In Cellulose - structure, modification and hydrolysis, eds. Young, R. A.; Rowell, R. M., John Wiley, New York.

Weber, C.J. (Ed.) (2000) Report on Biobased Packaging Materials for the Food Industry-Status and Perspectives ISBN 87-90504-07-0, November.

Weil, J., Sarikaya, A., Rau, S.-L., Goetz, J., Ladish, C.M., Brewer, W., Hendrickson, R. and Ladish, M.R. (1997) Appl. Biochem. Biotechnol. 68 (1-2): 21-40.

Willet, J.L. and Shogsen, R.L. (2002) Procesing and properties of extruded starch-polymer foams. Polymer, 43: 5935-5947.

Wood, T.M. and Saddler, J.N. (1998) Increasing the availability of cellulose in biomass materials. Meth. Enzymol. 160: 3-11.

Wood, T.M. and Garcia-Campayo, V. (1990) Enzymology of cellulose degradation. Biodegradation 1: 147-161.

Wool, R.P. and Srikant, N.K. (2002) Biobased resins and Natural fibers in American Society of Materials, 21: 184-193.

Wypych, F. (2004) Chemical modification of layered compounds surfaces in Clay Surfaces – Fundamentals and applications, ed. F. Wypych and K.G. Satyanrayana), Pub. Elsevier, USA (To be published in 2004).

Xu Quing (1999) Researching Development of fly ash silicate building products in China in the Proceedings of the International Workshop on Building Ceramics from Industrial Wastes, Organized by International Centre for Science and High Technology [UNIDO], Trieste-Italy and Regional Research Laboratory (CSIR), Thiruvananthapuram (India), March 22-26, 1999.

Yaman, S. (2003) Pyrolysis of biomass to produce fuels and chemical feed stocks, Energy Conversion & Management, 45 (5): 651-671.

Zimbardi, F., Ricci, E. and Braccio, G. (2002) Technoeconomic Study on Steam Explosion Application in Biomass Processing. Appl. Biochem. Biotechnol. 98: 89-100.

32

Reduce, Reuse, Recycle and Recover: Biotechnological Approach in Management of Solid Waste of Fruit and Vegetable Processing Industries

D. B. Satpute, Pratima N. Shastry and D. M. Dharmadhikari

Introduction

Fruit and vegetable processing industries contribute a major share in food processing industries. At global level, India is the largest producer of fruits and the second largest producer of vegetables in the world next to China. Its share in the world production of fruits is 11% and that of vegetables is 7%. Because of its varied agro-climatic conditions; temperate, sub-tropical and tropical, India produces a wide variety of fruits and vegetables. The production of fruits in India rose from 28.63 million tones in 1991-92 to 46.97 million tones in 1996-97 and vegetables from 58.33 million tones in 1991-92 to 110.62 million tones in the year 1997-98 (Singhal, 1999). India being one of the largest producer of fruit and vegetable, the extent of value addition is negligible compared to other developed countries like Korea and Malaysia.

In India, number of fruit and vegetable processing industries has increased from 3925 in the year 1991 to 5462 at the end of year 2002. These industries include 43% as home scale, 17% cottage scale, 16% small scale and 9% large-scale units. Most of these industries are not working to their full capacities. Among all the processed fruit and vegetable items, mango has 40% share followed by 24% for tomato, 17% for citrus, 9% for peas, 6% for guava, 4% for pineapple. The estimated installed capacity of fruits and vegetable processing industries has increased from 17.5 lakh tones in 1994-95 to 2.11 million tones in 2000 (MFPI-2001). In near future, there is a vast scope for expansion of post harvest processing and value addition in fruit and vegetable sector.

The food industry produces large volumes of wastes, both solids and liquids, generated from the production, preparation, and consumption of food. These wastes pose increasing disposal and potentially severe pollution problems and represent a loss of valuable biomass and nutrients (Laufenberg *et al.*, 2003). Due consideration to proper utilization and disposal of solid waste is the need of the hour for sustainable industrial development. Waste management improves the production efficiency and environmental performance. As such the waste needs to be renamed as an alternative resource rather than a matter to be disposed off. Much emphasis has been focused to the concept of clean technology, defined as manufacturing processes or product technologies that reduce pollution or waste, energy use, or material use in comparison to the technologies that they replace. In the last few years, however, it has become necessity for food processing industry to take into consideration aspects of preventing pollution for economic motives, and the need to conserve energy and materials. New processes and policies for waste handling and treatment have been introduced in the recovery of valuable by-products/constituents, bioconversion, and recycling for biomethanation and compost. In United States, key environment issues for industries are related to the treatment of wastewater and solid waste, and serious thought is being given to solid waste disposal, resource conservation and recovery act. Although several processes are being successfully used for reduction and treatment of wastewater, the treatment of solid waste has not received due consideration so far.

All the waste generation takes place at two major stages i) Preprocessing stage and ii) Production of processed material. Preprocessing waste is generated during harvesting, handling, storage, transport, and marketing. The post-harvest losses and consequent generation of wastes amount to 55.85 lakh tones equivalent to loss of worth Rs. 3000 crore. It is attributed due to various factors, such as, lack of infrastructure for storage and transport, seasonal fluctuations and unorganized marketing. During the processing, solid waste is generated at various stages of sorting, cutting, pulping, processing, filling, seaming and sealing. All the waste is biodegradable, and hence amenable for biotechnological approach. The quanta of solid waste generated from fruits and vegetables processing industries in India is presented in Table 1. The generated waste is normally collected from the various stages of processing and stored inside the factory premises or deposited in nearby municipal collection bins. Many a times, factory authorities arrange for waste carrier vehicle for which a definite charge is paid for lifting and taking away the waste. Presently, reduce, reuse, recycle and resource recovery is not practiced in these industries, may be due to the reason that there is no financial returns from the recycle and resource recovery (Satpute *et al.*, 2002).

To make the food industries sustainable, environmental aspects should be taken into consideration. The processes need to be focused on the development of cost effective technologies, the optimization of processes related to environmental issues, alternative processes for the reduction of wastes, optimization of the use of resources and improvement in production efficiency.

Industrial waste management techniques can be classified into three options :

- Source reduction by processing plant modification
- Waste recovery, recycle or waste treatment for value added products
- Eco-friendly detoxification or neutralization of the undesirable components

The first two options represent the most promising waste management strategies. Indeed, waste recovery is highly attractive option. Significant environmental and economic benefits can accrue, by separating fruit and vegetable industrial wastes with the objective of recycling, reusing these valuable components. Promising concepts include pervaporation in hybrid processes (Hausmanns *et al.*, 1999) or the upgrading of fruit and vegetable residues to create a secondary use for the waste products (Laufenberg *et al.*, 1999).

Table 1 : Quantities* of solid waste generated from fruit and vegetable processing industries in India

Region	1998	1999	2000
Eastern	226350	-	
Western	50200	33991	47923
Northern	-	68478	57744
Southern	149492	176447	175579

- Data not available

* The values are in metric tones estimated from fruits and vegetable processing units

Source: A Report-Management of waste from the food processing industries (Ministry of Food Processing Industries, Govt. of India)

The current practices of pollution control and waste management cannot completely meet the increasing stringent requirements for the reduction of environmental contamination. Therefore the manufacturing industry has to include the optimization of product-integrated environmental protection into strategic planning, research and development.

Holistic concept of food production

The concept of '3R' comprising of reduce, reuse and recycle has the ultimate goal to optimize utilization of material while minimizing pollution and environmental problems. The detailed review on this subject carried by Laufenberg *et al.*, (2003) has further brought holistic concept of food production which includes product quality, production efficiency and environmental protection. Apparent failure of current practice of waste management to satisfy the environmental issues has compelled the food processing industry to include optimization of product, integrated environmental protection in strategic planning and think of green productivity and has put forth the concept of co-product rather than by-product. The goal set in the operational programs is short term, medium term and long term to be followed by conducting waste audit, which includes:

- Assessment of the quality at source
- Composition of waste for each source
- Option to reduce, reuse and recycle
- Technical and economical feasibility
- Evaluate the result for waste reduction.

This approach tries to connect different goals, such as highest product quality and safety, highest production efficiency and the integration of environmental aspects into product development and food production. The recycling of residues is important to every manufacturing branch and includes high developing potential. A systematic reduction in product losses and emissions is profitable under both economical and ecological aspects. Within this concept, a special attention is needed for recovery of valuable substances or product losses and internal process of recycling. Fig. 1 illustrates holistic approach of food production.

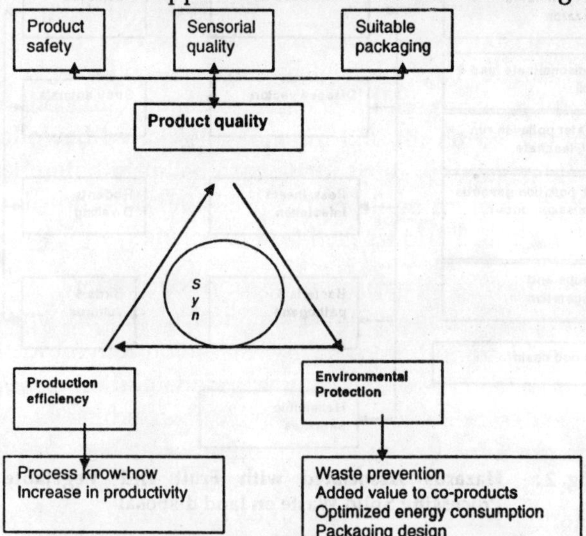

Source: Laufenberg *et al.*, (2003)

Fig. 1 : Holistic approach of food production

Kruszewska and Thorpe (1995) defined clean production in a similar way that the goal of clean production is to fulfill our need for products in a sustainable way i.e. using renewable, non-hazardous materials and energy efficient, while conserving biodiversity. Clean production systems are circular and use a few materials and less water and energy. It is new holistic and integrated approach to environmental issues centred on the product. This approach recognizes that most of our environmental problems related to global warming, toxic pollution, loss of biodiversity are caused by the way and rate at which we produce and consume resources. It also acknowledges the need for public participation in political and economic decision-making.

Hazards Associated with Solid Waste of Fruits and Vegetables

Problems related with solid waste management and minimizing the adverse environmental impacts are

- Seasonal availability of small quantity of solid waste of a fruit or vegetable
- Utilization of solid waste at the individual generating sources is not feasible due to small quantity of the waste
- Cost effective technologies are not practiced for value addition.

The most of the units are covered in small-scale sector without any organized method of solid waste disposal. Utilization practices for solid waste are negligible. The disposal of solid waste is associated with the problems of environmental degradation, poses health hazards to the residents as well as it has high impact over surrounding communities. Solid waste of fruits and vegetables is dumped on land mostly on the outskirts of the cities. The disposal of wastes on dumping sites shows that water, air and land component of the environment gets affected creating

- Environmental hazards
- Health hazards
- Aesthetics

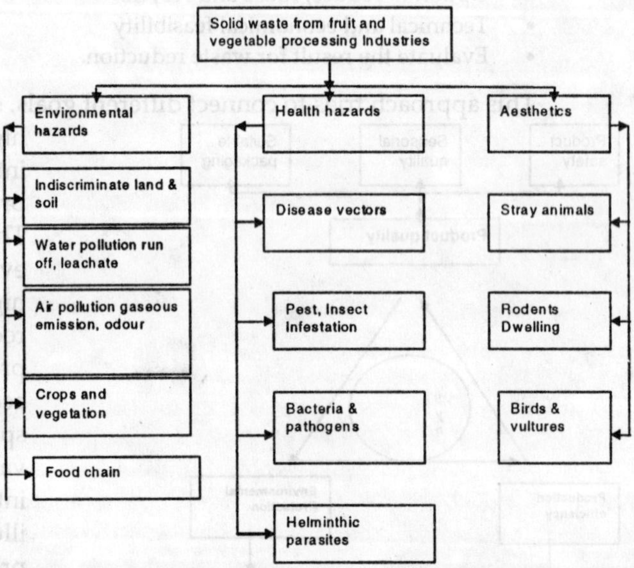

Fig. 2: Hazards Associated with Fruit and Vegetable Processing Solid Waste on land disposal

Solid waste has very high potential of producing undesirable effects through various biological and physico-chemical pathways. The postulated pathways for solid waste and disease relationship are indicated in Fig 2.

The leachate contaminate ground water and surface water bodies as percolation of water in the dumped waste which carries decomposition products, inorganic and organic chemicals, microbes etc. Disposal of solid waste on land contaminate surrounding air by emitting volatile organic, gases. And also it contaminates the land at interfacial layer. The problems associated with solid waste disposal are filthy odour, mosquitoes, flies, rodents, stray animals etc. Stray animals, rodents etc. play a role in dissemination of zoonoces. High level of bioaerosol is associated with compost operations at open disposal sites. The decomposition release spores of bacteria, fungi, actinomycetes. The study further revealed considerable concentration of suspended particulate matter (SPM), sulphur di-oxide, hydrogen sulphide and ammonia at the site (Sharma *et al.*, 2002).

Health studies in India indicate that solid waste disposal on land is at high risk of diseases like tuberculosis, bronchitis, asthma, pneumonia, typhoid fever, cholera, yellow fever, plague, dysentery, gastrointestinal parasites, worms and related organism. It is often difficult to trace the effects of such transmission to a specific population.

The impacts due to disposal of fruits and vegetables solid waste can be categorized as short term and long term. Short-term impacts are attributed to gaseous emission, leachate generation, odour nuisance, insect proliferation etc. These effects attribute to well-defined changes in clinical symptoms and lung functioning (Lawther *et al.*, 1977). Long-term impacts are related with leachate generation, gaseous decomposition products, and toxic inorganic chemicals. Assessment of the long-term effects need incorporation of mutagenicity, cytogenesis etc. considering the extent of exposure of toxic contaminants, which progresses through the food chain.

The solid waste has high density of microbial content. Donnelly and Scarprino(1984) observed that microbial population density in openly dumped solid waste of fruits and vegetables to that in undigested sewage sludge and hospital waste is quite comparable. The analysis of samples revealed the presence of coliform, salmonella species in waste samples, bottom soil and creek water.

Composition of fruit and vegetable solid waste

Most of the high amount of solid waste from fruit and vegetable is disposed untreated and has potentially severe pollution problems, however, fruit and vegetable waste represent a loss of valuable biomass and nutrients. The wastes contain cellulose, hemi-cellulose, lignin sugars, proteins, minerals, essential oils, pectin, pigments, enzymes and flavouring compounds, which offer tremendous technological potential to be used in isolation of byproducts/co-products, bioconversion processes and can be effectively exploited for bulk production of bio-chemicals.

Carbohydrates are one of the most important raw materials present in fruits and vegetables in the form of glucose, fructose, maltose, starch, cellulose, pectin etc. The nitrogen in the form of protein or nitrogenous compounds in fruit and vegetable wastes with an adequate level of fermentable carbohydrates and cellulose materials have an important role in solid substrate fermentation processes. The quality and quantity of nitrogen is important for the biotechnological processes. Favourable nitrogen sources are ammonium nitrate, ammonium chloride, oil cakes etc. The selection of initial concentration of nitrogen is very critical. However carbon to nitrogen(C:N) ratio plays an important role during fermentation processes. Besides carbon and nitrogen sources, magnesium, potassium, phosphate and sulphate are needed. Trace elements present in fruit and vegetable waste meet the requirement in fermentation processes.

The presence of cellulose, hemicellulose and lignin plays an essential role in solid state fermentation (SSF). The porous texture of cellulosic material has very good water holding capacity and oxygen retention capability, and are highly suitable substrate for SSF. The composition of some common wastes is summarized in Tables 2 and 3. The development of clean production processes can be achieved by;

- Waste reuse, reduction and recovery of valuable substances, byproducts and residues.
- Development and application of new and efficient production processes.
- Adding value to byproducts.
- Step-by-step implementation of environmentally benign manufacturing.

Table 2 : Chemical composition of solid waste from fruit & vegetable Processing

Type of waste	Organic matter	pH	Nitrogen (Total)	Phosphorus (as P_2O_5)	Potassium	Lignin	Pectin	Cellulose	Hemi Cellulose	Moisture
Mango waste										
Pulp	90.87	5.6	0.03	4.85	0.96	0.47	1.26	14.50	1.61	91.43
Peels	89.62	6.0	0.24	0.94	1.66	6.61	0.09	19.95	10.95	86.24
Kernel	88.67	6.0	0.30	1.95	1.47	11.27	0.62	12.30	3.59	84.63
Orange waste										
From pulper	91.11	4.5	0.16	0.77	3.97	8.55	1.80	47.26	1.45	84.51
Peels	90.13	4.5	0.41	1.11	1.47	1.69	1.47	52.36	1.23	89.61
Pineapple waste										
Pulper	94.77	5.0	0.14	0.94	6.26	4.12	1.65	18.21	1.95	87.36
Press	94.42	5.0	0.16	0.97	3.71	3.21	1.72	20.34	1.74	87.68
Peels	94.42	4.5	0.23	0.88	4.26	1.82	1.87	27.60	0.76	70.82
Tomato waste										
Peels	95.36	6.2	0.98	2.15	2.17	4.63	0.14	18.80	13.64	84.46
Seeds	34.49	6.0	1.26	1.11	2.26	1.41	6.16	20.94	17.28	81.75
Discard waste	93.32	6.0	1.47	1.26	1.35	4.1	0.4	39.08	25.9	85.55
Papaya waste										
Peels	90.30	5.0	0.56	0.71	0.89	4.15	13.06	35.72	14.98	89.48

Source: A Report-Management of waste from the food processing industries (Ministry of Food Processing Industries, Govt. of India 2003)

Table 3 : Composition of different fruit wastes (per 100 g)

Waste	Protein (g)	Fat (g)	Minerals (g)	Fibre (g)	Carbohydrate (g)
Apple pomace	2.99	1.71	1.65	16.16	17.35
Mango seed kernel	8.50	8.85	3.66	--	74.49
Jackfruit (Inner & outer portion)	7.50	11.82	6.50	30.77	14.16
Jack fruit seeds	6.60	0.40	1.20	1.50	25.80
Jackseed flour	2.64	0.28	0.71	1.02	18.12
Banana peel	0.83	0.78	2.11	1.72	5.00
Sweet orange seeds	15.80	36.90	4.00	14.00	--

Source: Anand and Maini (1997)

Options for fruit and vegetable waste management

Considering the chemical composition and the loss of valuable bio-chemicals with waste, an attempt to recover these products will be a strategic move for improving productivity and economy of the industry. The possible by-product recovery from fruit and vegetable is illustrated in Table 4.

Presently, solid waste management is focused on preventing pollution of the environment as well as for economic motives, i.e. the need to conserve energy and production of new materials. New innovative methods and policies for waste handling and treatment have been introduced in the recovery, bioconversion, and utilization of valuable constituents from food processing wastes. Besides its pollution and hazardous aspects, in many cases, food processing wastes have a potential for recycling raw materials or for conversion into useful products of higher value as a by-product, or even as raw material for other industries, or for the use as food or feed/fodder after biological treatment. The bioconversion of food processing residues is receiving increased attention, due to the fact that these residual matter represent a possible and utilizable resource for conversion to useful products (Martin,1998).

Table 4 : Possible by-product from fruit processing wastes

Fruit	% Waste	Nature of waste	By-products
Apple	20-30	Pomace	Juice, wine, vinegar, pectin cattle feed
Citrus			
Orange	50	Peels seeds and pulp	Essential oil, pectin, cattle feed, peel candy
Lime	60	Peels, seeds and pulp	Essential oil, pectin, cattle feed, peel candy
Grapes	20-30	Steams	Cream of tartar
		Pomace	Tartaric acid, cattle feed, jelly
		Seeds	Edible oil
		Skins	Food colourants
*Guava	40-45	Pomace, peel and seeds	Jam cheese, toffee, biscuits
Mango	40-60	Peel and pulp	Pectin, cattle feed, alcohol
Peel	12-15	Fibre	Wine, vinegar, juice
Pulper waste	5-10	Hull and kernel	Fat tannins, starch
Kernel	15-20		
**Pineapple	30-60	Mill juice	Bromellin, wine, sugar, syrup, citric acid, waxes, sterols, cattle feed
		Core	Conserve bulking agent
		Peels	Vinegar
		Stem	Bromellin, starch
		Crown	Cattle feed
		Leaves	Cattle feed, fibre, fabrics, paper

Fruit and vegetable processing industry's solid waste contains peels, pomace, seeds, roots, stem, leaves and rotten segment. For the fruit and vegetable processing industry, environmental issues could be sorted out by the usual approaches such as minimization, disposal, feeding, composting, closed loop production, or conversion. Several workers have reported the methodologies for the production of value added products utilizing industrial solid waste generated from fruit and vegetable industries. The options are listed in Fig.3.

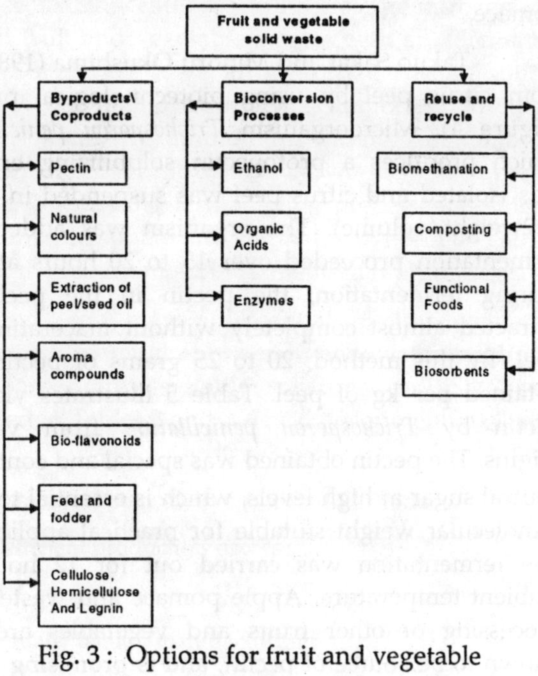

Fig. 3 : Options for fruit and vegetable processing solid waste management

Processes related to recovery of By-products/Co-products

Extraction of Pectin : Pectin is an important constituent of the cell walls, which help in the retention of the form and texture of fruits and vegetables. In plant tissues pectin exists in three different forms: water-soluble, oxalate soluble and acid soluble. Pectin is a purified carbohydrate product, which consists of partially methoxylated polygalacturonic acids and is capable of forming jelly with sugar and acids under suitable conditions (Chaliha, 1963). It acts as colloidal stabilizer and plays an important role in preventing flocculation and thus it is mainly used in pharmaceutical processes and as several other applications in various industrial processes. Majority of pectin produced in the world is extracted from dried citrus peels and apple pomace. A number of indigenous varieties of citrus, other than established sources have been examined which includes peel obtained from galgal (Himalayan lemon), varieties of mandarin limes, Assam lemons and oranges, sweet orange and grapefruit (Thamke,1997). The non-citrus fruit investigated for pectin source includes solid waste of guava, purple passion fruit, papaya, mango peel, jack fruit peel/core and other sources include sunflower heads could be exploited as viable technology for manufacture of pectin (Bhatia, 1959).

The presence of pectin up to 30% in dried residual matter like sugar beet pulp, carrot pomace, potato pulp or lemon peel and its availability in large quantities have made extraction worthwhile. Several other sources of pectin are reported; citrus peel (Widmmer and Montanari 1995), apple and sugar beet pulp (Grohmann and Bothast ,1994), cocoa husk (Nambudiri and Shivashankar ,1985), galgal (citrus fruit) peel (Attri and Maini,1996), hawthorn (Maxican fruit) peel (Higareda *et al.*,1995), mango peel (Srirangarajan and Shrikhade, 1976) or sunflower heads (Wang *et al.*, 1997).

However, no pectin has ever been extracted from fruit and vegetable residues with gel forming properties comparable to those of pectin extracted from apple pomace. Turqouis *et al.*, (1999) recently developed an alkaline extraction process for pectin from sugar-beet pulp and potato pulp with a gel forming properties comparable to those of pectin extracted from apple pomace.

Takuo Sakat and Minoru Okushima (1980) developed a process for production of pectin from citrus peel by using biotechnological process (Figure 4). Microorganism *Trichosporon penicillatum* which produces a protopectin solubilizing enzyme was isolated and citrus peel was suspended in water (1:2weight/volume). The organism was added and fermentation proceeded over 15 to 20 hours at 30°c. During fermentation, the pectin in the peel was extracted almost completely without macerating the peel. By this method, 20 to 25 grams of pectin was obtained per kg of peel. Table 5 illustrates yield of pectin by *Trichosporon penicillatum* from various origins. The pectin obtained was special and contained neutral sugar at high levels, which is essential to have a molecular weight suitable for practical application. The fermentation was carried out for 12 hours at ambient temperature. Apple pomace and waste from processing of other fruits and vegetables are also known to be source of pectin, and is promising in the

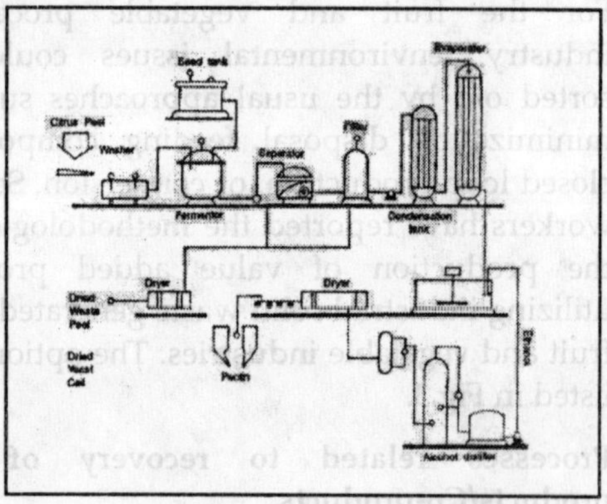

Fig. 4 Schematic diagram of manufacturing process of pectin from citrus peel
Ref: Takuo and Minoru (1980)

future as it will be an economical process. Table 6 presents comparison of some properties of pectin extracts.

Natural colours : Due to carcinogenic effects, natural colours or plants pigments are fast replacing coal-tar dyes in the processed food products. Natural colour of blue grape skin, Kokum (*Garcinia indica*), Phalsa (*Grewia subin acqualis*) and Jamun (*Syzygium cumini*) have been thoroughly investigated for its nature, concentration, extractability, stability and suitability as food colours. Technology of making natural colour concentrates and powders from these waste have been developed. A new fat-soluble pigment, garcinol has been isolated from ring of kokum (Krishnamurthy, 1981). Thus the recovery of natural colours from solid waste of these industries will be highly economical and eco-friendly process.

Table 5 : Yields of pectin extracted by *Trichosporon penicillatum* from various origins

Origin	Water soluble	Microbial soluble	Total
Savel Orange			
Peel	0.4	2.5	2.9
Segment cover	0	2.7	2.7
Grapefruit			
Peel	0.2	2.5	2.7
Segment cover	0	2.1	2.1
Mandarin			
Peel	0.3	2.6	2.9
Segment cover	0	1.8	1.8
Lemon			
Peel	0.9	1.7	2.6
Segment cover	0	1.6	1.6
Apple	0.1	0.1	0.2
Radish	0	0.2	0.2
Carrot	0	0.1	0.1

Source: Takuo and Minoru (1980)

Table 6 : A comparison of some properties of pectin extracts

Sr. No	Pectin Type	Ash %	Relative viscosity %	Methoxy l group %	Esterified Carboxyl group %	Galacturo nic Acid %	Neutral Sugar %	Mol. Wt. (x10³)
1.	Commercial product (from lemon)	2.6	1.53	9.24	63.1	85.0	5.7	102 (>200)
2.	From Mandarin peel extracted by fermentation	ND	1.46	8.58	73.8	68.2	23.2	105 (>200)
3.	Extracted by acid	3.5	1.23	9.13	66.1	80.3	10.5	50 (>200)

Source: Takuo and Minoru (1980)

Bio-colourants made of anthocyanins from olive cake (Clemente *et al.*, 1997), lycopenes from tomato skins (Al-Wanndawi *et al.*, 1985) or β-carotene from carrot pomace (Henn and Kunz, 1996) possess potential for pharmacological applications, as powerful antioxidants. Besides the blue-red colour the food will be enriched with healthy polyphenols (Anon, 1999).

Extraction of peel/seed oil : The aromatic essential oils are located inside the oil sacs in the flavedo area of the peel. The present method of oil recovery involves, distillation, solvent extraction or cold pressing neither highly efficient nor much economical for efficient oil recovery. The biotechnological approach by Sharma *et al.*, (1980) to treat orange peel with fungal xylanase enzyme prepared from *Monilea spp.* increased the major flavour component limonene several folds as compared to the above techniques of extraction.

Citrus wastes are rich resources of essential oils. Mandarin essential oil is extracted in small-scale units based on orange peels collected from fresh juice. Physico- chemical characteristics of large number of Indian mandarins and lime oils, both cold pressed and steam distilled, varies slightly. D-Limonine is a major by-product from citrus industry (Anand, 1997). Citrus peel oil extracted by the cold process fetches a better price than distilled oil, which is of inferior quality. Sud and Badyal (1993) extracted peel oil from various citrus cultivars, and the results are presented in Table 7.

Bhullar *et al.*, (2000) reported that tomato seed oil has physical and chemical characteristics comparable to any other edible oil and has potential to replace the conventional oils in food products. Wilds apricot seeds are extensively used for the extraction of both sweet and bitter oils (Bhatia, 1976).

In the process of canning, drying, and freezing processing of fruits such as peaches, apricots, and cherries, a fair percentage of the waste comes from their pits. The same has been utilized for the production of marketable oils. Apricots, cherries, prunes, peaches, and almonds are botanically closely related, and the oils obtained from these fruits are almost identical. The first step in product recovery from the kernel is to separate it from the shell. This is accomplished easily for apricots, almonds, cherries and prune pits by crushing them between heavy rollers adjusted so that the pits are broken, but the kernels are not crushed. The broken pits and kernels drop into a tank of brine which has the concentration adjusted, so that kernels and pits can be separated. The kernels are skimmed off and washed to remove excess salt, dried, and graded. Cherry pits are about 28% of the kernel, apricot about 24%, peach about 7%, and prune 15%. The oil is generally removed from the kernel by using high pressure technique however it is better to use solvent extraction. For pressure extraction, the kernels are coarsely ground, heated by steam injection, and pressed using a continuous or rack and frame press. Yields of about 33% oil should be obtained from apricot kernels, 25% from peach, and 30% form cherry and prune. The pressed oil usually contains some solid materials, such as particles of kernels, kernel skins, etc., which can be removed by screening or coarse filtration The raw oil seems to be usually high in free fatty acids and is dark coloured with a rancid odour. The oil can be refined by treating with sodium carbonate to neutralize the free fatty acids, adding a decolourizing agent such a activated carbon and steam heating in a vacuum to volatilize objectionable odour. These refined oils can be used in a multiple ways.

Table 7 : Recovery of peel oil from various citrus cultivars

Source of peel	Yield (ml/100g of peel)			
	Dried	Unruptured	Ruptured with needle	Cold press
Orange	0.45-0.50	1.20-1.25	1.66-1.89	1.68-2.10
Kinnow	0.35-1.12	1.00-0.41	1.30-1.42	1.42-1.60
Mandarin	0.42-0.48	0.78-0.90	1.42-1.58	1.50-1.80
Galgal	0.30-0.38	0.52-0.68	0.92-1.00	1.00-1.20
Kagzi lime	0.55-0.60	0.68-0.89	1.33-1.58	1.65-2.00

Source: Sud and Badyal (1993)

Aroma compounds : Natural flavours are chemical substance with aroma properties are produced from feedstock of plant or animal origin by means of physical, enzymatic or microbiological processing.

Flavour and aroma (including fragrances) compounds synthesized by biotechnological processes now a days is playing prime role in food, feed, cosmetics, chemical and pharmaceutical industries. This is mainly due to an increasing preference by the consumers for natural food and additives. Attempts are being made to produce such compounds by fermentation of simple nutrients such as sugars and amino acids. Specific aromas and flavours suitable for several food product like bread, fruity, spicy, pyrazines, terpenes, lactones and esters can be produced at industrial scale using microorganisms. Aroma chemicals and flavours are few high value products, which can be produced from fruit and vegetable waste like peel, pomace using biotechnology as a tool and are highly economical processes. Table 8 presents aroma compounds in various fruits.

The world market of aroma chemicals, fragrances and flavours has a growth rate of 4-5 % per year. Because of a higher consumer acceptance there is an increasing economic interest in natural flavours. Apple peel and pomace, the waste left after the juice operation, also can be used to produce natural flavourings. These compounds can be obtained by extraction with carbon-di-oxide, which is fractioned at two different temperatures obtaining a flavourless wax/water emulsion fraction and an intensely flavourful fraction. Flavour compounds extracted by this procedure have a broader flavour spectrum than did those prepared by distillation (Bundschich *et al.*, 1986).

Table 8 : Aroma compounds in various fruits

Flavour	Compound
Peach, fruity	γ - Deca lactone
Coconut, creamy	δ - deca lactone
Almond	Benzaldehyde
Fruity pineapple	Ethyl butyrate
Banana, fruity	Isomyl butyrate
Phenethyl alcohol	Roses
Vanilla	Vanillin

Microbial synthesis of natural flavours is generally carried out by submerge fermentation. Due to the high cost of currently used technology at an industrial scale there is a need of developing low cost processes. This could be achieved by exploitation of the metabolic pathways and by alternative technology such as solid state fermentation (Feron *et al.*, 1996). The production of bio-flavours is new technology offering opportunities of waste utilization.

Brarmorski *et al.*, (1998) compared fruity aroma production by *Ceratocystis fimbriata* in solid cultures from several agro-industrial wastes. According to them, apple pomace in combination with soybean have cassava produced a strong fruity aroma.

Bio-flavonoids : The single most important group of phenolics is the bioflavonoids. These compounds are highly diverse, in both their chemical structure and proposed biological functions (Macheix, 1990). Bioflavonoids are related to their diversity, biological significance as secondary plant metabolites and ecological role (Harborne, 1994), use as chemotasonomic markers (Kamiya, 1979), impact on fruit quality, physiological effects (Herrmann, 1970), and industrial applications (Ortuno, 1995) Some flavonoids may have an indirect effect on plant growth while other may protect the more vulnerable cell constituents from damaging radiation by virtue of their strong UV absorption. Bioflavonoids are potent antioxidants, free radical scavengers and metal chelators. They inhibit lipid peroxidation and exhibit various physiological activities, including anti-inflammatory, anti-allergic, anti-carcinogenic, anti-hypersensitive and anti-arthritic activities. The biological properties of quercetin, the most frequently studied bioflavonoid, are certainly consistent with these activities.

Citrus waste streams are used in horticulture (Widmer and Montanari, 1995). Mandarin peel flavonoids are interesting due to their fungistatic activity (Chkhikvishvili and Gogiya, 1995), which may be applied to naturally protect vegetables and fruits from moulding. The limonoid compound in citrus peel and seeds have recently been found to have important pharmacological properties as well as potential in the use as an insect antifeedant for agriculture crops (Manthey and Grohmann, 1996).

Bioflavonoids like hesperidin, naringin or rutin are able to normalize capillary permeability and vascular brittleness. Hesperidin is applied in vein dedication, acts antiviral in the flue therapy and owns artificial sweetener properties; hydrated naringin from citrus is about 300 times sweeter than saccharose, neohesperidin almost 2000 times. The major flavonoids found in selected fruits have been highlighted in Table 9.

Table 9 : Major flavonoids in selected fruits

Fruit	Flavonols	Anthocyanins	Flavan-3-ols*/flavanones	Chalcones	Others
Apple	Quercetin, quercetin glycosides including rutin, kaempferol	Cyanidin glycosides including acylated derivatives	(+)Catechin, (-)epicatechin	Phloretin derivatives, notably phloridzin	
Citrus –Sweet orange		Glycosides of pelargonidin, peonidin, delphinidin, petunidin	Hesperidin, narirutin, eriocitrin, narirutin-4'-glucoside		Flavones:sinensetin, nobiletin, tangeretin, isosinesitin
Grapefruit	Quercetin, kempferol, glycosides of quercetin kaempferol myricetin, isorhamnetin including rutin		Naringin, narirutin, hesperidin, neohesperidin,		Flavones: tangeretin, polymethoxylated flavones
Lemon	Rutin, limocitrol, limocitrin, isolimocitrol		Hesperidin, eriocitrin		Flavones: diosmin, luteloin-7-rutinoside
Grape	Quercetin, Kaempferol Glycosides Quercetin, Kaempferolmyricotin Isorhamnetin Rutin	Glycosides of cynanidin,peonidin,delphinidin,petunidin,malvidin	(+)Catechin (-)epicatechin (+) gallocatechin, (-)epigallocatechin		Flavanonols:glycosides of dihydroquercetin and dihydrokaempferol
Pear	Quercetin, glycosides of quercertin ,isoquercetin kaempferol and isorhamnetin	Cyanidin glycosides	(+)Catechin (-)epicatechin	Arbutin, phloretin glucoside	
Stone fruit					
Peach	Myricetin, quercetin, kaempferol,quercetin and kaempferol glycosides including rutin	Cyanidin glycosides	(+)Catechin, (-)epicatechin		
Plum	Glycosides of kaempferol and quercetin	Glycosides of Cyanidin,peonidin	(+)Catechin (-)epicatechin		
Apricot	quercetin ,quercetin3-O-glucoside , rutin ,kaempferol glycoside				
Sweet Cherry	Glycosides of kaempferol and quercetin	Cyanidin glycosides	(+)Catechin (-)epicatechin		
Tomato	Kaempferol and quercetin glycosides		Naringenin, Naringenin glycosides	Chalconaringenin	

*Associated mainly skin and seeds

Ref: Spanos *et al.*, 1992 and Fernandez *et al.*, 1992

Feed and fodder : Regarding processing of mangoes discard, it has been reported that mango seed kernel is good source of carbohydrate and can replace around 20% of corn in growing and laying rations of poultry feed. (Anand, 1997).

Orange waste in the form of peels, rags and seeds are dried and are being used as a livestock feed. It is low in crude proteins, fibre and fat but high in carbohydrates. The liquid concentrate made from crude citrus waste is called 'molasses'. Its use is similar to cane molasses and is widely used for feeding dairy and beef cattle.

Pineapple waste obtained from peels is rich in sugar and fibre and poor in protein content and hence needs supplementation with high protein feeds. Silage can also be prepared from waste. Chopped pineapple tops when fed to cows results in better milk yield (Anand, 1997).

An extrusion process for transformation of tomato pomace into a value-added product has been described and product composition has been evaluated by Tadeu Pontes *et al.,* (1996). Tomato pomace containing skin, seed, some fibre and adhering pulp of tomato after processing has been found to serve as a good protein feed source for poultry and livestock (Anon, 2000). Product obtained by utilizing this pomace contains about 20-25% crude protein and 13% more lysine compared to soybean–meal protein. Besides, it is a very good source of vitamin B, vitamin A and xanthophyl pigment. This pigment has well-established role in improving colour and quality of egg yolk.

The practice of feeding unprocessed wastes to animals is a potential health hazard as excreta may contain agents harmful to human and animal health. Solid-state fermentation is the most appropriate technology for the production of protein-enriched feed. Wastes from citrus, banana, potato, apple pomace obtained from processing have been converted into feed supplements of high nutritional value. The use of fruit and vegetable wastes has been thoroughly investigated by several workers (Pande, 2000). Cultivation of GRAS microbes (generally regarded as safe), such as yeast and basidiomycetes, namely *Pllyporus sp, Pleurotus sp., Trichoderma sp.,* etc. on ligno-cellulosics should be considered for animal feed and fodder production. Microorganisms like *kloceckera apiculata* and *candida utilisis* can transform apple pomace in to an improved feedstock by solid state fermentation (Rahmat *et al.*, 1995). An animal feed byproduct produced from apple pomace by solid state fermentation and was further evaluated by Joshi and Sandhu (1996). Joshi *et al.*, (2000) have recommended to look into the bio-safety aspects of the process and optimse/analyze the conditions adequately of the process.

Onodera *et al.,* (1996) have reported a silage preparation technique from residual juice of mandarin orange and carrots in production of shochu, a fermented product. Mahadevswamy and Venkataraman(1990) have developed an integrated approach for utilization of mango processing waste . Leftover residue of biogas production from mango processing waste can be utilized as food for fresh water fishes like a major carp Rohu (*Labeo Rohita*) and a common carp (*Cyprinus carpio*). Cull grapefruit has been examined as potential feedstock. Gandhi *et al.*, (1997) have carried out a study on nutritional and toxicological evaluation of wild apricot pomace. Studies on the utilization of orange waste as a substrate for protein production by the fungus *Sporotrichum pulverulentum* were performed and the biomass containing about 32% protein was obtained (Karapinar and Okuyan, 1982).

Cellulose, hemicellulose and lignin : Cellulose is a linear polymer of anhydroglucose units linked at C1- and C4- atoms by a β-glucoside bond. In addition to cellulose other major cell wall constituents are lignin and hemicellulose. Lignin is a complex of 3-D polymer of phenolic and enolic origin where as hemicellulose consists of cellulose and polyuronide hemicellulose (Norman, 1954; Whistler and Richards, 1970). Goring (1971) and Higuchi (1980) have elucidated structure and size of lignin. Chemically hemicellulose has hardly any resemblance with cellulose but due to its close association with cellulose it is named so. All properties of cellulose are determined by its chemical structure but the forces between the cellulose polymer chains produce super molecule texture that profoundly influences most properties of material (Mark, 1954). Cellulose itself is among the least degradable natural polymers due to its high molecular weight, high degree of structural order, insolubility, low surface area and its association with lignin and hemicellulose. Several research findings in this area firmly established that the potentiality of cellulose waste for production of foods and feeds or their principle component is fairly good. Principle constraints of commercial exploitation of cellulose wastes for production of food and feeds are lack of rapid breakdown of cellulose by enzymes, rapid growth and proliferation of microorganism on celluloses, exorbitantly high cost of pretreatment to improve degradability or digestibility and close association of celluloses with some inhibitory substances (Ghose *et al.*, 1975, Mandels *et al.*, 1975, Brown *et al.*, 1975).

A large number of fungi have been screened for their ability to grow on cellulose (Chahal and Gray, 1968; Chang, 1983). Several lignocellulosic and other foods and agricultural wastes have been utilized for production of edible mushrooms. The growth of biomass of edible Basidiomycetes has been supported by cellulosic wastes obtained form citrus processing (Elisashvili *et al.*, 1992) and tea production unit (Kokhzeidre *et al.*, 1993). A good production of edible mushrooms such as *Agaricus bisporus* and *Morchelia crasipes* on citrus waste has been observed by Labeneish (1960). Nancib *et al.*, (1997) has reported a successful utilization of date wastes as substrates to the production of yeast (*Saccharomyces cerevisiae* biomass with a maximum yield of 0.62 g of dry weight/lit. Sturion and Oetterer(1995) have established that banana leaf waste from banana plantation is a more efficient substrate than wheat straw for cultivation of edible fungi, such as *Pleurotus sp. Subsp. florida, Pleurotus obstreatoreseus* and *Pleurotus sajor-caju.*

Bioconversion processes

Production of Ethanol : Substantial progress has been made over the past ten years with respect to the conversion economics for yeast-based ethanol production resulting in three fold reduction of projected selling price (Blaschek, 1992). Biotechnological processes and various solid state fermentation processes for ethanol production using sugar rich substrates have been studied (Pande, 2000). Solid waste of fruits and vegetables contain substantial amount of carbohydrates. Ethanol as fuel to replace gasoline is being synthesized on commercial level by bioconversion of the waste (Pande, 2000).

Fermentation of apple pomace by solid state fermentation at 30° c for 72 hours with or without yeast addition showed that the natural microflora could induce fermentation, and it was accelerated by the yeast addition (Shankaranand and Lonsane, 1993). In solid state fermentation of apple pomace for ethanol production addition of nitrogen, phosporus and trace elements increased the fermentation efficiency (Gupta *et al.*, 1989). Out of various nitrogen sources tried, ammonium sulphate gave the highest stimulation with *Saccharomyces cerevisiae* giving the higher ethanol productivity in solid state fermentation of apple pomace (Joshi *et al.,,* 1994)

Orange waste coming from food industries when was subjected to continuous fermentation in a fixed bed immobilized cell reactor showed satisfactory ethanol productivity values (Gera, *et al.*, 1969). The initial saccharification of polysaccharids by commercial cellulase and polygalactouronase followed by removal of inhibitory compounds by filtration and pH adjustment of the hydrolase was necessary for a successful fermentation (Convert, *et al.,,* 1989).

Citrus molasses prepared by evaporation, concentrate of the press liquor and molasses mixed with the citrus pulp have also been used by distilleries as an alcohol feed stock (Gera *et al.,,*1969). In the process moisture content of the solid has been shown to be a limiting factor for maximum ethanol production (Nadgi *et.al.*, 1992).

Ethanol yield from normal banana was found to be as ripe whole fruits 0.091, pulp 0.082 and pea 0.0061 /kg of ripe whole fruits .the green fruits gave 0.090, normal ripe 0.082 and overripe 0.069 /kg of ethanol. Enzymatic hydrolysis was necessary for higher ethanol yield while dilution with water was not essential for effective fermentation.

The current technology for the conversion of cellulose to ethanol has the potential for significant improvements in the areas of pretreatment of the cellulose, improving enzyme production, increasing the efficiency of fermentation, and enhancing product recovery. Genetically engineered microorganisms that produce their own cellulase have also been suggested (Surbriggen *et al.*, 1990).

Organic Acids : De lima *et al.*, (1995) used peels and pomace of pineapple for citric acid by solid state fermentation using *Aspergillus niger*. The highest production of citric acid to the tune of 290 gm/kg of the solid waste was obtained after 6 days fermentation when the medium was supplemented with 4 % of methanol. Apple pomace as a substrate has also been employed in the production of citric acid and 250 gm of citric acid per kg of waste was obtained. Addition of methanol in fermentation of apple pomace increased the citric acid production (Hang, *et al.,,* 1988). A rotting disc immobilized fermenter has been developed for waste wherein the mycelium could be grown on the surface with concomitant production of citric acid and during waste treatment. Synthesis of citric acid has been reported by Wang (1998) by using agro-industrial residues such as sugar beet molasses.

The amount of tartarates present, in grapes varies form 0.8-1.2 percent (Giridhar, 2000). The pomace is extracted with water and the water filtered to remove any solid materials. Precise amounts of calcium hydroxide and calcium chloride are added, the amount depending upon the pH of the solution and the amounts of tartarates expected. After stirring and settling, the supernant is discarded and the calcium tartarate sludge washed and dried. The calcium tartarate can be further purified by re-dissolving in water and re-crystalline with calcium hydroxide (Marsh, 1944).

Production of Enzymes

Bromelin : Papain and bromelin are proteolytic enzymes. Crude papain is collected from matured papaya before harvesting. Bromelin is isolated from peel and waste of pineapple. The recent process developed for bromelin production consist of extraction of the enzyme from the fresh pineapple or pineapple waste containing 10-30% fruit pomace with Dibasic Sodium or Potassium Phosphate Buffer (pH 5.5-7.5) followed by precipitation by ethyl alcohol by ethyl alcohol or acetone or n-butanol at a temperature of 5°C to 15°C and drying at reduced pressure (Ksheersagar, 2002). Fig 5 represents flow sheet for production of bromolin.

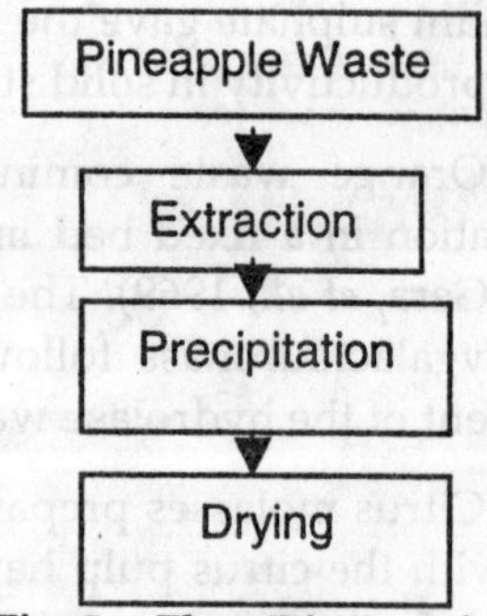

Fig. 5 : Flow Diagram for Bromelin Production

The proteolytic activity of bromelin from mature and green pineapple and from edible portion (slices, pineapple juices and crush) and corresponding waste (skin, top and bottom) have been studied. Bromelin was isolated by cold acetone precipitation (-20°C) from samples of pineapple.

A simplified method has also been developed for extracting bromelin. The method involves enzymatic clarification of juice, precipitated by ammonium sulfate at final concentration of 70% for one hour at room temperature without pH control and recovery by precipitation and centrifugation.

Other Enzymes

Amylase production from banana waste by solid state fermentation with *Bacillus subtilis* has been optimized. The organism produced very high titers of X- amylase. The enzyme level was 2.55 times higher in banana fruit stalk than that obtained with medium having wheat bran (Krishna *et al.*,, 1996, Pandey, 1991).

Cellulase has been produced from cabbage waste. Production of cellulase enzymes is important in hydrolysis of cellulose into fermentable sugar for production of ethanol as an alternative energy source . Cellulase are produced by microorganism *Pseudomonis bisillus* and *Cellulamonas*. Among *the molsa trichoderma Aspergiuulus* and *Penicillium* are known to produce high titers of this enzyme (Nirmal *et al.*,,1994).

A novel stream of *Vibrio costicola* (CB TK 108) isolated from the banana stalk produced under solid state fermentation 146 Ku/g amylase and 1.92 ku/g cellulase. Enzyme production from citrus waste employing A*spergillus foetidus* in solid state fermentation has shown that up to 16,000-17,000 μg/l isometric readings could be obtained after supplementing the waste with yeast extract and mineral salts. Higher yield of pectinase was achieved with apple pomace as a substrate. Production of pectinase enzyme from apple pomace and the factors affecting solid state fermentation have also been reported (Garzan, et.al,1992).

There exist an ample opportunity for the production of number of enzymes from pretreated solid waste as substrates. These include pectinase, amylases, alpha and beta galactosidases. Solid state fermentation of waste can be an alternative methodology for synthesis of these products and will be an economical solution (Leda, 2000)

Recycle

Biomethanation : Anaerobic digestion has been considered as a simple and economically viable technology for energy production. Along with energy in the form of methane it also yields spent solids and effluents. Spent solids can be utilized for soil enrichment and effluents after solid recovery would be a good feed for aqua culture. Besides, it reduces the level of pollution, and conserves the wastes. Utilization of fruit and vegetable wastes for methane generation by anaerobic digestion is somewhat different from the conventional method of treating sewage or manure or cattle dung. These wastes are rich in cellulose, hemi-cellulose, lignin and pectin and these are difficult to hydrolyze as compared to partially hydrolyzed cattle dung, human excreta, sludge etc. Mango peel rich in carbohydrates has been found to be deficient in proteins and therefore, organic and inorganic nitrogen is to be supplemented in order to correct the carbon and nitrogen ratio of 30-20:1. Similarly citrus contain citrus oil, which inhibits the growth of methanogenic and non- methanogenic microorganism. Tomato peel and seeds are rich in protein (23.3%), whereas lemon (15.65%)and deoiled orange peel (32.6) contain very high percentage of pectin on dry weight basis. It has been reported that the improved rate and yield of biogas was achieved in the digester when the waste was microbiologically pretreated with selected strains of *Sporotrichum, Aspergillus, Penicillium* and *Fusarium* (Krishna Nand 1995). Table 10 represents biogas generation from fruits and vegetables waste.

Table 10 : Biogas generation potential from fruit and vegetable processing wastes

Parameters	Pineapple	Mango	Orange deoiled peel	Lemon	Tomato
Hydraulic retention time (days)	30	30	32	30	32
Loading rate (Kg/day)	4.0	4.0	4.0	4.0	4.5
Volatile solids (%)	93.79	96.4	94.2	92.27	95.73
C/N ratio	101.9	185.0	23.0	19.0	22.0
PH	7.0	7.0	7.0	7.0	7.0
Biogas yield (m^3/Kg Vs added)	0.4	0.207	0.633	0.720	0.512
Methane content (%)	59.7	53.0	62.0	53.0	70.0

Source : Krishna Nand 1995

Composting : Composting is decomposition and stabilization of organic matter under controlled conditions. Bacteria, fungi etc. feed upon the organic matter and in the process stabilize the waste. Compost can be used as a soil conditioner. Composting can be done both aerobically and anaerobically.

i) Aerobic composting: During aerobic composting, aerobic microorganisms oxidize organic matter into carbon dioxide and oxides of nitrogen. Carbon is the source of energy to microorganisms while nitrogen is recycled.

ii) Anaerobic composting: Anaerobic microorganisms while metabolizing nutrients break down the organic compounds by the process of reduction. The gases evolved are mainly methane and carbon dioxide.

Depending on the physical state and composition, organic waste is used for composting. For example solid waste containing excessive large fraction such as pineapple solid waste, it should be reduced in size. Wastes which are heterogeneous mixture of inert (recalcitrant) and non-biodegradable, minimization of inert material is essential. Normally, solid wastes of bioorganic origins contain nutrients essential for microorganisms. Limiting factors for composting are carbon to nitrogen ratio and oxygen availability. Optimum carbon to nitrogen ratio is 35-45% for composting. Waste such as lignocellulosics and solid waste from fruit and vegetables processing operations are deficient in nitrogen content. Under such conditions, supplementing nitrogen from outside sources is essential.

Solid waste except agricultural residues and waste from processing commodities normally contain high proportion (80-90%) of moisture. Such waste while composting needs organic amendment to distribute the moisture thereby controlling moisture for combined waste in optimum (30-45%) range. In composting of fruit and vegetable solid waste, it is preferable to add compost in desirable proportion.

Windrows composting : Windrows of height 1.8 to 2.5 m depending on the provision of air supply are prepared preferably on the cemented platform. Windrows can be turned to provide required oxygen to the decomposing waste mass. Initially, frequency of turning is more i.e. 2-3 turnings for initial 15 days and later on weekly turning is sufficient. Monitoring the process is essential to control the moisture level for optimum degradation. Temperature is prime parameter for monitoring the performance of the composting process. Increase in temperature above 60° c should be controlled. When the temperature declines and approaches to ambient temperature, composting is normally considered as stabilized. Advantages of composting include pathogen destruction due to thermal inactivation and antagonistic effects. Minimization of nitrogen loss as ammonia (volatilization of ammonia) can be controlled by adding surface soil for compost covering. Finally at the stabilization level, the compost needs screening through 1 cm mesh opening and residue is recycled back to fresh material. Fig 6 illustrates schematic flow-sheet for windrows composting.

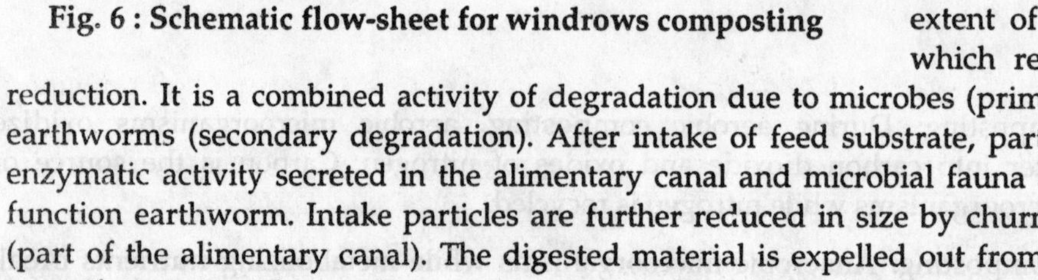

Fig. 6 : Schematic flow-sheet for windrows composting

Vermi-composting : Compost prepared by adopting vermi-compost technology produces the end product where extent of stabilization is more which results in particle size reduction. It is a combined activity of degradation due to microbes (primary degradation) and earthworms (secondary degradation). After intake of feed substrate, particles are subjected to enzymatic activity secreted in the alimentary canal and microbial fauna act as indigenous gut function earthworm. Intake particles are further reduced in size by churning action in gizzard (part of the alimentary canal). The digested material is expelled out from the anus, commonly termed as casting. Conversion to casting is actual process of a vermicompost.

Vermi-composting is influenced by physico-chemical characteristics of the waste, viz. temperature, aeration, toxicity etc. Earthworms prefer stable waste for conversion. Pre-stabilization of the waste minimizes toxic effects of degradation products. Temperature exceeding 30ºç is lethal to earthworms. To reduce the lethal effects of temperature increase over 30ºc is avoided by initial phase of pre-stabilization of the waste by sprinkling water.

Indigenous earthworm species have limitations for use in vermiconversion as it is active only in the rainy season. In the post-rainy season period, microclimatic changes influence the earthworm activity and earthworms normally undergo "hibernation" or "aestivation". Exotic earthworm species i.e. *Eisenia* and *Eudrillus* are suitable for vermiconversion. The earthworms are very sensitive to toxic effect of chemicals and succumb to parasitic activity of *Eimeria sp.* (protozoa). Therefore, stock of vermiculture is essential for carrying out vermicomposting. Cow dung is preferred media for culturing earthworms.

To start with the vermiconversion, heaps of waste not exceeding 10 m x10 m size using pre-established waste are used. Surface of the heaps should better be covered with wet gunny bags to minimize the moisture loss. Earthworms are added to the composting heaps at the rate of 50 - 100 earthworms per square meter. As the worms avoid direct exposure to sun, it is preferable to install the heaps under the shade. Castings are normally deposited at the surface of the heaps and hence it needs frequent separation using screening (4-5 mm mesh opening). The screening separates earthworm from the castings and the separated earthworms are transferred back to stock culture. A schematic flow-sheet of vermiconversion is presented in Fig 7.

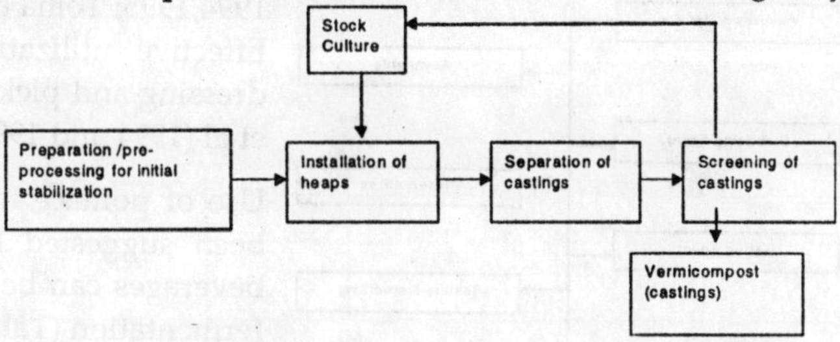

Fig. 7 : Schematic Flowsheet for Vermicomposting

Functional foods

Upgrading of fruit and vegetable residues for the production of novel types of multifunctional food ingredients in fruit juices, bakery goods and health foods is promising application in food industry. Currently there are great variety of raw materials from which dietary fibres are obtained such as wheat or rice, etc. Useful substitutes could be obtained from fruit and vegetables residues like citrus peels, mango peels, and fruit pomace (Larrauri *et al.*, 1999). Various treatments could be performed to improve the functionality of insoluble fibre, which is the main component of the residual matter:

- Partial removal of lignin
- Incorporation of fruit and vegetable fibre to produce a product with superior sensory properties
- Incorporation of fibre for functional properties of extrusion products
- Enzymatic modification to improve sensorial and nutritional properties

Dietary fibres from cereals have better properties than fruit fibres due to higher total and soluble fibre contents, lower phytic acid content and soluble fibre contents, fermentation ability, and water and oil holding capacities. However, fibre in fruit and vegetable residues is of particular interest for the carrier function of dietary fibres, which has been already used as antioxidants (Saura-calixto, 1998), encapsulation of flavours (Zeller, 1999) or dyes (Henn, 1998). There is a need to develop processes for fruit fibres, which have natural bioactive compounds and higher health promoting effects.

Besides the application as a texturing or gelling agent the fat replacement function in diet food is a important advantage of fibres. Recently new food additives have been developed on the basis of fruit and vegetable residues (Broughton *et al.*, 1995; Inglett, 1998). In bread and bakery goods, as well as in pastry cereals and diary products, the investigated carrot pomace works as a stabilizer. Besides crude fibre it is rich in provitamins, colours and natural acids. Addition as substitutes of sour dough in bread functions as acidifying agent, preservative or antioxidants in several food products (Filipini and Hogg, 1997; Lucas *et al.*, 1997; Ohsawa *et al.*, 1994,1995; Toma *et al.*, 1979; Masoodi and Chauhan, 1998). Effective utilization of carrot pomace to bread, cake dressing and pickle has been studied in detail by Ohsawa et al (1994 and 1995).

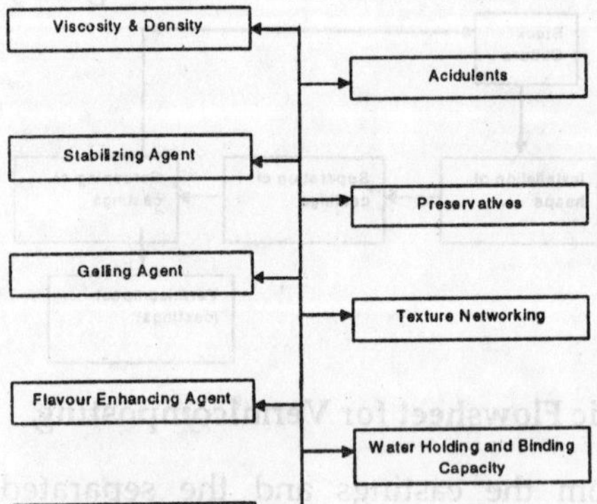

Fig. 8 Multifunctional ingredients for food

Use of pomace for manufacture of functional drinks has been suggested by Henn and Kunz (1996). Sour fruit beverages can be produced from apple pomace by lactic fermentation (Takanami *et al.*,, 1996). The mango kernel fat is good quality and can be considered as substitute for cocoa butter (Anand,1997). A sauce of acceptable quality has been prepared from apple pomace by the technique standardized (Joshi *et al.*,, 1996). The product had favorable economic return and could be stored for six months without deterioration in physico-chemical and sensory qualities. Narottam(1992) reported that apple pomace toffees of high physico-chemical character could be prepared by using sugar and pulp in the ratio of 1:1.5 along with glucose (15g), skimmed milk (30g), vegetable ghee (30g), and citric acid (3.5g) per kg of pulp. The multifunctional ingredient of food has been shown in Fig. 8. Acceptable quality of apple pomace lather can be obtained by using sugar (10%) with citric acid (0.2%) and sodium benzoate (200ppm) in apple pomace pulp followed by drying in mechanical dehydrator at 60+2°C for 10-12 hours.

Fruit and Vegetable residues as biosorbents

Fruit and vegetable solid waste could be used as a biosorbent for the pretreatment of waste water. Activated carbon being used so far is relatively expensive. Activated carbon has non-polar surfaces that are used to adsorb non-polar molecules, especially hydrocarbons. These are manufactured from both organic and inorganic sources and can be used to recover solvents. Overall carbons make a broad and important class of adsorbent (Cussler, 1997).

Classical methods for treating wastewater containing dyes, aromatics compounds or heavy metals are coagulation, flocculation, reverse osmosis (Paul et al, 1998). Activated carbon adsorption, combined with membrane process like nano filtration or ultra filtration has significant efficiency and economic advantages (Lenggenhager et al, 1997).

In order to obtain cheaper adsorbent, biosorbents are being attempted lignocellulosic materials have been studied as biosorbents. A number of low cost adsorbent have been tried for wastewater treatment as microbial biosorbents (Xie *et al.*,1996), pillared clays (Baksh *et al.*, 1992), banana pith (Namasivayam and Kanchana, 1992), orange peel (Namasivayam *et al.*, 1996), Onion skin (Bankar,1982). Apple fibre is most suitable because of its binding capacity for cadmium and lead (Borycka and Zuchowski, 1998). Laufenberg and Filipini (2000) determined the adsorption rate and is carrot pomace >corncob>sugar beet pulp. Carrot pomace adsorbs vanillin at pH 4, corncob at pH 7 and sugar beet pulp adsorbs vanillin at pH 10.

The efficiency of biosorbent depends on the rich quantity of lignin and phenols. Therefore effective biosorbent material would be lignin-rich grape pomace(45%), combined with polyphenol rich mango, orange peel (Manthey and Grohman, 1996). Dronnet *et al.*, (1997) reported that positively charged metals are attracted to negative charges on the adsorbing pomace, the effect becomes stronger in alkaline solutions. Torre et al (1995) investigated the adsorption behavior of grape and olive pomace and lemon peel towards Fe^{3+}, Fe^{2+} and Ca^{2+}. They determined that the quantity of polyvalent cations bound by dietary fibre materials increases with a higher level of mineral addition, which is in compliance with other results.

Table 11 : Fruit & Vegetable waste as bio-adsorbants

Sr. No.	Material/ Adsorbents	Removed components	Adsorbent dosage (G/L)	Reference
1.	Apple/Black current pomace	Cd, Pb, Ca , low 30 - 40 %	10 and 20	Borycha and Zuchowski, 1998
2.	Banana pith	Textile dyes, upto 80 %	0.5 - 6	Namasivayam and Kanchana,1992
3.	Carrot pomace and sugar beet pulp	Vanillin, humic acid	10 and 20	Laufenberg and Filipini, 2002
4.	Onion skins	Ca^{2+} , Mg^{2+}	40	Bankar and Dara, 1982
5.	Orange peel	Dyes, 73 - 92 %	2- 10	Namasivayam *et al.*,, 1996

Other byproducts

Apple as well as grape processing waste can be utilized as fuel source. The waste can be converted to carbon products by low temperature pyrolysis, with definite heat of combustion. The pits, waste shells and hulls from fruits and also nuts can be made into charcoal for use in briquettes. This charcoal is produced by dry-distilling the raw product at high temperature. The carbon thus produced can be marketed for numerous uses besides just briquettes (Walter, 1976).

Conclusion

Clean production concept shows a good utilization potential for solid waste of fruits and vegetables. It could achieve a reduction of investment and raw material costs and can contribute to a waste minimized food production. Unless the industry considers the waste disposal as an asset, rather than liability, the situation can not be expected to improve. Isolation of value added products and bioconversion of solid waste has significant potential. It is an environmentally sound application for sustainable development. On one hand it helps in solving the disposal problem of fruit and vegetable processing industry waste, and at the other hand it provides clean up process with better economic opportunities.

With the application of above discussed recent technologies on utilization of fruit and vegetable processing waste, there will be tremendous prospects for food industries with highly favorable, socioeconomic and positive environmental impacts.

References:

Al-wanndawi, H. (1985) Tomato processing waste as essential raw source. Journal of Agriculture and Food Chemistry 33, 804-807

Anand, J. C. & Maini, S. B. (1997) Utilization fruit and vegetable wastes, Indian food packer; March-April 1997:45-64

Anon. (1999).Gesunde Rottone .Lebensmitteltechnik 7-8, 38

Anon. (2000) Tomato pomace, a protein source for livestock and poultry. Processed Food Industry (India). Jan 2000, Vol. 3,33.

Attri,.B..L., Maini ,S. B.(1996) Pectin from galgal (*Citrus pseudolimon* tan) peel. Bioresource Technology 55 (1), 89 – 91

Bankar, D. B., Dara, S. S. (1982) Binding of calcium and magnesium by modified onion skins. Journal of applied Polymer Science 27, 1727 – 1733

Baksh, M. S., Kikkinides, E. S., Yang, R.T. (1992) Characterization by physisorption of a new class of microporous adsorbents : pillared clays. Industrial Engineering and Chemical Research 31 (9), 2181-2189.

Bhatia, B. S., Siddappa, G. S. and Lal, G. (1959) Preparation of pectin, pectin extract and syrup from Jack fruit rind. Ind. Jr. Aarici. Sci. 29 :75

Bhatia S. K. (1976) Apricots of Ladakh, status and problems, Ind. Food Packer 30(2]: 35-37

Bhullar, J. K., Sogi, D. S. (2000) Shelf life studies and refining of tomato seed oil. Jr. Food science and Tech. 37, 5:

Borycka, B. Zuchowski, J. (1998) Metal sorption capacity of fibre preparations from fruit pomace. Polish. Journal of Food and Nutritional Sciences 7 (48),67-76

Bramorski, A., Soccol C. R,, Christen P. Revah 'S. rebl 9B) (1998) Fruity aroma production by *Ceratocystis firnbriata* in solid culture from agro industrial waste. Rev. Microbiol. 29[3] 208-212

Broughton, N. W., Dalton, C. C. Jones, G.C. Williams, E.L., (1995a) Adding values to sugar beet pulp. Hochera Erloese aus Zuckerruebenschnitzein. Zuckerindustrie 120(5), 411-416.

Broughton, N. W., Dalton, C. C. Jones, G.C. Williams, E.L., (1995b) Adding values to sugar beet pulp. International Sugar Journal 97(1154), 57-60, 93-95.

Brown, D. E., Halsted, D. J. and Howard P. (1975) P.137, In; Symposium on enzymatic hydrosis of cellulose, Bailey, M. Ehari, T.M. and Linko, M., Eds., SITRA, Aulanko, Finland.

Bundschish, E., Tylla, M. Baumann, G., and Glerschner, K. (1986) Isolation of natural flavours from food industry byproducts by high pressure CO_2 extraction. Leben. Wiss. Technol. 19(6) – 493

Chadha, M. 1993 Nutritional evaluation and utilization of guava pomace. M. Sc. Thesis CCSHAU, Hissar.

Chahal, D. S. and Gray, W. D. (1968). In Biodeterioration of Material. Microbiological and allied and aspects. Walter, A.H. and Elphic, J. S., Eds., Elsevier Publ. Co., England.

Chaliha, B. P., Barua, A. D., Gohain, A., Siddapa, G. S. (1963) Assam lemon as a source of pectin Part I Effect of method of extraction, drying and storage of peels and pomace on the recovery and quality of pectin. Ind. Food Packer 17(3]: 8

Chang, S.T., Khar, G.L., Ng, C.L., Ong, K.C., Quimio, T. H., Stanton, WR & Wang, W.C. (1983) Mushrooms:producing cell (microbial) protein on lignocellulosic or other food & agricultural waste p573-604, In ; Handbook of lignocellulosic waste or other food & indigenous fermented foods, steinkrus, K.H., Ed, Macrel Dekker, New york.

Chkhikvishvili,I..D., Gogiya N.N., (1995). Flavonoids of mandarin fruit wastes and their fungi static effect on the fungus *Phoma tracheiphila* .Applied Biochemistry and Microbiology 31 (3) , 292-296

Clemente, A., Sançhez-Vioque, R., Vioque, J.,Bautista, J.,Millaio, F. (1997). Chemical composition of extracted dried olive pomace containing two and three phases. Food Biotechnology 11(3), 273-291

Convert, A. Perego, P., Borghl M.D. and Ferraito G. (1989) Pretreatment Operations and Alcohol Fermentation of orange wastes. Journal of Fermentation Bioengg , 68:42-77)

Cowling, E.B. and Brown, W. 1969 Celluloses and applications. Advances in chemistry series. Gould R.,F. ED.Vol.95. P. 152.American Chemical Soc. Publications, Washington, D.C.

Cussler, E. L. (1997). Diffusion. Second ed. Cambridge University Press, New York

De Lima, V.A.G. , Stamford, T.M. and Salgueiro, A.A. (1995). Citric acid production from pineapple waste by solid state fermentation using Aspergillus niger. Arg. Biol. Technol.,38(3):777.

Donnelly, J. A. and Scerpino P. V. (1984) Isolation, Characterization and Identification of Microorganisms from Laboratory and full scale lysimules EPA-600/2-84-199 Mucicipal Environmental Research Lab. Cinconnal

Dronnet, V. M., Axelos, M.A.V., Renard, C. M., Thibault, J. F.(1997).Binding of divalent metal cations by sugar-beet pulp. Carbohydrate Polymers 34, 73- 82

Elisashvili, V. I. , Glonli, N. M. , Kachilishvill, E. T., Kiknodze, M. O., Tusishvili, K. A. (1992). Selection of higher basiolomycetes-protein and enzyme producers. Applied Biochemistry and Microbiology, 28(3):271-274.

Emert, G. H., Gum, E.K., Lang, Y.A., Liu, T.H. and Brown, R.D. Jr. (1974) Food related enzymes. Advances in Shem. Ser., Vol. 126 P. 79 Amer. Chem. Soc. Public., Washington, D.C.

Feron, G., Bonnar,E, P., Durand, A. (1996) Prospects for the microbial production of food flavours. Trends in food science and technology 7, 285-293.

Filipini,M., Hogg,T. (1997) Upgrading of vegetable wastes and applications in the food industry, pp.1329-1331. In: 11 Forum for applied Biotechnology. Gent (Belgium).25-26 September 1997. Mededelingen-Facultiei-Landbouwkundige-en-Toegepaste-Biolog-ische-Wetensch appen-Universiteit -Gent (Belgium) .62 (4a).

Gandhi, V. M., Mukherjee, B., Iyer, V. J. and Cherrian, K. M. (1997). Nutritional and toxicological evaluation of wild apricot pomace. J. of Food Sc. & ech 34(2) 132-135.

Garzon, G..G. and Hours, R. A. (1992) .Citrus waste: An alternative substrate for pectinase production in solid state culture. Bioresource Technology 39 (1) :93

Ghose, T. K., Pathak, A.N. and Bisaria, V. S. (1975) In: Symposium on Enzymatic Hydrolysis of cellulose. Bailey, M. Enaxi, T.-M., Linko, M., Eds. P111, SITRA, Auklanko, Finland.

Gera, I.B. and Kramer, A. (1969). The utilization of Food Indusstries Wastes, In: Advances in Food Research. C. O. Chichester, E.M. Mkak and G. I. Stewart, cds. P.28. Academic press, New York- London.

Gerg , I.B. and Kramer , A. (1969) The utilization of Food industries wastes, p.28, In : Advances in Food research c.o. Chichester,.E. M. Mkak and G.F. stewart eds , Academic Press, New York-London

Giridhar, N. and Satyanarayana A. (2000) Grape waste as source of tartarates. Indian Food Packer 54(4) 59-61

Goring, D. A. I. (1971). Polymer properties of lignin and lignin derivatives. P.695-768, In: lignins, occurance, formation, structure and reactions, Sankanen, K. V. and Ludwig, C.H. Eds., Chap-17. Wiley- Inter- science, New York.

Grohmann, K. Bothast, R.J. (1994) Pectin rich residues generated by processing of citrus fruits, apples and sugar beets :enzymatic hydrolysis and biological conversion to value added products. ACS - symp - ser, vol. 566. American Chemical society, Washington, DC pp. 372 - 390

Gupta, L.K. ,Pathak, G. and Tiwari, R. P. (1989) Effect of nutrition variables on solid state fermentation of apple pomace by yeast. Journal of Food Science Agriculture;50 (1) : 55

Hans, P. Blaschek (1992) Approaches to making the food processing industry more environmentally friendly. Trends in food Science & Technology 3:107-110

Hang ,X.D. (1988) improvement of the nutritional value of apple pomace by fermentation. Nutritional report International, 38(1) :207

Hang, Y. D. and Woodams, F. F. (1988) Production citric acid from corncobs by Aspergillus niger. Bioresource Technology 65(3), 251-254

Harborne, J. B. Phytochemical methods, Chapman & Hall. London, 2nd edn, 1984.

Hausmanns, S., Laufenberg, G. Lipnitzki, F. Field, R. (1999). Prospects and performance of hydrophobic pervaporation in the concept of clean production In: 2nd Asia –specific cleaner production Round table and Trade Expo. Brisbane April 21/24 1999.

Henn, T.and Kunz, B. (1996) Use of plant pomace for manufacture of functional drinks. Flussiges Obst ,63(12):715-719

Henn, T., Kunz B. (1996) Zum Wegwerfrn zu schade ZFL 47 (1/2),21-23

Henn, T. 1998. Untersuchugen zur Entwicklung and Bewertung Funktioneller Labensmitte Lzataten dus Reststoffen arn Beispiel von Mohrentrestern and inrer Anwendung in Getranken(Thesis Bonn/D1998) Cuviller Verlag Gottingen.

Herrmann, K. Z. (1970) Lebensm-Unters. Forsch. 1998, 186I

Higareda, R.A. *et al.*, (1995) Commercial pectin concentrate from hawthorn pulp. Revista Chapingo , Serie Horticultura 1(4),155-157.

Higuchi, T. (1980). Lignin structure and morphological distribution in plant cell walls. In lignin biodegradation, Kirk, T. K., Higuchi, T. Chang, H.M., Eds., Vol. 1, Chap-1, P19, CRC Press, Boca Raton, F.L.

Inglett, G. E., (1998). New cereal products with health benefits. Food Ingredients Europe, Conference proceedings 17-19.

Joseph, G., and Mahadeviah, M. (1988) Utilization of waste from pineapple processing industry, Indian Food Packer 42(1): 46-58

Joshi, V.K. and Sandhu, D.K. (1994) Solid state fermentation of apple pomace for production of ethnol and animal feed in solid state fermentation, Ashok Pandey ed.,p 93 , New Age Publisher , New Delhi

Joshi, V.K., Gupta, K., Devrajan, A., Lal, B. B., and Arya, S. P. (2000) Production and evaluation of fermented apple pomace in the feed of broilers. J. of Food Science and Tech. 37. 6 : 609-12

Joshi,V. K. and Sandhu, D. K. (1996) Preparation and evaluation of an animal feed by product produced by solid state fermentation of apple pomace. Bioresource Tech , 56 (2/3):251-255.

Joshi, V.K., Kausal, N. K. and Thakur, N. S. (1996) Apple pomace sauce-Development and quality of fresh and stored products. J. Fd. Sci. Technol. 33(5):414-417.

Kamiya(1979) S. Esake. G. Agric. Biol. Chem.1979, 161

Karapinar, M. and Okuyan, M. (1982) The utilization of citrus waste as substrate for microbial protein production by fungus *Sporotrichum pulrerutentum.* J. Chem. Technol. Biotechnol. 32, 1055-58.

Katzen, R. and Fowler, D. E. (1994) Ethanol from lignocellulose waste with utilization of recombinant bacteria Application of Biochemistry, Biotechnology 45/46, 697

Kokhzeidre, N.G. and Elisashvili, V. I. (1993). Ligno cellulolytic activity of Pl. ostreatus IBK 191 in solid phase fermentation of the waste of tea production. Appl. Biocheem and microbiolo., 29(2): 169-173.

Krishan C. and Chandrasekaran ,M. (1995) Banana waste as substrate for amylase production by *Vibrio costicola* under solid state fermentation in Micon held at Mysore abst. No. 1MB - 14

Krishna, C. and Chandrasakaran , M. (1996). Banana waste as substrate for alpha-amylase production by *Bacillus substillis* under solid substrate fermentation. Applied Microbiol. Biotechnol.: 46(2):106.

Krishna Nand, (1995) Evaluation of fruit and vegetable processing wastes for bioconversion to methane by anaerobic digestion, In Wealth from waste, Tata Energy Research Institute.

Krishnamurthy, N., Lewis, Y. S. and Ravindranath, B. (1981) On the structures of Garcinol, Isogarcinol and Camboginol. Tetra Hedron Letters, 22:793-96.

Kruszeaska, I. and Thorape, B. (1995) What is clean production? HTML version of Greenpeace briefing. Htpp://www.rec.org/Poland/wpa/cpb.htm.

Ksheersagar, P. and Kawadkar, D. K. (2002) Report on Value added products, Department of Food Technology, L.I.T., Nagpur, Nagpur University (Unpublished)

Laufenberg, Kunz B., Nystroem, M. (2003) Transformation of vegetable wate into value added products (A) The upgrading concept; (B) Practical implementations. Bioresource Technology 87 (2003) 167-198

Laufenberg,, G,, Hausmanns, S., Kunz, B.. Nystroem, M., (1999). Green productivity concept for the utilization of residual products from food industry—trends and performance. Lecture at the 2nd Asia-Pacific Cleaner Production Roundtable and Trade Expo, Brisbane 21-24 April. 1999.

Lawther, P. J. (1977) Air pollution and pulmonary airways resistance: A six year study with three individuals environmental research 13, 478-492

Larrauri, J. A. *et al.*, (1999). New approaches in the preparation of high dieiary fibre powders from food by-products. Trends in Food Science and Technology 10 (1), 3-8.

Leda R. Castilho, Ffcardo A. Medronho, Tito L.M. Alves (2000) Production and extraction of pectinases obtained by solid state fermentation of agro industrial residues with *Aspergillus niger.* Bioresource Technology 71; 45-50.

Lenggenhager... T., Lyndon, R. (1997). PrilHI-generaling benefits of ultra filtration and adsorber technology. Fruit Processing 7; 250-256

Macheix, J. J., Fleuriet, A. and Billot, J. (1990) Fruit phenols. CRC press, Boca Raton FL

Mahadevswami, M., and Venkataraman, L.V., (1990) Integrated utilization fruit processing wastes for biogas and fish production. Biological wastes 32(4) 243-251

Mandels, M., sternberg, D. and Andreotti, R.E. (1975). Growth and cellulose production by *Trichoderma.* P. 81-109, In Symp. Enzymatic Hydrolysis of cellulose, Bailey M., Enari, T. – M. and Linko, M. Eds., SITRA, Helsinki.

Manthey. J.A.. Grohiiiann. K., (1996). Concentration of Hesperidin and other orange peel Flavonoids in citric processing byproducts. Journal of Agriculture and Food Chemistry 44,811-814.

Mark, H. (1954). Interaction and arrangement of cellulose chains. P. 217-230, In; High polymers V5, Pl., Cellulose and cellulose derivatives. Emil, O. H. Spurlin, H.M. and Graffin, M.W., Eds., Inter-science Publishers, New York.

Marsh, G. L. (1944) Recovery of tartartates from winery wastes. Wine Rev. August.

Martin. A.M. (Ed.). (1998). Bioconversion of waste materials to industrial products, second ed. Blackie Academic & Professional,London.

Masoodi ,F.A.,, Chauhan, G.S. (1999) Use of apple pomace as a source of dietary fibre in wheat bread. Journal of Food Processing and Preservation 22:255-263.

Ministry of Food Processing Industries (mfpi), New Delhi, Annual report (2000)

Narottam, K. (1992) Utilization apple pomace for edible product preparation. M. Sc. Thesis submitted to YSP UHF, Nauni (Solan) , India

Namasivayam, C., and Kanchana, N. (1992) Waste Banana Pith as Adsorbent for colour removal from waste waters. Chemosphere 25 (11), 1691-1705.

Namasivayam, C., Muniasamy, N., Gayatri. K., Rani, M., and Ranganathan, K. (1996) Removal of dyes from aqueous solutions by cellulosic waste orange peel. Bioresource Technology 57, 37-43.

Nambudiri & Shivshankar (1985) Cocoa waste and its utilization. Indian Cocoa. Arecanut and Spices J. 8(3). 78-80.

Nancib, N., Nancib, A. and Boudrant, J. (1997). Use of waste date products in the fermentative formation of bakers yeast biomass by S. cerevisiae. Bioresource Technology, 60(1):67-71.

Norman, A.G. (1954) Noncellulosic carbrohydrates in cellulose and cellose derivatives. In P 459, Emil, O., Spurlin, M. H. and Graffin, W. Eds., Part - !, Inter science Publshers Inc., New York

Nagdi, M. D. and Correia, L. R. (1992) Solid state ethanol fermentation of apple pomace as affected by moisture and bioreactor mixing speed. Journal of Food Science 57 (3): 667

Nirmala, P.J. and Ramakrishna, M. (1994) . Production of cellulases by solid state fermentation. Recent development in solid state fermentation by Ashok Pandey , ed. P. 145 .New Age publisher, New Delhi

Ohsawa, L. K.. Chinen, C..Takanami, S. Kuribayashi. T., and Kurokouchi, K. (1994) Studies on effective utilization of carrot pomace.2:Effective utilization to bread. Research Report of the Napano State laboratory of Food Technology Japan (22). 24-28.

Ohsawa, K., Chinen, C., Takanami, S,, Kuribavashi, T., Kurokouchi, K., (1995). Studies on effective utilization of carrot pomace, 2: Effective utilization to cake dressing and pickles. Research Report of Food Technology Research Institute of Nagano Prefecture, Japan 4 (23), 15-18.

Ohsawa, K., Chinen, C., Takanami, S., Kuribayashi, T and Kurokouchi, K. (1994). Studies on effective utilization of carrot pomace I. Effective utilization to bread. Res. Rep. of the Nagano State Lab. of Food Tech. No. 22:24-28

Ohsawa, K., Chinen, C., Takanami, S., Kuribayashi,T and Kurokouchi, K. (1995) Studies on effective utilization of carrot pomace II. Effective utilization of cake dressing and pickles, Res. Rep. of the Nagano State Lab. of Food Tech. No. 23:15-18

Onodera, R., Kawamura, O., Inazawa,A., Izumi, T., Okuda, M., Katayama, H. and Yokoyama, M. (1996). Preparation of lilages consisting of barley shochu distillery by product and pulps of mandarin oranges and carrots. Bull. of Fac. of Agri, Miyazaki University 43(2):145-150.

Pandey, A. (1991) Effect of Particle size substrate on enzyme production in solid state fermentation, Bioresource Technology 37:169

Pandey, Ashok., Cerlos R Soccol, Poonam Nigam, Vanete T. Soccol (2000) Biotechnological Potential of agro-industrial residues: Sugarcane bagasse; Bioresource Technology 74; 69-80.

Paul, D., Ohirogge, K., (1998) Membrane separation processes for clean production. Environmental progress 17 (3), 137-141.

Pintado & Lonsane (199B) On line monitoring of citric acid production in solid state fermentation by respirometry process, Biochem, 33(5), 513-517

Rahmat, H., Hodge, R.A. ,Manderson, G. J. .and Yu, P. L. (1995). Solid substrate fermetnation of *Kloeckera apiculata* and *Candida utilis* on apple pomace to produce an improved stock feed. World. J. of Microbiology and Biotechnology, 11(2) : 168 - 170 .

Satpute, D. .B., Pratima Shastri, and Dharmadhiakri, D. M., (2002) Economic Utilization of Fruit and Vegetable solid waste for value addition-An environmental Approach; Proceedings of National Conference on Pollution Prevention and Control in India . IAEM , 2- 3 March 2002 : VRCE , Nagpur & Indian Association for Environmental Management (IAEM) , Nagpur

Saura-Calixto, F. (1998). Antioxidant dietary fiber product: a new concept and an potential food ingredient. Journal of Agricultural and Food Chemistry 46(10), 4303-4306.

Shankaranand, V. S., and Lonsane, B. K. (1993) World Journal of Microbiology Biotechnology 9:337

Sharma, A., Joseph, R., and Sreenivasan, M. V. (1980) Improved method of extraction of oil from orange peel using fungal xylanase. Proceedings of the symposium on wastes from food industries utilization and disposal p. 18 Association of food scientists and technologist CFTRI, Mysore, India.

Sharma, V. J., Ambulkar, A. .R.,.and Bhoyar, R. V. (2002) Potential Health Hazards associated with solid waste management in India; Proceedings of National Conference on Pollution Prevention and Control in India . IAEM, 2-3 March 2002: VRCE, Nagpur & Indian Association for Environmental Management (IAEM), Nagpur

Singhal, Vikas. Indian Agriculture (1999). Published by Indian Economic Data Research Centre, New Delhi

Srirangarajan, A.N. and Shrikhade, A,J. (1976) Mango peel waste as a source of pectin. Current Science 45 (17), 620-621

Stamford V.A.G.,T.M. and Salgueiro, A. A. (1995). Citric acid production from pineapple waste by solid state fermentation using *Aspergillus niger*, Biological technology :38 (3) :777

Sturion, G.L. and Oetterer, M. (1995). Utillization of banana leaves as substrate on edible mushroom cultivation (Pleurotus spp.) Cienciae Technologia de Alimentos, 15(2): 194-200.

Sud, R. G. and Badyal, J. M. (1993) Utilization of citrus rind for peel oil: Stage of fruit & Evaluation of methods. Indian Food Packer 47(4) 45-47.

Surbriggen, B., Bailey, M. J. Penttia, M. E., Poutanen, K. and Linko, M. (1990) J. Biotechnol. 13, 267-270.

Tadeu -Pontes, M. A., Carvalheiro, F., Roseiro ,J. C. and Amaralcollaco, M. T. (1996) .Evaluation of product composition profile during an extrusion based process of tomato pomace transformation .Agro food Industry Hi. Tech ,7(3):39-40

Takanami, S., Ohsawa, K., Kuwabara, H., and Kurokouchi, K. (1996) ,Studies on sour fruit beverages by malolactic fermentation. IX production of sour beverage from apple pomace .Research report of the Nagano State Lab of Food tech .,No. 24: 12-15.

Takuo Sakai and Minoru Okushima. (1980) Microbial Production of Pectin from Citrus Peel ,Applied and Environmental Microbiology , 39 (4) : 908 –912

Thamke Ajay and M. G. Bhotmange (1997) Report on pectin from fruits and vegetables processing waste, Nagpur University, Maharashtra

Toma, R.B. *et al.*, (1979). Physical and chemical properties of potato peel as a source of dietary fibre in bread. Journal of Food science 44:1403-1407.

Torre, M., Rodriguez, A. R. ,Saura-Calixto, F., (1995). Interactions of Fe(II), CA(II), and Fe(II) with high eietary fibre materials : a physicochemical approach, Food Chemistry 54(1), 23-31.

Turqouis, T., Rinaudo. M., Taravel. F.R., Heyraud, A. (1999) Extraction of highly gelling pectic substances From sugar beet pulp and potato pulp: influence of extrinsic parameters on their gelling properties. Food Hydrocolloids 13. 255-262.

Wang, J. *et al.*, (1997) Continuous countercurrent extraction of pectin from sunflower heads . Transactions of the ASAE 40 (6):1649 - 1654 .

Wang, H. (1998) Improvement of citric acid production by *Aspergillus niger* with addition of phytase to beet molasses. Bioresource Technoloov 65:243-247.

Walter, R. H., and Sherman, R. M. (1976). Fuel value of grape and apple processing waste. Agriculture Food Chem. 24(6): 1244.

Widmmer, W., Montanari, A. M. (1995) Citrus waste streams as a source of phytochemicals in: 107 the Annual meeting of the Florida State Horticultural society, Orlondo /Florida USA, vol. 107 pp. 284 - 288

Whistler, R. L. and Rechards, E. L. (1970) Hemicelluloses, the carbohydrates: chemistry and biochemistry, Pigman, W. and Harton, D. Eds., P. 447, Academic Press. New York.

Widmer, W. and Montanari, A. M. (1995) Citrus waste stream as a source of phytochemicals. In: 107th Annual Meeting of the Florida State Horticultural Society. Orlando/florida, USA, vol. 107 p. 284-288.

Xie J. Z., Chang, H. L., Kilbanell, J. J. (1996) Removal and recovery of metal ions from wastewater using biosorbents and chemically modified biosorbents. Bioresource Technology 57, 127-136.

Zeller, B. L. (1999). Development of porous carbohydrate food ingredients for the use in flavour incapsulation. Trends in Food Science and Technology 9(11-12), 389-394.

33

Biotechnological Application for an Effective Management of Liquid and Solid Wastes

A. G. Murugesan and John Ruby

Introduction

Biotechnology is the exploitation of versatile metabolic machinery or components of living organisms to produce valuable metabolites from renewable resources (Becker *et al.*, 1990). Increasing emphasis on protecting environmental degradation and sustainability of ecosystems in creating favourable climate for biochemical processes. Technological advances in biological sciences are now being exploited to produce commercial commodities and the newly growing technologies are creating novel efficient processes for producing old products. Biotechnological strategies are applied at varied frontiers. The ultimate mission behind all these ventures is to replace energy-intensive and pollution-producing chemical synthesis by low energy consuming and pollution-free microbial technology. This ultimately paves way for the microbial machineries to trans form waste and toxic pollutants into complex chemical structures that are difficult to assemble chemically. Biotechnological strategies therefore serve as viable tools in environmental clean up. In some approaches, biotech is being used in the manipulation of microorganisms with application in the introduction of new bio-based products (Dorsh, 2002).

Bioreactors for Waste Water Management

The treatment of domestic and industrial wastes has a long history and in many industrialized countries sewage systems and treatment plants were installed initially some 100 years ago. Most effluents are heterogeneous and contain both dissolved and suspended matters, be it organic or inorganic Treatment of such effluent or waste water involves the removal of contaminants and is done to prevent any adverse effects on the receiving water or allow its reuse. The conventional system comprises two mains processes i.e fluidized or suspended microbial culture and film flow system. The former includes trickling filter and rotating biological contactor.

Activated sludge process

This process involves the agglomeration of the microbes in the form of floods each having a diameter of 0.1 mm or larger. This is necessary to separate cells from the treated liquid. The essential features are aeration tank, solid/liquid separation and sludge recycle (Rao, 1991). The effluent form aeration tank is separated in a secondary clarifier. There the separated sludge is without contact with organic matter and becomes activated. A part of this is recycled in aeration tank and the rest discarded and disposed later. Development of protozoa in sludge indicates a healthy condition. The sludge disposal is a problem, however they may be lagooned (Lowries and Wells, 1991).

Trickling filter

Conventional trickling filter consists of a rock bed (1-3m depth) with an opening for aeration the influent is sprinkled bed. As the liquid trickle through the biofilm, microbial metabolism enhances the production of CO_2 and NO_2. As the slime gets thickened it is sloughed and settles in the settling tank. The microfilm contains bacteria such as *Flavobacterium*, *Pseudomonas, Ascaligenes* and algae such as *Ulothrix, Phormidium, Stigcoclonium, Chlorella* and fungi and yeast. When the slime thickness is about 70-100 Mm maximum removal occurs.

Rotating biological contactor (RBC)

In this type the microbial slime film is built upon a partly submerged support medium, which consists of a series of light weight disks of polyethylene, styrofoam or polyvinyl chloride mounted on a horizontal shaft in a tank through which wastewater is allowed to flow. A low speed motor rotates the shaft slowly. This type of plant is widely used in the treatment of municipal waste and industrial ones like pulp, textile, meat and vegetables (Banerjee, 1994).

Fluidized beds

These are column reactors in which wastewater is pumped upwards through a particulate bed of material upon which biofilms have developed. The main advantage of these fluidized beds are high surface area of biofilm per unit of reactor.

Anaerobic biological treatment

Anaerobic digestion has long been used for sludge digestion and stabilization in sewage treatment process. Here microbial fermentation occurs wherein organic matter is converted to CO_2 and methane in the absence of high levels of nitrates and sulfates.

Contact digesters

This is an anaerobic equivalent of activates sludge process, which consists of a stirred tank and a tank under anaerobic conditions. The output of completely stirred tank is settled under anaerobic conditions and apart of settled sludge is returned to the digester. This results in the concentration of sludge and longer retention time, which enables retaining of methanogenic organisms over a wide range of loading. Contact digesters are used for treating wastes from sugar processing, distilleries, citric acid and yeast production, canned vegetables, farm slurries etc.

Packed bed reactors

The microorganisms are contained within the packing medium in an enclosed vessel and liquid wastes pass upwards. No slime formation occurs. The treatment rate is directly related to surface area of packing.

Anaerobic baffled digesters

These reactors have walls across the tank built form the top to the bottom, so that wastewater flowing along the tank has to go alternately under and over the baffles. The baffles tend to keep the bacteria in the tank and also help to prevent problems with the floating solids.

Upflow anaerobic sludge blanket reactor

The UASBR system is a process that works under suspended growth systems. The substrate passes through a bed of active sludge granules in the upward direction and gets converted into methane and carbondioxide. Part of carbondioxide further reacts with excess hydrogen prevailing in the high hydrogen partial pressure zone in the lower part of the granular bed producing more methane. The sludge is immobilized by natural mechanism, that is the formation of highly settable aggregates of microoganisms known as granules (Jayantha and Ramanujam, 1995). This is the most widely applied treatment method for the municipal sewage and industrial effluents. The main advantage of this process is that no support material is needed for immobilization of microorganisms (Jayantha and Ramanujam, 1996). UASB system has been made use for the treatment of distillery spent wash and using cowdung slurry as a startup material . A well-granulated biomass has been observed after 12 weeks. The hydraulic loading ranged from 60 hours to 80 hours. High COD reduction has been achieved, accompanied with a higher quantity of biogas generation. UASB reactors have been also employed for the treatment of pulp and paper mill effluent. The COD reduction rate was 70% The amount of biogas generation gradually increased with the increase of organic loading rate (Murugesan and Rajakumari,1997).

Phytoremediation

Phytoremediation is a concept involving the use of plants to remove chemical contaminants in a biomatrix, sometimes referred to as a biofilter. Understanding plant use of nutrients, and the use of appropriate soil additives can accelerate plant root growth, mobilize metals and hydrocarbons, and improve plant toughness and growth rates. Design of the biofilter may include tiers using different types of grasses and other plants to encourage different nutrient uptake and/or to accelerate evapotranspiration. Design of phyto-remediation systems consists of several simple steps to encourage the plants to maximize their uptake, and ongoing maintenance activities to insure that the uptake is properly realized. These steps include identifying the contaminants of concern, appropriate cultivars or plants for the locale and for the contaminants, needed soil nutrients to mobilize and/or breakdown the contaminants and appropriate aeration capacity. Wastewater products may include reclaimed water, storm water or industrial wastes. Storm water may contain many nutrients required by plants, as well as hydrocarbons and less desirable constituents.

Plant Nutrient Uptake

To understand the implications of plant nutrient uptake, it is important to understand the nature of plants and their interactions with soils on a macro-scale. During daylight hours, plants appear to absorb carbon dioxide and respirate oxygen by the process of photosynthesis, using the sun's energy (light) (Bedell, 1992). From a photochemistry perspective, what happens is that the plant's surface fluorescence causes the bonds between the carbon and oxygen to break, thereby allowing the plant to use the oxygen for respiration and the carbon for cell building, while the oxygen is needed of respiration. The excess oxygen is released as apart of a photochemical reaction.

At night, photosynthesis cannot occur because the sun is absent, but the plant still must respirate and build cells. Therefore, at night respiration occurs via the roots using oxygen instead of carbon dioxide. The plant brings oxygen into the roots via the soil, along with water, carbon and other nutrients. The water, oxygen and nutrients travel up the xylem to the leaves to continue building of the plant. The plant releases carbon dioxide into the soil as a waste product. Plants generally need carbon, nitrogen, phosphorous, potassium, selenium in a ratio of 100:10:1:0.1:0.001, respectively. Calcium, magnesium, iron and silica are also incorporated via this method, albeit at lower ratios. Carbon and water are readily available in most soils so the processes of bringing nutrients in through the roots becomes the primary route of nutrient transfer as well as respiration.

Phytoremediation mechanisms of organic contaminants

Plants and trees remove organic contaminants utilizing two major mechanisms: direct uptake of contaminants and subsequent accumulation of nonphytotoxic metabolites into the plant tissue, and release of exudates and enzymes that stimulate microbial activity and the resulting enhancement of microbial transformations in the rhizosphere (the root zone).

Direct uptake

Not all organic compounds are equally accessible to plant roots in the soil environment. The inherent ability of the roots to take up organic compounds can be described by the hydrophobicity (or lipophilicity) of the target compounds. This parameter is often expressed as the log of the octanol–water-partitioning coefficient, Kow. Direct uptake of organics by plants is a surprisingly efficient removal mechanism for moderately hydrophobic organic compounds. There are some differences between the roots of different plants and under different soil conditions, but generally the higher a compound's log *K* ow, the greater the root uptake. Hydrophobicity also implies an equal propensity to partition into soil organic matter and onto soil surfaces. Root absorption may become difficult with heavily textured soils and soils with high native organic matter. There are several reported values available in the literature regarding the optimum log *K*ow value for a compound to be a good candidate for phytoremediation.

The analogous resin in this system is wood, the vascular tissue of the tree, and this "resin" is replenished each year by normal growth. Wood is composed of thousands of hollow tubes, like the bed of a hollow fiber chromatography column, with transpirational water serving as the moving phase. The hollow tubes are actually dead cells, whose death is carefully programmed by the tree to produce a water-conducting tissue that also functions in mechanical support. A complex, cross-linked, polymeric matrix of cellulose, pectins, and proteins embedded in lignin forms the walls of the tubes.

The cell wall matrix is chemically inert, insoluble in the majority of solvents, and stable across a wide range of pH. Once an organic chemical is taken up, a plant can store (sequestration) the chemical and its fragments in new plant structures via lignification, or it can volatilize, metabolize, or mineralize the chemical all the way to carbon dioxide, water, and chlorides. Detoxification mechanisms may transform the parent chemical to nonphytotoxic metabolites, including lignin, that are stored in various places in plant cells. Many of these metabolic capacities tend to be enzymatically and chemically similar to those processes that occur in mammalian livers, and one report equated plants to "green livers" due to the similarities of the detoxification process. Different plants exhibit different metabolic capacities. This is evident during the application of herbicides to weeds and crops alike. The vast majority of herbicidal compounds have been selected so that the crop species are capable of metabolizing the pesticide to nontoxic compounds, whereas the weed species either lack this capacity or metabolize it at too slow a rate. The result is the death of the weed species without the metabolic capacity to rid itself of the toxin. The shear volume and porous structure of a tree's wood provides an enormous surface area for exchange or biochemical reactions. Some researchers are attempting to augment the inherent metabolic capacity of plants by incorporating bacterial, fungal, insect, and even mammalian genes into the plant genome.

Phytoremediation mechanisms of heavy metals

Most heavy metals have multiple chemical and physical forms in the soil environment. All forms are not equally hazardous, nor are all forms equally amenable to uptake by plants. The chemistry of the metal and its mobility will inherently impact the toxicity in the environment. Metal fractionation or sequential extraction schemes (such as TCLP) sometimes are used to describe metal behavior in soils. Most metals interact with the inorganic and organic matter that is present in the root-soil environment. Potential forms of metals include those dissolved in the soil solution, adsorbed to the vegetation's root system, adsorbed to insoluble organic matter, bonded to ion exchange sites on inorganic soil constituents, precipitated or co precipitated as solids, and attached to or inside the soil biomass. Phytoremediation of heavy metal contaminated soils can be divided into phytostabilization and phytoextraction approaches.

Phytostabilization of heavy metals

Implementation of phytostabilization involves the reduction in the mobility of heavy metals by minimizing soil erodibility, decreasing the potential for wind-blown dust, and reduction in contaminant solubility by the addition of soil amendments. Erosion leads to the concentration of heavy metals because of the selective sorting and deposition of different size fractions of the soil. Eroded material is often transported over long distances, thus selectively extending the effects of contamination and increasing the risk to the environment. Erosion can, therefore, cause the buildup of concentrations of normally nontoxic contaminants to toxic levels at locations where transported material is deposited Planting of vegetation at contaminated sites will significantly reduce the erodibility of the soils both by water and wind. Density of vegetation will effectively hold the soil and provide a stable cover against erosion. Another element of phytostabilization is to supplement the system with a variety of alkalizing agents, phosphates, organic matter, and biosolids to render the metals insoluble and unavailable to leaching. Materials with a calcareous character or a high pH can be added to influence the acidity: compounds such as lime and gypsum. Specific binding conditions can be influenced by adding concentrated Fe, Mn, or Al compounds. To maintain or raise the organic matter content in the soils, various materials such as humus or peat materials, manure or mulch can be added. This chemical alteration should be quickly followed by establishing a plant cover and maximizing plant growth. The amendments sequester the metals into the soil matrix and plants keep the stabilized matrix in place, minimizing wind and water erosion.

Phytoextraction of heavy metals

The use of unusual plants that have the ability to accumulate very high (2 to 5%) concentrations of metals from contaminated soils in their biomass provides the basis for this phytoremediation technique. The metals are translocated to the shoot and tissue via the roots. These plants are called hyper accumulator plants and they exhibit the ability to tolerate high concentrations of toxic metals in above-ground plant tissues; these species contain toxic element levels in the leaf and stalk biomass (LSB) of about 100 times those of nonaccumulator plants growing in the same soil, with some species and metal combinations exceeding conventional plant levels by a factor of more than a thousand many hyper accumulator plants that are nonwoody (not a tree) have been identified to have the capacity to accumulate metals. *Thlaspi caerulescens* was found to accumulate Zn up to 2000 to 4000 mg/kg. The Indian mustard plant *Brassica juncea,* grown throughout the world for its oil seed, was found to accumulate a significant amount of lead. One planting of mustard in a hectare of contaminated land was found to soak up 2 t of lead. If three plantings could be squeezed in per year, 6 t of lead per hectare can be extracted. Both hemp dogbane (*Apocynum* sp.) and common ragweed also have been observed to accumulate significant levels of lead. *Aeollanthus subcaulis* var. *lineris* and *Paspalum notatum* are other hyper accumulator plants known to accumulate Cu and Cs, respectively. Hyper accumulator plants can address contamination in the shallow soils only up to 24 in. in depth. If contamination is deeper, 6 to 10 ft, deep-rooted poplar trees can be used for phytoextraction of heavy metals. These trees can accumulate the heavy metals by sequestration. However, there are concerns specifically for trees that leaf litter and associated toxic residues will be blown off site. This concern may be tested in the laboratory to see whether uptake and translocation of the metals into the leaves exceed the standards. Hyperaccumulators have the metal-accumulating characteristics that are desirable, but lack the biomass production, adaptation to current agronomic techniques, and physiological adaptations to the climatic conditions at many contaminated sites. It has been reported that harvesting done at different seasons in a year had pronounced differences in accumulation levels.

In the future, genetic manipulation techniques may provide better hyperaccumulator species. The success of phytoextraction depends on the use of an integrated approach to soil and plant management. The disciplines of soil chemistry, soil fertility, agronomy, plant physiology, and plant genetic engineering are currently being used to increase both the rate and efficiency of heavy metal phytoextraction. Translocation from the root to the shoot must occur efficiently for the ease of harvesting. After harvesting, a biomass processing step or disposal method that meets regulatory requirements should be implemented.

Phytosorption and phytofiltration of heavy metals

Aquatic plants and algae are known to accumulate metals and other toxic elements from solution. There are large differences in bioremoval rates due to species and strain differences, cultivation methodology, and process control techniques. In the past, commercial systems have used immobilized algal biomass for removing radionuclides and other heavy metals in the aqueous phase.

Plants that are naturally immobilized, such as attached algae and rooted plants, and those that could be easily separated from suspension, such as filamentous micro algae, macro algae, and floating plants, have been found to have high adsorption capacities. In a recent study, one blue-green filamentous algae of the genus *Phormidium* and one aquatic rooted plant, water milfoil (*Myriophyllum spicatum*) exhibited high specific adsorption for Cd, Zn, Ph, Ni, and Cu. It has been reported that porous beads containing immobilized biological materials such as sphagnum peat moss can be used for extracting metals dissolved in the aqueous phase. The beads designated as BIO-FIX beads readily adsorbed Cd, Pb, and other toxic metals from dilute waters. In a recent study, it was reported that *Saccharomyces cerevisiae* yeast biomass, when treated with a hot alkali, exhibited an increase in its biosorption capacity for heavy metals.

Phytoremediation of nitrogen compounds

Nitrogen is an important nutrient for plant growth. When nitrates or other inorganic nitrogen species are present above the allowable limits in groundwater, phytoremediation is a viable remediation alternative under shallow water table conditions. Corn and deep-rooted poplar trees have been found to be very effective in reducing nitrate concentrations in groundwater. Poplar trees were most effective at reducing the nitrate concentrations, and the perennial nature of the trees ensures prolonged protection throughout the year. Deeper planting of the trees, closer to the water table, resulted in better growth and greater nitrogen accumulation in leaves. Munitions wastes such as 2,4,6-trinitrotoluene are of considerable interest as candidates for phytoremediation. An excreted nitroreductase activity in the root zone has been identified as the catalyst for degradation of this compound.

Many successful field-scale applications of phytoremediation have been reported so far. Each successful application reported has site-specific conditions unique to that particular site as well as different remediation objectives. Although phytoremediation is not a panacea for hazardous waste problems, it has proven effective in several applications for remediation of shallow contaminated sites. Before the technology can mature, we need a better understanding of the role of metabolites, enzymes, and the selection of plant and tree systems for various wastes and hydrogeologic conditions. Trees potentially are the lowest-cost plant type to be used for phytoremediation. A number of trees can grow on land of marginal quality, and this will allow establishment of trees on.

Sites with low fertility and poor soil structure, thus keeping costs low for plant establishment. Since trees are perennial plants with long life spans, site remediation can continue for decades with little or no maintenance costs. Trees of the Salicaceae family (willow and poplar) have been planted at several locations because of their flood tolerance and fast growth. Various grass species such as prairie grass, crested wheat grass, and cattails have been used in phytoremediation. Alfalfa has been used widely for its deep rooting and root zone metabolic activity. Parrot feather and Eurasian water milfoil have been applied to break down ammunition wastes.

Root systems are known to facilitate the uptake of contaminated groundwater by hybrid poplar trees up to 30 ft below land surface. In addition, the hybrid poplar trees can transpire a quantity of water sufficient to impact the flow of groundwater when planted in sufficient density. Such hydraulic control in shallow groundwater conditions is valuable in that the migration of the contaminant plume can be reduced or eliminated.

One recent study demonstrated that poplar trees, which possess cytochrome P-450s analogous to the oxygenases responsible for transformation of compounds such as TCE in the mammalian liver, exposed to 100 mg/l of TCE did uptake and chemically alter this contaminant. TCE and its metabolites were found in the roots and tissue of the study trees, but not in control trees or in the soil used for potting the trees.

In a subsequent study, popular seedlings exposed to C-labeled TCE were found to generate C-labeled carbon dioxide. Intermediate compounds generated during oxidation are thought to be 2,2,2-trichloroethanol, and di- and trichloroacetic acid. Similar studies have shown positive results for toluene and benzene. Poplars are phreatophytic, capable of extending their roots into aerobic water tables. For example, the roots of poplars growing alongside streams can easily be observed intertwined in the stream bottom. The degree to which poplar roots would penetrate the saturated zone cannot be easily estimated. However, if their access to soil moisture from precipitation is limited, poplars will draw large amounts of water from the top of the saturated aquifer. Evapotranspiration will draw down the water table below the trees similar to a pump and treat system. Under optimal conditions, a hybrid poplar occupying 4 m of ground can cycle 100 l of groundwater per day; this translates to approximately 30,000 gal of water per acre of trees per day.

Poplar trees have been reported to grow 6 to 8 ft per year. Simulations of a proposed design can be carried out based on extent of contamination, hydrogeological data, and past precipitation and infiltration records. A big advantage of phytoremediation over conventional pump and treat systems is the ability of the roots to penetrate the microscopic scale pores in the soil matrix. Contaminants adsorbed or trapped in these micropores are minimally or not impacted by the pump and treat system. In the case of phytoremediation, the roots can penetrate these micropores for contaminant removal. Another reported application of phytoremediation is the planting of 10,000 poplar trees per hectare as the final cap on a landfill in Oregon.

Treatment of organic wastes is not the main goal at this site, but rather to keep the site natural and free from infiltration. This proprietary technique is known as Ecolotree™ cap, and installation involves the placement of a thick cover with soil amendments for storage of water within the root zone. Densely planted poplar trees uptake water through their extensive root systems as a pump and transpire the water back to the atmosphere. Advanced moisture control systems using time domain reflectometry (TDR) technique have been installed at this site. Although hybrid poplars seem to tolerate organic chemicals quite well, high concentrations of metals, salts, and ammonia are toxic.

For long-term closure of landfills in temperate climates, it might be desirable to use higher value hardwood trees such as walnut or pecan, in addition to fast-growing poplars, unless there is a ready market for poplar products. Walnuts or pecans under good fertility conditions use large amounts of water and produce both nuts and wood. In subtropical regions, eucalyptus would be more appropriate than poplars. One would use alfalfa only in a situation where a short-lived perennial is satisfactory. Replanting would be necessary after a few years, if the remediation was not completed in that time. It is essential to choose climatically adapted species from among those with desired growth characteristics. Degradation of petroleum contaminants in the rhizosphere has been reported in the literature. Microbial numbers were substantially greater in soil with plants when compared to soil containing no plants, indicating that plant roots enhance microbial populations in contaminated soil for enhanced biodegradation of contaminant.

Another study investigated the uptake of two pesticides, alachlor and atrazine, present in the soil. It was found that plant uptake is an influential process on the fate of alachlor and atrazine. A recent report considers some strategies for engineering plants to improve bioremediation in the root zone.

One of the simpler approaches is to make use of the organism *Agrobacterium rhizogenes* to induce a state called hairy root disease. Predicted groundwater flow conditions at maturation stage of Tree Mediation system. virulence of the strain used, the extent of root production is variable, but generally infection leads to a significant enhancement of rooting without obvious detrimental effects on the host plant. Increased root mass has the apparent advantage of increasing the surface area available for microbial colonization. Root exudation may be increased in proportion to increase in root area. Such rhizosphere enhancements could improve bioremediation potential of the plant–microbial system. It is suggested that when water is not freely available in unlimited quantities, increased root mass could lead to greater water uptake, and hence greater contaminant mobilization and potential degradation. Genetic engineering of plants by insertion of genes for chlorinated phenolics catabolism is in progress. These enzymes may also allow metabolism of TCE. A number of companies have introduced genes for degradation of herbicides into crop plants. Some of these approaches could prove useful for remediation of other contaminants.

Economical status of phytoremediation

Phytoremediation is very competitive with other treatment alternatives. It is aesthetically pleasing and its public acceptability is high. Darlene Bader of the U.S. Army Environmental Center at Aberdeen Proving Ground, reports that two anaerobic wetlands cells followed by two aerobic cells with canary grass were successful in removing TNT at 30% of the cost of granular activated carbon treatment phytoremediation is less than half the cost of the pump and treat technology. The estimated cost advantage of phytoextraction for metals compared to in situ fixation, excavation and landfilling in a RCRA approved hazardous waste facility, and soil extraction. Phytoremediation is far less expensive, but it requires five years rather than shorter periods for the competing technologies. The advantage of phytoremediation on petrochemical wastes is shown relative to competing technologies. Once again, phytoremediation offers cost advantages, but the trade off is the amount of time that is required to achieve treatment to action levels. Phytoremediation is most comparable to in situ bioremediation and natural attenuation. In these technologies, mathematical modeling and monitoring are necessary to demonstrate the effectiveness of the technology to regulatory agencies. The same will be true of phytoremediation.

Phytoremediation is an emerging technology for contaminated sites that is attractive due to its low cost and versatility. It is not a panacea for hazardous waste problems, but it shows tremendous potential in several applications for treatment of metals and organics at sites where contamination is shallow. The role of enzymes, metabolites, and the selection of plant systems for various wastes must be better understood. Plants have the ability to withstand relatively high concentrations of pollutants; they can sometimes take-up the chemicals and convert them to less toxic products, and they are known to stimulate degradation of organics in the rhizosphere. The technology has not been demonstrated conclusively at many sites to date, and it remains to be seen if it is effective at full scale. The importance of industrial waste discharges in relation to water supply and water use is not the same in different part of the country, depending upon the distribution and type of industry. Reduction of waste may frequently be accomplished by closer supervision of the processes or good house keeping.

Aquatic Macrophyte – Based Wastewater Treatment

Aquatic macrophytes take an important role in wastewater treatment. Each species contributed its special function and co-operates in the purification process. Some nutrient pollutants were absorbed by the macrophytes and were removed from the effluents. The macrophytes used for BOD removal seemed to function as fixed film reactors with the submerged plant structures acting as a substrate for bacteria. Vascular structures of the macrophytes can transport atmospheric oxygen from the foliage into the root. Oxygen not required for root respiration may diffuse into the wastewater and be utilized by bacteria for the oxidation of BOD (Howes and Teal, 1994; Flessa, 1994). Floating aquatic macrophytes are also capable of assimilating large quantities of trace elements, some of which are essential for plant growth (Reddy and Smith, 1987; Zaranyika *et al.*, 1994).

Since algae were among the most important purifying organisms in the oxidation ponds, there often existed high BOD and TSS levels in the effluent of these ponds. Regardless of the type of aquatic macrophytes treatment system (AMATS) whether a natural wetland or an artificial system planted with a monoculture or polyculture of either floating or emergent plants, the processes thought to operate are essentially the same. In addition to the direct uptake and accumulation of contaminates, pollutant removal may be achieved by a complex range of chemical and physical reactions, occurring at the water-sediment, root-sediment and plant-water interfaces. Although the basic biological processes responsible for the removal of pollutants by AMATS are well established, there has been little standardization of their design or methods of construction. Constructed wetlands are one of the brighter new water treatment technologies that offer low maintenance, require little energy are quite stable and provide very good secondary or tertiary treatment. They do require considerable land area and this will limit their applications in some locations. However, based largely on the root-zone concept, a body of general design criteria and principles for the construction of reed bed systems is beginning to emerge.

Artificial wetlands may well have a role to play in the control of nonpoint pollutions. It is conceivable that small strips of wetland could, for example, be employed to contain pesticide and fertilizer run off from agriculture systems and heavy metal run off from roadways. The potential of wetlands to accumulate and contain synthetic organic pollutants is largely unexplored, as is the possibility of improving the performance of AMATS by selectively introducing desirable bacterial strains capable of specific pollutant decomposition.

Aquatic macrophytes, particularly floating species, such as the water hyacinth and pennywort are exhibiting very high rates of growth, e.g. 10gm/sq.m/day. Such growth rates are associated with high levels of nutrient uptake and demand, particularly for nitrogen and phosphorus. It has been estimated that water hyacinth plants are capable of removing from the water, by direct uptake, 5850 kg N/ha/year and of storing within their biomass between 300 and 900 kg N/ha. Significant quantities of phosphorus can also be removed (350-1125 Kg P/ha/year) and accumulated (20-57 kg P/ha). Direct uptake and accumulation by plants can also be significantly reduce the concentration of other contaminants in wastewater. Natural and artificial wetland systems have been shown to be effective in lowering effluent metal content (e.g. Cu, Zn, Pb, Hg, and Ni), partly as a result of both active and passive plant uptake and accumulation.

The conventional engineering methods of industrial effluent treatments are not fulfilling the requirements of the pollution control standards and so alternative methods are essential. Aquatic macrophytes are efficient in removal of pollutants from industrial effluents (Murugesan and Rajakumari, 2000). The ability of *Eichhornia*, *Trapa*, *Salvinia* and *Pistia* in absorbing several pollutants has raised their value for pollution abatement and so it has been employed to treat rubber factory and integrated liquid hydrogen plant effluent and also the pulp and paper mill effluent.

Table 1 : Efficiency of aquatic weeds in treating rubber factory effluent (percent reduction in 7days)

Macrophyte	Chloride	Sulphate	BOD	COD
Eichhornia	96.5	78.5	80	86
Trapa	39	59.5	71	75
Pistia	59.5	56	66.6	72
Salvinia	68.5	89	92	78

Values are in mg/l

In the rubber factory effluent chloride and sulphate were reduced to 96.5 & 78.5% respectively in 7 days by *Eichhornia crassipes*, 39 and 59.5 % respectively by *Trapa* in 7 days, 68.5 and 89% in seven days by *Salvinia* (Table 1). More than 80% BOD and 86% COD were reduced by water hyacinth, while 71% BOD and 75%COD were reduced by *Trapa*. Aquatic macrophytes have been tested for their effiency in treating textile mill effluent and the results obtained showed that the macrophytes are highly efficient in treating the effluent (Table 2).

Table 2 : Percent decrease of pollutants from integrated textile mill effluent treated with floating aquatic macrophytes (Retention time – 7 days)

Characteristics	Floating Macrophytes			
	Eichhornia	*Pistia*	*Salvinia*	*Trapa*
TS	88.06	86.42	84.10	73.65
TDS	87.93	86.38	84.48	73.45
TSS	90.00	87.00	80.10	76.50
BOD	90.12	88.00	79.00	73.20
COD	86.96	87.77	72.39	64.96
Phosphate	73.13	67.19	64.71	60.80
Sulphate	41.03	40.61	67.76	36.75
Chloride	55.89	53.46	51.11	50.13

Values are in mg/l

Table 3 : Efficiency of floating aquatic macrophytes in treating paper mill effluent (values except pH are in mg/l) Retention time (8 days) mean ± SD

pH	8.8 ± 0.4	8.7 ± 0.2	8.6 ± 0.1
TS	3941 ± 818	2509 ± 713	2618 ± 613
TSS	2819 ± 136	1908 ± 96	2196 ± 142
BOD	425 ± 42	289 ± 53	316 ± 48
COD	3100 ± 112	2322 ± 96	2459 ± 110
Nitrate	42.7 ± 2.1	39.7 ± 0.9	40.3 ± 1.2
Phosphate	22.4 ± 1.3	16.5 ± 1.1	17.1 ± 0.9
Phenol	2.3 ± 0.3	1.3 ± 0.2	1.8 ± 0.2

Cyperus exaltatus, *Scirpus articulatus* and *Typha angustata* demonstrated their ability in rapidly in removing the pollutants from integrated textile mill effluent. Addition of sewage to the effluent stirred the microbial activity and thereby increased the rate of removal of pollutants. Effluent with 50% sewage showed a higher pollutant removal rate when compared with 75% and 100% effluent. There was a marked reduction in BOD, COD, alkalinity, hardness, sulphates, nitrates and total phosphates, chlorides in effluents treated with aquatic macrophytes. The colour of the effluent has also removed significantly by these reeds with the retention period of 5 days. The efficiency of colour removal was most significant in *Scirpus articulatus* and *Typha angustata* when compared with *Cyperus exaltatus*. Murugesan and his team have documented that reeds are efficient in treating the pulp and paper mill effluent (Table 3).

Application of Cyanobacteria in Industrial Wastewater Treatment

In recent years, the importance of employing cyanobacteria for wastewater treatment has attracted the attention of the world. Growth of different species of cyanobacteria to different types of wastewater shows a positive response and the cyanobacteria can grow well even in higher concentration of the waste water. Along with the growth of cyanobacteria there is significant reduction in nutrients and other categories of toxic pollutants like (zinc, iron, copper, lead, chromium) and the pesticides (cytrolane, furaden, aldrin, methoxychlor, diazinon, parathion). Cyanobacteria can take up and bioconcentrate variety of environmental contaminants present in the wastewater. Uma and Subramanian (1990) studied the effective use of cyanobacteria in ossein industry effluent, which has very high levels of TDS, Ca, and chloride content. This waste water was treated with a novel biological treatment systems using a halophilic bacterium, *Halobacterium* sp. US 101, a marine filamentous cyanobacterium *Oscillatoria* sp. BDU 10142 and a calcicolous unicellular cyanobacteria *Aphanocapsa* sp. BDU 16 both by lab and field studies. Attention on algae as indicators of pollution their usage for removal of toxicants from wastewater treatment and as bioassay organism for testing the toxicity of chemicals has been drawn in recent years. From the works of Murugesan it is evident that cyanobacteria such as *Spirulina* and Chlorella grow well in distillery and sago effluent which contain a high organic load (Table 4) and it has also been documented that the cyanobacteria

Table 4 : Efficiency of reeds in treating combined partially treated pulp and paper mill effluent. Retention time – 5 days

Characteristics	Aquatic Reeds		
	Scirpus	*Typha*	*Cyperus*
BOD	95.40	93.80	94.60
COD	90.80	90.69	90.30
Chloride	57.80	66.59	56.82
Phosphate	47.29	76.50	69.43
Nitrate	82.51	72.50	83.80
Nitrite	66.09	61.90	65.63
Sulphite	86.60	70.60	65.00
Sulphate	80.60	70.16	65.28
TSS	57.79	56.86	56.42
TS	26.07	24.90	25.50

Values are in mg/l

Table 5 : Efficiency of *Spirulina* and *Chlorella* in treating distillery and sago effluents (values are in mg/l) Retention time – 10 days.

Effluent	Characteristics (mg/l)	Initial	Final	
			Spirulina	*Chlorella*
Distillery	BOD	982	246	364
	COD	2648	640	850
	TS	5654	2310	2406
	TDS	2036	942	864
	TSS	3618	1368	1542
Sago	BOD	1100	284	320
	COD	3256	718	943
	TS	5738	2050	2384
	TDS	2020	787	948
	TSS	3718	1263	1436

are capable of uptaking the organic material from seafood and gherkins processing effluents and grow well in them (Table 5).

Biodegradation of Xenobiotics

Many microorganisms can metabolize particular aromatic compounds. Natural biodegradation is often a slow process and depends upon a complexity of waste, concentration of individual pollutants, toxic intermediates produced, pH, temperature, oxygen availability, redox potential, types of microorganisms and their numbers, competition by easily degradable substrate etc. Still natural biodegradation does occur and considerable disappearance of xenobiotics is possible. The use of added microorganisms for the treatment of hazardous wastes has been attempted and has a promising future. Bioaugmentation with parachlorophenol degrading bacteria achieved a 96 % removal in 9 hour as compared with a control that exhibited 57 % removal after 58 hours (Kennedy *et al.*, 1990). *Candida tropicalis* cells also have been used to remove high concentrations of phenol in wastewater. *Desulfomonile tiedjei* , when added to a methanogenic upflow anaerobic granular – sludge blanket increased the ability of the bioreactor to dechlorinate 3 – chlorobenzoate.

In most cases, xenobiotic degradation is brought about by mixed consortia in a sequential mechanism rather than individual species alone. The consortial microbial species enhances the degradative process by allowing better adaptation, more tolerance and sby supplementing nutrient and cofactor needs of each other in the community. Inducible enzymes of many types are better produced when a community of different species acts on a mixture of pollutants.

Microbial dehalogenses

The biochemical pathways of *Pseudomonas* sp. lead to the spontaneous dehalogenation of the haloaromatic compounds. There are enzymes called 'dehalogenases' (Jensen, 1960), which specifically catalyze the cleavage of carbon-hydrogen bond. Microbial dehalogenases cleaves the carbon-hydrogen bonds through catalysis of oxygenolytic, reductive and hydrolytic reactions and the enzymes are active towards substituted aliphatic and aromatic compounds.

Oxygenolytic dehalogenases

These enzymes have also been seen to act as dehalogenases by actively catalyzing the halogenation step. Under anaerobic conditions, the metabolism of halogenated hydrocarbons is mediated by the strictly anaerobic methanogenic bacteria (Belay and Daniels, 1987). These microbial forms utilize CO_2, H_2 or formate, acetate or methanol for growth and co-metabolize haloalkanes leading to complete mineralization. Methane monoxygenases from the various methanogens catalyze different reactions when haloalkanes are the enzymes substrates. It has been found that methane-oxidizing bacterium strain 46-1 (Little *et al.*, 1988) oxidized trichloroethanes to the corresponding acids and halogens. *Methylococcus capsulatus* converted bromomethane to formaldehyde via the putative intermediate bromoethanol (Stirling and Dalton, 1980). The involvement of cytochrome monooxygenases has been suspected since mammalian systems were shown to mediate detoxification of halogenated hydrocarbons. Growth of a strain of *Pseudomonas putida* on camphor has been seen to induce high levels of cytochrome P_{450} (Lam and Vilker, 1987). *Nitrosomonas europaea* has been shown to be capable of the degradation of a range of aliphatic compounds with the degradation apparently catalyzed by an ammonia monooxygenase (Vannelli *et al.*, 1990).

Dioxygenase activities have also been implicated in the microbial metabolism of polychlorinated biphenyls (PCBs). The principal metabolic pathways catalyze the hydroxylation of unsubstituted carbons at positions 2 and 3 or 3 and 4. Degradation pf PCBs by many bacterial genera has been reported. Most of the aerobic PCB-degrading bacteria are capable of metabolizing mono, di and trichlorobiphenyls although some can metabolize congeners containing upto five chlorine sustituents (Clark *et al.*, 1979). The white rot fungus *Phanerochaete chrysosporium* can go one step further and degrade hexachlorobiphenyls.

Reductive dehalogenation

Reductive dehalogenation results in the direct substitution of the halogen substituent by hydrogen and has been reported in both aerobic and anaerobic organisms. Anaerobic degradation of meta-chlorobenzoates was first reported for an anaerobic microbial community isolated from sewage sludge and subsequently reductive dechlorination was the first step in degradation of several aromatic compounds (Quensen *et al.*, 1988). Methanogenic communities isolated from lake sediments and sewage sludge have been shown to possess the ability to mineralize halobenzoates after they were isolated by enrichment for growth on 3-chlorobenzoate. With these microbial communities meta-halogens were more susceptible than the ortho-or para-isomers.

Hydrolytic dehalogenation

There are two groups of enzyme, which catalyze the hydrolytic dedhalogenation of haloorganic compounds, namely the glutathione-dependent dehalogenases and the halodohydrolases (Hardman, 1991). A glutathione-dependent conversion of dichloromethane to formaldehyde was catalyzed by cell free extracts of *Hyphomicrobium* sp. Halidohydrolase-type dehalogenases have been isolated as being active towards both aliphatic and aromatic compounds. The physiology of the microbes producing the 2-haloacid halidohydrolases have been extensively studied. The importance of the haloacid halidohydrolases as enzymes forming part of a catabolic pathway for the degradation of environmental chemicals is demonstrated by the degradation of 1,2-dichloroethane by *Xanthobacter autotrophicus* (Janssen *et al.*, 1985). A second example is provided by the degradation products of TCE by the methane monooxygenases present in a methanotrophic community (Little *et al.*, 1988).

The widespread use of chlorophenols as herbicides and fungicides and the associated environmental contamination has made it necessary to study the biodegradation of such compounds. A number of bacteria and bacterial communities have been shown to be able to mineralize pentachlorophenol (PCP). The actinomycetes, *Rhodococcus chlorophenolicus* PCP-11 (Apajalahti *et al.*, 1986) has been isolated as an organism capable of mineralizing 11 different chlorophenols.

Haloalcohol degradation

Haloalkane halidohydrolases are not only active towards haloalkanes, but also catalyze the dehalogenation of haloalcohols. Bacterium belonging to the *Flavobacterium, Pseudomonas, Arthrobacter* and *Corynebacterium* genera can utilize compounds such as 1,3-dichloro-2-propanol,

3-chloro-1,2-propanediol and 1-chloro-2-propanol and in some instances epichlorohydrin as carbon and energy sources (Nagasawa *et al.*, 1992).

Ligninases

Ligninases are a group of extracellular enzymes which catalyze the degradation of lignin and include lignin peroxidases, aryl methoxy demethylase and phenol oxidase. Lignin peroxidase is the enzyme produced by the most studied white rot fungi *Phanerochaete chrysosporium*. Ligninases have been shown to oxidize polyaromatic hydrocarbons such as 1) biphenyls like pyrene, anthracene and dibenzo (p) dioxin, 2) chlorinated aromatics like 4-chlorobenzoate, pentachlorophenol and 2,4,5-trichlorophenoxyacetic acid 3) polycyclic chlorinated aromatics 4) chlorinated alkylhalides such as Lindane, Chlordane and a range of triphenyl methane dyes.

Biodegradation of Lignin and Tannin

Chlorinated lignin derivatives in the bleach plant effluents can be degraded by white rot fungus *Trametes versicolor* when glucose is provided as co-substrate 90% reduction in colour is obtained in 5 days Simultaneously concentration of chloroorganic compounds decreases by 45% efficient colour removal from such effluents can be achieved by immobilizing the fungus in calcium alginate beads. White rot fungi have a great potential in treating effluents from paper mills. Using *P. chrysosporium* on rotating biological contactor 70% of organic chlorine in bleach plant effluent can be removed in less than 48 (Jogdun, 1995).

Microbial desulflonation has been attempted by using mixed culture of anaerobic sulphate reducing bacteria and *Pseudomonas* sp. growing yeasts or fungi like *Paecilomycetes variotii*, *Sporotrichum pulverulentum* and *Chaetomium cellulolyticum* is used for SCP on sulphate waste.

The outstanding microbiological characteristic of lignin is its resistance to enzymatic degradation. It seems probable that the first attack in lignin decomposition is due to the exoenzymes. Because lignin-decomposing fungi usually excrete phenolic oxidases they have been thought responsible for the cleavage of aromatic compounds. However it is not to demonstrate the function of phenolic oxidase in lignin degradation especially in the cleavage of c-c or ether bonds.

The rate and extent of lignin decomposition is affected by temperature, availability of nitrogen anaerobiosis and by constituents of plant residue undergoing decay. Lignin is depolymerised to give simple aromatic substances such as vanillin, and vanillic acid or possibly other methylated aromatic structures. Among the chemical changes that occurred in the molecule are removal of some of the side chains of aromatic nucleus and decreases in the number of methoxyl groups. In the final stage the remaining methoxyls are removed with the formation of hydroxy benzene derivatives that are ultimately cleaved to give low molecular organic acids. On the other hand a partial or complete methoxyl loss may precede liberation of simple aromatic products.

Simple compound formed during lignin metabolism the fungi that primarily attack lignin include *Polystictus versicolor* and *Stereum hirsatum*. Other fungi such as *Pleusotus ostreatus*, *Ganoderma applanatum*, *Polyporus adustus*, *Amillaria mellea* attack lignin and cellulose simultaneously. Lignin decomposition can also be demonstrated in number of the following genera *Pholiota, Clitocybe, Lenzites, Pomus , Poria, Trametes* and others.

Microorganism such as *Aspergillus fumigatus* and *Pseudomonas* species when grown on chrome waste from tannery waste can leach out chromium. Fungi can be used to control pollution from tannery effluents. *Aspergillus niger, Chaetomium cupreum* can degrade tannin in the presence of utilzable carbon source like glucose.

Heavy Metal Biosorption by Microbial Cells

Aqueous effluent streams originating from production processes can contain heavy metals that due to their chemical and toxicological properties cannot be discharged directly into the environment. Heavy metals are of great concern when found in high concentrations in soils, surface water and ground water, due to their toxic nature. The non-biodegradability of these elements is responsible for their accumulation in the biosphere and translocation through food chain. To reduce the concentration of these metals conventional physico-chemical procedures ie, precipitation, reduction and oxidation can be applied. However, because such processes may not be very suitable for some concentrations of heavy metals, methods using microorganisms for these purposes are getting more attention (Shumate *et al.*, 1978). Some microorganisms have developed different and efficient mechanisms for tolerating high levels of metal content and exhibit a number of enzymatic or non-enzymatic processes for the uptake and accumulation of heavy metals (Gadd, 1990; Silver and Mira, 1988; Silver and Walderhaugh, 1992).

Microbial biomass appears to involve the combination of different processes like adsorption and absorption (Mchale and Mchale 1994). The metals thus bound can be recovered by eluting with an acidic solution without affecting the metal ion absorption capacity of the bacterial cells (Nakajima and Sakaguchi, 1996).

Role of Bacteria in Biosorption

A lead resistant strain of *Brevibacterium iodinum* has been found to grow on concentration of Pb $(NO3)^2$ as high as 7.2 mm. *Brevibacterium iodinum* immobilized about 4.0mg of lead per gram of dry cells. The uptake was accomplished after 1 hour (Andreoni *et al.* 1994). The microorganism besides being resistant to lead, chromate and arsenate and tolerant other heavy metals such as chromium, zinc, cobalt, mercury, aluminium and nickel was found to be multiple antibiotic resistant. Pseudomonas fluorescence immobilized on polyvinyl chloride granules and packed into columns at the rate of 588 g/lit column volume has been capable of removing metals such as, lead and zinc (Jogdund, 1995). *Pseudomonas aeruginosa* on an oxygen plasma treated poly propylene web were able to biosorb plutonium, however, no attempt has been made to recover plutonium. *Citrobacter* 'restingcells' were capable of removing cadmium, copper, lead and uranium. *Brevibacterium* immobilized cells are also able to accumulate cadmium. When lead and cadmium were fed in pairs into the column, the presence of lead influenced the cadmium removal, and so cadmium was removed to a lesser extent (Vecchio *et al.*, 1998).

Researchers at Miyazaki medical college have succeeded in recovering precious metals like gold and palladium present in industrial or mining wastewater. Microbes can act as adsorbents to remove metals, even in ppm amounts, thus trapped by adsorption, and be easily recovered further by using thiourea as an intermediary. The recovery rate is 98.6% by

Pseudomonas, Microccocus leutus, Streptomyces phacermogenes which are capable of acting as adsorbents. Both free cells as well as immobilized forms can be used. *S. phareotiles* natants have been used in the accumulation of iron , copper, magnesium, cobalt and cadmium. A bacterium *Proteus mirabilis* was able to grow in the presence of Cd^{2+} upto 300 mg/l. Accumulation of cadmium is reported for growing and non growing cells of the organism and the metal is was shown to be associated with the cell envelope (80%) as also accumulation in the cytoplasm (20%) (Andreoni *et al.*, 1991).

Application of fungi in biosorption

Active research is going on where removal of heavy metals and radioactive substance from effluents is done using free and immobilized fungal mycelium cells walls of fungi contain many anionic sites such as amines, carboxyl, hydroxyl and phosphate groups.

Rhizopus arrhizus can take up 170 mg radionuclide per gram dry weight. They also have the ability to adsorb between 30-130 mg cadmium per gram dry matter depending on the treatment applied and ph (Jogdund, 1995). *Aspergillus niger* waste mycelium can selectively remove silver from solution. Adsorption is intensive at extremes of pH and temperature. However, desorption of silver cannot be done easily. *Aspergillus niger* was grown in a leach liquor containing metals and high cyanide concentrations. The strain was capable of removing 64.42 % of gold, 100 %silver, 58.74 %of copper, 80.34 % of iron and 73.67 % of zinc by solution. The highest removal of metals occurred in the period between 60 h and 72h of incubation (Gomes *et al.*, 1998).

Many fungal species such as *Aspergillus niger, Pencillium chrysogenum* and *Claviceps paspali* have been used to remove zinc from effluents in the fermentation industry. Two patents exist for the use of growing heat-killed biomass of fungi for recovery of platinum, gold, copper, and zinc. The process is uneconomical. *Trichoderma viride* immobilized in china clay in columns can be used to remove copper from effluent buffered at pH 5.5. The possible range of removal is 5.94 - 127.5 mcg/ml. *Aspergillus oryzae* immobilized on reticulated foam can be used for removal of cadmium with 99%efficiency. Silver biosorption has been possible using *Saccharomyces cerevisiae* (Singleton and Simmons, 1996). Three strains of *Saccharomyces cerevisiae* and one strain of a *Candida* sp. have been found to uptake silver and copper. Considerable differences in metal uptake capacities were found between the different strains of *Saccharomyces cerevisiae* and between *Saccharomyces cerevisiae* and the *Candida* sp. used. Copper uptake capacities ranged from 0.05 mol/g dry wt to 0.184 mol/g dry weight while values of 0.034 mol/Ag/g dry weight and 0.193 mmolAg/g dry weight biomass were observed (Simmons *et al.*, 1995). Removal of nickel from the effluent of electroplating industry by *Saccharomyces cerevisiae* has been studied, the maximum removal has been found to be at 2.5 g of biomass wither five hours of treatment. The efficiency was observed to be influence by biomass, time, pH, and temperature, stationary and shaking conditions. The bound metal was recovered by 0.1 N HCl and the biomass was changed to be reusable. In the regeneration cycle, the efficiency was not affected in first two cycles, however it decreased insignificantly in the third cycle (Gill *et al.*, 1996). The brewer' yeast strain of *S. cerevisiae* have been reported to possess the most pronounced ability to remove uranyl ions from solution in both it living and nonliving form. The *S. cerevisiae,* which contains glucan and mannan and traces of chitin in its cell wall, sequestered 0.4 mmmol u/g cells at an equilibrium concentration of 0.7 mm (Volesky and Phillips, 1996).

Role of algae in biosorption

A large number of algae have been involved in the bioaccumulation or biosorption of heavy metals and other metal ions from polluted waters. The effective uptake of 50%calcium by *Oscillatoria salina* BDU 10244 and *Aphanocapsa* BDU 16 from hard-to-treat ossein effluent has been well dernonstrated. The increased glutamine synthetase activity reported in a marine cyanobacterium *Nostoc calcicola* indicates the non-toxicity of the lead being actively taken by this organism (Srivalli, 1994). Petterson *et al.* (1986) have demonstrated the uptake of aluminium by passive diffusion in *Anabaena cylindrica*. It has been demonstrated that non-viable cells remove quantities of metals comparable to viable cells by physical adsorption. Report was started to account to the suitability of *Spirulina* to uptake chromium form the electroplating effluent. The columns were treated with phosphate (8.0 pH) to strip the bound chromium and 83% of the Cr+ was recovered. The ability of a marine cyanobacterium *P. valderianum* to reduce iron by 25% in a chlor -alkali effluent possibly indicates the siderophore production, which could be commercially, exploited metal recovery.

Spirulina maxima live cells and dry cells have been found to uptake cadmium (Augusto da Costo and De Franca, 1998). Reports have been documented on the biosorption of cadmium, copper and lead by cells of *Chlorella fusca* (Wehrheium and Wettern, 1994). The uptake of mercury and its regulation in a cyanobacteriuym *Nostoc calcicola* Breb. has been studied by Pandey and Singh (1993). Mercury uptake pattern consisted of two distinct phases, one rapid binding of the cation to the negatively charged cell surface and its subsequent metabolism dependent intracellular import. The biosorbent prepared form *Sargassum* algal biomass have been found to bind approximately 2.3 meq / g metal cations from water by ion exchange. The values of ion exchange equilibrium constants showed that the affinities of metals towards the biosorbent decreased in the following order Cu > Ca > Fe (Kratochil and Volesky, 1998). Biosorption of the nonliving brown algae *Sargassum fluitans* has been found particularly effective in binding metals ions of cadmium. High sorption capacity (exceeding 100 mg / g), easy regeneration and low costs make this biomass of special interest for the purification of high volumes of low concentration waste water (Volesky and Holan,1995).

Live and immobilized cells of *Saccharomyces cerevisiae* and *Corynebacterium* sp. have been used in the biosorption of lead and they have been efficiently found to uptake the heavy metals from the aqueous solutions (Table 6). Bagirath *et al.* (1999) report that cadium is biosorped from aqueous solution by *Spirulina platensis*. Biosorption technology is a very effective technology for tackling heavy metal pollution (Murugesan *et al.*, 2002).

Biopolymer Production Employing Microorganisms

Most of the food processing industries generate large amount of wastes. Some of the wastes are cellulosic in nature, containing surplus amounts of cellulose and lignin. These waste materials serve as good source of carbon for the production of biopolymers. In nature, bacteria most often synthesize molecules belonging to a class of polymers, known as the PHA (poly-beta-hydroalkanoate) family. The general structure of a PHA monomer has a three or four carbon structure with a single-bonded oxygen and a double-bonded oxygen on a one terminal carbon. One of the most commonly produced plastic polymer in bacteria is poly - β - hydroxybutyrate or PHB.

Biodegradable plastics such as PHA emerge as potential substitutes for overcoming the environmental problem created by the massive use of plastics. Furthermore, PHA get easily complexed by hydrogen bonding with other polymer like nylon polyethylene oxide polyvinyl acetate, polyvinyl chloride and polysaccharides forming copolymers of varied properties. Such polyesters can be used as biodegradable matrix for the sustained release of plant growth regulators or pesticides, or in controlled drug release, surgical sutures and bone plates. Commercially produced PHA, Biopol is a copolymer of p (3HB – CO – 3HV). The initial production of Biopol was started using *Azotobacter* sp., but the excessive uncontrolled production of polysaccharide led to use of methanol – utilizing organisms, methyl bacteria. Due to slow growth and low polymer content the commercial production of PHA was shifted to *Alcaligenes eutrophus*, glucose – utilizing mutant from original H16 strain (Bryom, 1987). Novel microbes with desirable physical properties and high productivity are still being discovered in the quest of biodegradable plastics.

Coir industrial waste is the recalcitrant lignocellulosic waste. This waste also contains 8 – 12 % soluble tannins like phenols that inhibit the microbial growth and activity. However, it is recorded that native microflora generally comprises fungi, bacteria and actinomycetes. These locally available forms can be used for the biotransformation of lignin and cellulose to a biopolymer. Reena *et al.* (2003) report that microbes such as *Aspergillus niger, A. flavus, Penicillium* sp., *Fusarium* sp., *Streptomyces* sp. and *Bacillus* sp. can be employed for the bioconversion of coir waste into a compound similar to PHB. The starchy waste, Thippi from the cassava industry is an ideal substrate for the biopolymer production. From the works of Rathiga and Murugesan (2002) it is evident that microbes such as *Ralstonia eutrophus, Alcaligenes* sp., *Arthrobacter viscosus* and *Azotobacter beijernicksii* can be utilized for the production of biopolymer from cassava waste.

Organic Waste Management

In India, at least 60 % of the solid wastes are organic in nature. Being rich in plant nutrients these wastes are a major source of pollution and disease where they decay. The ideal solution to this problem is to turn them into products of economic value. Some potent techniques available to bioprocess organic solid wastes into useful products are composting and vermicomposting.

Composting

This process use microorganisms to degrade the waste organic matter. To carry out this process it is necessary to prepare a bed. This involves the separation of the waste into the organic and inorganic part. It is also necessary to add nitrogen, if deficient in the raw material. Digestion involves most of the microbiological action. The waste is piled in windrows and regularly turned to ensure adequate aeration, preferably by a machine. It takes about two to three weeks for complete digestion (Abassi and Ramasami, 1999). The process that follows next is termed as curing, which allows further decomposition of the cellulose and lignin, which has not been digested. It takes about three months for curing. This is followed by screening, which is to remove the remaining inorganics, drying, grinding and pelletization of the compost and make it ready for marketing.

Vermicomposting

Role of earthworms in the improvement of soil fertility and of plants in absorbing gaseous pollutants and aquatic weeds in absorbing and concentrating pollutants has been known since several decades. However, these agents were not commercially utilized for pollution control. But in the last two decades, vermicomposting had found commercial applications. Pretreated non-toxic solid and liquid organic wastes, sewage sludge, agricultural wastes, animal wastes, dairy wastes, wastes from distilleries and hatcheries etc. can be effectively tackled by vermicomposting. Vermicomposting is a small scale, low cost technology approach and uses locally available raw material.

Certain species of earthworms can consume organic residuals very rapidly and fragment them into much finer particles by passing them through their gizzard. These organisms derive their nourishment from the microbes that grow upon the organic matter. They also promote further microbial activity in the residuals so that the faecal matter or 'casts' that they produce are much more fragmented and microbially active than what the earthworms consume. During this process the important plant nutrients in the organic material, particularly nitrogen, phosphorus, potassium and calcium are released and converted through microbial action into forms that are much more soluble and available to plants than those in the parent compounds (Edwards, 1995).

There are several observations, based on which the earthworms have been subjected as potential tools for waste management.

a) Earthworms dominate the soil faunal biomass in many ecosystems. Studies on energetics show that they assimilate at least 13% of the net annual primary production in grasslands in India (Dash, 1978).

b) Earthworms break up the organic matter, thereby decreasing particle size and increasing the surface area for microbial action. Earthworms also mix humified material into the soil and produce about 10 - 89 tonnes of worm castes/ha/year (Satchel, 1967).

c) By means of nucleus production, nephridial excretion and from their dead tissues, earthworms contribute about 99-198 kg of nitrogen to the soil/ha/year (Dash, 1978).

d) Earthworms posses the ability to lower the ratio of the litter. The C/N ratio of freshly fallen litter ranged from 38.2 : 1 to 90.6 : 1 (Wittich, 1953). Observations made by Dash (1978) in his studies on vermicomposting revealed that earthworms could lower the C/N ratio of the soil from 22.15 : 1 to 15.13 : 1 within 25 days.

Recombinant DNA Technology for Waste Treatment

Genetically engineered microbes can be useful in several areas of waste treatment, among which are biomass production, biodegradation of recalcitrant molecules, fermentation (methane and organic acid production), enhancement of enzyme activity, and increased resistance to toxic inhibition. Waste treatment processes can be improved by selection of novel microorganisms that can thrive in domestic wastewater and perform a desirable function. There are two steps involved in the application of GEMS for waste treatment. The first step consists of finding a microorganism that has the desirable function (e.g. the ability to degrade a pesticide).

The second is the transferring of the desirable function to the suitable host preferably a microorganism with some relevance from an environmental viewpoint.

Potential applications include the use of genetically engineered multi – plasmid *Pseudomonas* strains for degrading several components found in crude oils. GEMS have now become a handy tool to the biotechnologists. Kellog *et al.* (1981) by their plasmid assisted molecular breeding could isolate *Pseudomonas ceparia* strain AC1100 that harbored two plasmid and could finally degrade 2,4,5-T. It has been possible to genetically transfer TOC plasmid and PCB degrading gives from *Pseudomonas putida* strain LB400 to *E. coli.* Multiplasmid system has also been constructed through hybridization with plasmids encoding other genes like CAM, HAA, OCT etc, so that the resultant strain with broad range degradative power could grow in crude oil several other potential GEM are listed below.

Substrate	Gem
Alkane	*Pseudomonas oleovorans*
4-chlorobenzene	*Actinetobacter* sp.
Mono and dichloro aromatics	*Pseudomonas putida*
Parathion	*P. diminuta*
2,4.d	*Alcaligenes* sp.
2,4,5-t	*P. cepacia*
SDS(detergents)	Pseudomonas sp.

Most of these degradative genes are found in the microorganisms of waste sites and it is essential to isolate than for the development of efficient biodegraders through the process of gene cloning (Labeda, 1970) pseudomonas is a versatile microbial species in providing a lot of degradative genes that can be cloned to others. Geneticallly EMS may also be structured in a way to either synthesize extracellular polymers as in zooglea to bird with toxic metals such as, chromium, cobalt, copper, nickel etc. and help precipitate them from polluted water or synthesize phosphatase enzyme on cell surface to cause metal precipitation from solution as in *Citrobacter*. So GEMS may prove to be 'biotechnological wonders' in the forthcoming years including environmental clean up and management (Fry and Day, 1993).

Summary

Biotechnological strategies are applied at varied levels. In recent years the microbial machineries have been utilized for the transformation of waste and toxic pollutants into complex chemical structures that are difficult to assemble chemically. Bioremediation and biosorption techniques serve as potential tools in environmental clean up. The nuisance aquatic weeds have the ability to absorb toxic pollutants and hence these macrophytes and several other aquatic reeds also play a vital role in environmental clean up. Biotechnological tools aid in enhancing the activity of the microbes by employing advanced tools such as genetic engineering in choosing the gene of interest and deleting the unwanted ones. Such altered microbes are efficient candidates in environment management.

References

Abassi, S. A. and Ramasami, E. (1999). Solid waste management with vermicomposting. In: Biotechnological methods of pollution control. University Press, Hyderabad, pp. 168.

Andreoni, V., Fasoli, R., Sorlini, C. and Finoli, C. (1994). Investigations of lead resistance in a *Brevibacterium* strain. Ann. Microbiol. Enzimol, 44, 153.

Andreoni, V., Finoli., Manfrin, P., Pelosi, M. and Vecchio, A. (1991). Studies on the accumulation of cadmium by a strain of *Proteus mirabilis.* FEMS Microbiology Ecology, 85: 183 – 192.

Apajalahti, J. H. A., Karpanoja, P. and Salkinoja-Salonen, M. S. (1986). *Rhodococcus chlorophenolicus* sp. nov., a chlorophenol-mineralizing Actinomycete. Int. J. System. Bact, 32 (2): 246 – 251.

Augusto da Costa, A. C. and De Franca, F. P. (1998). Cadmium uptake by *Spirulina maxima* : toxicity and mechanism. World Journal of Microbiology, 14: 579 – 581.

Banerji, G. (1994). Ph.D thesis,IIT, Kharagpur. India.

Becker, J. M., Caldwell, G. A. and Zachyo, E. A. (1990). Biotechnology. A Laboratory Course. Academic Press, Orlando, FL.

Bedell, G.W. (1992). Phytoremediation. The World & I, 12: 260-263.

Belay, N. and Daniels, l. (1987). Production of ethane, ethylene and acetylene from halogenated hydrocarbons by methanogenic bacteria. Appl. Environ. Microbiol., 53: 1604 – 1620.

Bryom, D .1987. TIBTECH. 5: 246 – 250.

Clark, R. R., Chian, E. S. K. and Griffen, R. A. (1979). Degradation of polychlorinated biphenyls by mixed microbial cultures. Appl. Environ. Microbiol., 37 (4): 680 – 685.

Dash, C. M. (1978). Role of earthworms in the decomposer system. J. S. Singh and B. Gopal (eds). Glimpses of ecology. International Scientific Publications. 399 – 406.

Dorsh, R. (2002). Sustainable materials and chemicals for the next generation. In 'The Biobased Economy of the Twenty – first Century', National Agricultural Biotechnology Council, Repotr 12, pp. 69 – 74.

Edwards, C. A. (1995). Historical overview of vermicomposting. Biocycle. 56 – 58.

Flessa, H. (1994). Plant-induced changes in the redox potential of rhizospheres of the submerged vascular macrophytes *Myriophyllum verticillatum* L. and *Ranunculus circinatus* L. Aquatic Botany. 47: 119 – 129.

Fry, J. C. and Day, M. (1993). Release of Genetically Engineered and other microorganisms. Cambridge University Press, Cambridge.

Gadd, G, M. (1990). Heavy metal accumulation by bacteria and other microorganisms. Experientia, 46, 834 – 840.

Gomes, N. C. M., Camargos, E. R. S., Dias, J. C. T. and Linardi, V. R. (1998). Gold and silver accumulation by *Aspergillus niger* from cyanide-containing solution obtained from gold mining industry. World Journal of Microbiology and Biotechnology, 14: 149.

Hardman, D. J. (1991). Biotransformation of halogenated compounds, Crit. Rev. Biotechnol., 11 (1): 1 – 40.

Howes, B. L. and Teal, J. M. (1994). Oxygen loss from *Spartina alterniflora* and its relationships to salt marsh oxygen balance. Oecologia. 97: 431 – 438.

Janssen, D. D., Scheper, A., Dijhuizen, I. and Witholt, B. (1985). Degradation of halogenated aliphatic compounds by *Xanthobacter autotrophicus* GJ 10. Appl. Environ. Microbiol., 49: 673 – 677.

Jayantha, K. S. and Ramanujam, T. K. (1995). Applicability of upflow anaerobic sludge blanket (UASB) reactor for treatment of distillery spentwash. Indian J. Environmental Protection, 15 (2): 106 – 110.

Jensen, H. L. (1960). Decomposition of chloroacetates and chloropropionates by bacteria. Acta Agricult. Scand, 10: 83 – 103.

Jogdund, S. N. (1995). Envrionmental Biotechnology. Industrial Pollution Management. Himalaya Publishing House, Bombay, pp. 235.

Kellog, S. T., Chatterji, D. K. and Chakraborty, A. M. (1981). Plasmid assisted molecular breeding: a new technique for enhanced biodegradation of persistant toxic chemicals. Science, 214: 1133 – 1135.

Kennedy, M. S., Grammas, J. and Arbuckle, W. B. (1990). Parachlorophenol degradation using bioaugmentation. J. Water Pollut. Control Fed. 62: 227 – 233.

Kratochvil, D. and Volesky, B. (1998). Biosorption of Cu from ferruginous wastewater by algal biomass. Biosorption Ferruginous Mine Drainage. 1 – 6.

Labeda, D. P. (1990). Isolation of Biotechnological organisms from Nature. Mc Graw – Hill, New York.

Lam, T. and Vilker, V. L. (1987). Biodehalogenation of bromotrichloromethane and 1,2-dibromo-3-chloropropane by *Pseudomonas putida* PpG-786. Biotech Bioeng, 24: 151 – 159.

Little, C. D., Palumbo, A. V., Herbes, S. E., Lidstrom, M. E., Tyndall, R. L. and Gilmer, P. J. (1988). Trichloro ethylene biodegradation by a methane-oxidizing bacterium. Appl. Environ. Microbiol., 54: 951 – 956.

Lowries, P. and Wells, S. (1991). Microorganisms, Biotechnology and Disease, Cambridge University Press, Cambridge.

Mchale, A. P. and Mchale, S. (1994). Microbial biosorption of metals. Potential on the treatment of metal pollution. Biotech, Advances. 12: 632 – 667.

Murugesan, A. G. and C. Rajakaumari. 1997. Application of Upward Anaerobic Sludge Blanket Reactor for treating pulp and paper mill effluent. Proc. Vth Sem. Tamil Nadu State Council. Sci. Tech. 104 – 106.

Murugesan, A.G., Navaneetha, N. and Pratheeba, V. (2002). Microbial biosorption of heavy metals a potential technology for effective pollution. Environics 2002, Proc. National Conf. Poll. Contr. Indust. Ecol., 95-98.

Nagasawa, T., Nakamusa, T., Yu, F., Watanbe, I. and Yamada, H. (1992). Purification and characterization of halohydrin hydrogen halidelyase from a recombinant *Escherichia coli* containing the gene from a *Corynebacterium* sp. Appl. Environ. Microbiol., 36: 478 – 482.

Nakajima, A. and Sakaguchi, T. (1986). Accumulation of Ni^{2+} by bacteria, yeast and fungi. Appl. Microbiol. Biotechnol. 24: 59 – 64.

Pandey, K. and Singh, S. P. (1993). Hg^{2+} uptake in a cyanobacteria. Current Microbiology, 26: 155 – 159.

Petterson, A., Hallbon, l. and Bergmann, B. (1986). Aluminium uptake by *Anabaena cylindrical,* J. Gen. Microbiol, 132.

Quensen, j. F. I., Tiedje, J. M. and Boyd, S. A. (1988). Reductive dechlorination of polychlorinated biphenyls by anaerobic microorganisms from sediments. Science, 242: 752 – 754.

Rao, C. S. (1991). Environmental pollution control design, Wiley Eastern Ltd., New Delhi.

Reddy, K. R. and Smith, W. H. (1987). Aquatic plants for water treatment and resource recovery. Magnolia Publishing Inc., Orlando. Pp. 1032.

Reena Mol, Murugesan, A. G. and Potty, V. P. (2003). Microbial production of biopolymers from coir industrial waste. Proc. Sem. Tamil Nadu State Council for Sci. and Tech. 312 – 314.

Satchell, J. E. (1967). Lumbricidae. A. Burgers and F. Raw (eds). Soil biology. Academic Press, London and New York. 259 – 322.

Shumate, S. E., Strandberg, G. W. and Parraot, J. K. (1978). Biological removal of metal ions from aqueous process streams. Biotechnology and Bioengineering Symp. No. 8, 13 - 20.

Silver, S. and Mishra, T. K. (1988). Plasmid – mediated heavy metal resistances. Ann. Rev. Microbiol. 42, 717 – 743.

Silver, S. and Walderhaugh, M. (1992). Gene regulation of plasmid and chromosome – determined inorganic ion transport in bacteria. Microbiol. Rev. 56, 195 – 228.

Simmons, P., Tobin, J. M. and Singleton, I. (1995). Considerations on the use of commercially available yeast biomass for the treatment of metal containing effluents. Journal of Industrial Microbiology, 14: 240 – 246.

Srivalli, B. (1994). Effect of lead on nitrogen metabolism of the marine cyanobacteria *Nostoc calcicola* BDU 40302, M.Sc dissertation,Bharathidasan University, Tiruchirapalli.

Stirling, D. I. and Dalton, H. (1980). Oxidation of dimethyl ether, methyl formate and bromomethane by *Methylococcus capsulatus* (Bath). J. Gen. Microbiol., 116: 277 – 283.

Uma, L. and Subramanian, G. (1990). Effect of cyanobacteria in effluent treatment, Proceedings of the National Symposium Cyanobacterial Nitrogen Fixation, Indian Agricultural Research Institute, New Delhi, 437 – 444.

Vannelli, T., Logan, M., Arciero, D. M. and Hooper, A. B. (1990). Degradation of halogenated aliphatic compounds by the ammonia oxidizing bacterium *Nitrosomonas europaea*. Appl. Environ. Microbiol., 56: 1169 – 1171.

Vecchio, A., Finoli, C., Simine, D, D. and Andreoni, V. (1998). Heavy metal biosorption by bacterial cells. Presenius J Anal. Chem. 361: 338 – 342.

Volesky, B. and Holan, Z. R. (1995). Biotechnol. Prog, 11: 235 – 250.

Volesky, B. and Phillips, H. A. M. (1995). Biosorption of heavy metals by *Saccharomyces cerevisiae.* Appl. Microbiol. Biotechnol. 42: 797 – 806.

Wehrheim, B. and Wettern, M. 1994. Biosorption of cadmium, copper and lead by isolated mother cell walls and whole cells of *Chlorella fusca.* Microbiol. Biotechnol, 41: 725 – 728.

Wittich, W. (1953). Untersuchungen uber den verlauf der strenzersetzung auf. Linem. Boden mit Regenwarntatigkeit. Schriff Reiche. Fak. Univ. Gottingen, 9: 7 – 33.

Zaranyika, M. F., Mutoko, F. and Murahwa, H. (1994). Uptake of zinc, cobalt, iron and chromium by water hyacinth (*Eichhornia crassipes)* in lake chiveroh, Zimbabwe. Science of the Total Environment, 153: 117 – 121.

34

Biotechnological Practices for Sustainable Livestock Production and Waste Management

Ajoy Mandal, P.K.Sahoo and T.K.Dutta

Introduction

Livestock is an important driving force and national resource for development and sustaining livelihood in all the developing countries and this sector is predicted to become the most important agricultural sector. Domestic animals are a crucial component in sustainable livestock production in meeting future global food requirements. There is a growing world demand for livestock products with important implications for rural poverty, hunger, national trade balances and the environment. Human population, increasing urbanization and rising incomes are fuelling a massive demand for food of animal origin in developing country including India. Globally, livestock production is growing faster than any other agricultural sector. In view of its substantial dynamics, this process has been referred as to as the "Livestock Revolution". Animal husbandry practices are cohesively associated with the livelihood and social-ecosystem either in rural or urbanized condition. They contribute to the basic inputs of the family in terms of money, work or social status maintenance. In India, the value of the output from livestock and fisheries sector together accounts for 27% of the value of the output from agricultural and allied fields. At present, India is the largest producer of milk in the world reaching 86 million tones. Meat production is estimated as 4.5 million tones (FAO, 2002). In spite of this production potential of livestock sector, the country is unable to maintain proper protein-nutrition balance to our growing population. Thus, the country is facing with the challenge to rapidly increase agricultural productivity to feed their growing populations without depleting the natural resource base. Besides this, animal wastes (such as urine, faeces or slaughterhouse waste) from farms and livestock/poultry and dairy production are increasing day by day with the growing livestock population in the country. These waste materials can pollute the environment by contributing excess nutrients, pathogens, organic matter, solids, and odorous compounds to the environment. These may be hazardous, if it is not used judicially and scientifically for the benefit of human population. Therefore, concentrated effort in terms of latest scientific approach is required to enhance the productivity of animal origin and to reduce the environmental pollution, caused by animal wastes.

Biotechnology is regarded as a means to meet all objectives through addressing the production constraints of small scale or resource poor farmers and managing the waste materials. The term "biotechnology" or "biologically based technology" can be defined as *any technique that uses living organisms or substances from such organisms to make or modify a product, to improve animals or plants or to develop microorganisms for a specific purpose.* It generally refers to heterogeneous body of knowledge that falls into three general areas, namely genetic engineering, fermentation technology and enzyme technology. Biotechnology, therefore, is not a new development and despite of its current momentum, it has a deep-rooted history in human activities. Man has used it for thousand of years to manufacture products such as beer, wine and bread etc. The conventional plant and animal breeding, which involves selection and mating of phenotypically preferred individuals, is a good example of age-old application of biotechnology. Biotechnology has achieved some dramatic advances in recent years in both crop and livestock production. These include transferring a specific gene from one species to another to create a transgenic organism; the production of genetically uniform plants and animals (clone); embryo manipulation technology and the fusing of different types of cells to produce beneficial medical products such as monoclonal antibodies. These have enhanced the possibilities for manipulating biological systems for the benefit of mankind. Thus, biotechnology offers considerable potential for improvements in sustainable livestock production and agro-industrial processing, particularly for more environmental friendly and energy-efficient processes.

Today, biotechnology has a number of applications in livestock production. It is being used to hasten animal growth, enhance reproductive capacity, improve the animal health and develop the new animal products. The role of biotechnology to offer new opportunities to enhance the productivity and efficiency of animals through nutrition, reproduction, health and welfare is now well recognized (Robinson and McEvoy, 1993; Bonneau and Laarveld, 1999). Among agricultural and allied fields, animal production and heath have probably benefited the most from biotechnology. Biotechnology can play a crucial role to convert the wastes of livestock origin into useful products. But successful application of biotechnology for the improvement of livestock production in developing countries has lagged behind which needs special attention.

Status of Available Biotechnology Research

Biotechnologies cover a vast field of application in animal production. The priority areas identified in the developing countries where the biotechnology has potential application are: (i) production of monoclonal antibodies for disease diagnosis, (ii) biotechnological approach to improve low quality forages and agro-industrial by-products as livestock feed, (iii) genetic and non-genetic manipulation of rumen micro-flora, (iv) embryo transfer technology, and (v) gene cloning and gene transfer. The application of modern genetic technologies in livestock such as genome mapping, marker assisted selection and transgenesis has lagged behind its application in the plant sector. While genetically modified livestock, in contrast to genetically modified crops, are not likely to play a major role in developing countries in near future, considerable potential exists for the application of biotechnologies in the use of bio-engineered inputs covering the entire food production chain from animal feed to product proceeding, importantly also including animal health.

Biotechnology in Livestock Production

The overview of available biotechnologies with current and/or potential application in the areas of genetics and animal breeding, animal nutrition, reproductive physiology, physiology of lactation & growth and animal health are highlighted.

Animal genetics and breeding

Genetic improvement of farm livestock is a powerful technology, which can influence changes to livestock production systems. Each species has its own optimal balance, which depends upon variables such as genetics of the trait involved, the reproductive rate and the technology available to adjust the rate and particular economics of the situation. The potential application of new biotechnologies for the acceleration of genetic change has been an important factor in encouraging their detailed study in livestock and for this reason, some general points about the nature of such change may be helpful. Genetic improvement of livestock depends on access to genetic variations and effective method for exploiting this variation. In other words, genetic progress of any trait of economic importance primarily depend on four main factors, its rate increasing with the amount of genetic variations, with the accuracy and intensity of selection and with the genetic intervals. Genetic diversity constitutes a buffer against changes in the environment and is a key in selection and breeding for adaptability and production on a range of environments.

In developed countries, breeding programs are based upon the performance recording and this led to substantial improvement in animal production. Developing countries have distinct disadvantages for setting up successful breeding programs due to lacking of infrastructure facility needed for performance testing of animal. This problem mainly arises because of small herd size and large variability exists between farms, farming systems and seasons, low reproductive efficiency due to poor nutrition and grazing, and animal health programs. A number of techniques that are being developed at present have the potential to contribute greatly to animal genetics and breeding.

Multiple ovulation and embryo transfer (MOET) scheme and open nucleus breeding system

Various reproductive technologies such as super-ovulation, embryo recovery from donors and embryo transfer to recipients has opened the possibilities for increasing the reproductive rate of the females to a significant degree in cattle, sheep, goat and pig. Multiple ovulation embryo transfer (MOET) is a composite technology, which includes super-ovulation, fertilization, embryo recovery, short-term *in vitro* culture of embryos, embryo freezing and embryo transfer. This technique allows the acceleration of genetic progress through increased selection intensity of females and freezing of embryos enables low cost transport of genetic material across continents. Other benefits from MOET also include increasing the number of offspring produced by valuable superior females, increasing the population base of rare or endangered breeds or species, *ex-situ* preservation of endangered populations and progeny testing of females. In several countries over the past decades multiple ovulation and embryo transfer has been used in nuclear breeding schemes to accelerate genetic changes in dairy animals. However, problems arise in practice: variation in embryo recovery following super-ovulation extends genetic intervals and can reduce selection differentials; furthermore, evaluation and selection within the nucleus will encourage inbreeding. The nucleus herds established can also serve as a focus for the application of further reproductive technologies. A more exact counterpart to AI than super-ovulation and embryo transfer are the *in vitro* techniques of maturation (IVM) and fertilization (IVF) of oocytes (Gordon and Lu, 1990). They may be useful in the production of crossbred beef embryos for transfer to the dairy herd or as recipient cells during nuclear transfer, but not in selection schemes.

Nicholas (1979) was the first to examine the possible impact of a MOET nucleus scheme on dairy cattle development. He showed that rates of gain equal to those possible from an efficient national progeny testing scheme could be made in a 500-cow herd. The idea was later elaborated (Nicholas and Smith, 1983) and since then, MOET nucleus breeding scheme have been subject of much interest. In MOET scheme as envisaged by Nicholas and Smith (1983) increment of annual genetic gain by 30-80% is possible. Some workers concluded that the earlier figures were over predictions. The realistic average number of live progeny per donor flushed is in the range of 2-3 in sheep and cattle and 6-8 in goats (Macmillan and Tervit, 1990).

Consideration of these figures suggests that MOET could increase the annual genetic gains by 10-20% in large breeding schemes. However, costs of operating such schemes in developing countries need to be evaluated before they can be recommended.

Genetic improvement of ruminants in developed countries has made much progress through the use of large-scale progeny testing of males. But the progeny testing schemes in developing countries has been implemented without much success due to general failure of use of AI. The small herd/flock and uncontrolled breeding in communal grazing situations preclude implementation of progeny testing. Smith (1988a) suggested that the Open Nucleus Breeding System (ONBS) might be especially valuable for the developing countries.

The ONBS is a breeding scheme in which a "nucleus" herd/flock is established under controlled conditions to facilitate selection. The nucleus is created from the "best" animals by screening the base (farmers') population for outstanding females. These animals are then recorded individually and the best individuals chosen to form the elite herd/flock of the nucleus. The "nucleus" herd/flock would be open in the sense that the lowest yielding females (say about 20% to 30%) are culled every year and are replaced by procurement of that many high yielding females from the population, i.e., farmers' herd/flock. The nucleus herd is utilized entirely for production of males for breeding purpose in the population. With the availability of the technology of MOET, the elite female herd is used through MOET with superior sires to produce embryos, which are carried by recipient females from the base population. The resulting offspring are reared and recorded and the males among them are evaluated either based on the performance of their sibs and paternal half sibs or their own performance. The elite group of males with high breeding values for the specific trait is selected from this herd/flock and used in the base population for genetic improvement through natural service or AI. The level of the genetic response depends on the size of the scheme (i.e., number of participating herds/flock and the total number of animals) and the selection intensity. Though MOET improves the genetic progress substantially, it has been tried to operate MOET without ET technology in sheep in West Asia by FAO (Jasiorowski, 1990). However, availability of AI and ET, in addition to increasing rates of genetic gain, enhances the flexibility of the system. For example, germplasm from other population can be introduced easily through semen and/or embryos. Under ONBS-MOET scheme, the evaluation and the selection of males and females are conducted within the nuclear herd/flock, so there would be greater degree of control on the determinants of genetic change, i.e., intensity of selection, generation interval and estimation of changes in levels of inbreeding. The scheme also provides the opportunity to record information on more traits accurately than is possible in a decentralized progeny-testing scheme. The ONBS can be used for the improvement of an indigenous or exotic breed. It can also be used to improve a stabilized crossbred population.

Physiological indicator traits

Genetic improvement for dairy traits has been hampered by the fact that they can only be measured in the female. If it is possible to measure a trait significantly correlated with milk yield in both sexes before reproductive age, this would be advantageous. Indicator traits are characteristics, which are genetically correlated to traits of economic importance and are easier to measure than the latter. Such traits are usually not the target of genetic improvement but provide an indirect means of improving a targeted trait. Indicator traits can improve genetic response by increasing accuracy of selection and reducing generation interval. The value of an indicator trait will depend largely on the magnitude of co-heritability (square root of the product of heritability of the indicator and of the target trait) and the genetic correlation between the two traits (Woolliams and Smith, 1988). Some physiological and/or metabolic characteristics may be used as potential indicator traits. Traits such as testicular size in rams or bulls or FSH in ewe lambs (Bodin *et al.*, 1986) have potential as indirect predictors of fertility. Woolliams and Smith (1988) studied the effect of an indicator trait, such as blood urea nitrogen level, on genetic response for milk yield. They showed that the possible increase in genetic gain due to using indicator trait was greater with MOET nucleus schemes than with progeny testing scheme.

Packed cell volume (PCV), an indication of the extent of anemia, is widely used as indicator trait (Baker, 1998) for pathological conditions associated with anemia. As for example, PCV is used as an indicator of the effect of trypanosomiasis and hence of trypanotolerance and also as an indicator of effect of the endoparasite *Haemonchus contortus* and hence as an indicator of resistance to the parasite.

Genetic marker and marker-assisted selection

Genetic progress of any trait through traditional selection procedure involves the increase in frequencies of favorable alleles through selection on the phenotype. The mapping of genes controlling traits of value (quantitative trait loci or QTL) provides the possibilities of increasing the frequencies of favorable alleles directly by selection of marker alleles known to be linked to the favorable QTL allele. A genetic marker for a trait is a DNA segment, which is associated with, and hence segregates, in a predictable pattern as, the trait. Genetic markers facilitate the "tagging" of individual genes or small chromosome segments containing genes, which influence the trait of interest. The likelihood of detection of major genes influencing quantitative traits has enhanced due to availability of large numbers of such markers. The method involves screening the genome for genes with a large effect on traits of economic importance through a procedure known as linkage analysis. The process of selection for a particular trait using genetic markers is called marker assisted selection (MAS). MAS can also accelerate the rate of genetic progress by increasing accuracy of selection and by reducing the generation interval. Ruane and Colleau (1995) found possible increase in selection response of only 0.2 to 1% in six generations using a single marker. However, the benefit of MAS is greatest for traits with low heritability and when the marker explains a larger proportion of the genetic variance than does the economic trait. Lande and Thompson (1990) suggested that about 50% additional genetic gain could be obtained if the marker explains 20% of the additive genetic variance and the economic trait has a heritability of 0.20. Ruane and Colleau (1996) found an increase of 6 to 15% from MAS in the selection response for milk production in cattle nucleus that used MOET in the first six generations of selection. Marker assisted selection could be used to enhance the introgression of alleles of value to one line to another, as an adjunct to traditional within-line selection or to allow new methods of selection.

Gene introgression : Having located a desired gene and genetic markers associated with it in an inferior breed, the aim of the introgression phase is to fix the favourable alleles in the commercial population with as little as possible of remainder of the genome from the inferior breed. The use of markers to help the introgression of one or a few QTL of values from one line to another line is probably where marker-assisted selection will reflect its great impact on animal breeding. The efficiency of the marker assisted introgression (MAI) programme depends on 1) the frequency of the introgressed allele in the final population and 2) the genetic progress of the traits of economic importance. The marker-assisted introgression (MAI) can enhance future prospects for breeding for resistance to environmental stresses, including diseases. Transfer of disease resistance genes could also have substantial impact in animal sciences because many of the efficient breeds cannot survive because of their susceptibility to diseases to which local, less efficient, breeds are resistance. Already, identification of carriers of genes for resistance and introduction of such genes into a population seems feasible for resistance against *Trichostrongylus colubriformis* and *Haemonchus* contortus (Gogolin-Ewens *et al.*, 1990). It should also be possible to eliminate factors predisposing sheep to Listeriosis or Salmonellosis (Blancou, 1990). There is also evidence for a major gene for resistance to the cattle tick (*Boophilus microplus*) in a Hereford x Shorthorn cattle line called Belmont Adaptaur (Kerr *et al.*, 1994). The mapping of alleles responsible for resistance to trypanosomiasis in West African N'Dama for introgression into susceptible zebu breeds (Soller, 1990) has already been made. Research is underway at International Livestock Research Institute (ILRI) in Kenya to identify genetic markers for resistance to endoparasites in Red Maasai sheep.

Within population selection : Genetic markers may provide information, which can be used as an adjunct to phenotypic information in within-line or breed selection programme. The detection and utilization of QTL is very difficult in an out-breeding population as the phase of linkage between a gene of value and marker is unknown and variation exists between individuals. Lande and Thomson (1990) suggested selection based upon an index of markers using the overall linkage disequilibrium between markers and QTL detected by multiple regressions. But this approach will only be effective for a few generations of selection in a population newly established from a line cross. An advantage of marker-assisted selection is that it allows within family selection and thus it may be used to reduce the inbreeding.

Recombinant DNA Technology

Transfer of gene or gene construct allows for the manipulation of individual genes rather than the entire genome. There have been dramatic advances in gene transfer technology in the last three decades. The first few reports of successful gene transfer in laboratory animals (Gordon *et al.*, 1980; Costantini and Lacey, 1981) stimulated some livestock scientists to become actively engaged in the transfer of genes in farm animals in 1985 and 1986 (Hammer *et al.*, 1985; Brem *et al.*, 1986). There are four requirements for an effective scheme which exploits gene transfer: it must be possible (1) to make the change, (2) to regulate expression of the gene in the desired manner, (3) to identify genes that are able to have a significant effect, and (4) to disseminate the changes into the target population. The technique has now become routine in the mouse and resulting transgenic mice are able to transmit their transgenes to their offspring thereby allowing a large number of transgenic animals to be produced. A transgenic animal is an animal whose hereditary DNA has been augmented by addition of DNA from a source other than parental germplasm through recombinant DNA techniques. Successful production of transgenic livestock has been reported for pigs, sheep, rabbits and cattle. The majority of gene transfer studies in livestock have, however, been carried out in the pig. Although transgenic cattle and sheep have been successfully produced, the procedure is still inefficient in these species.

The transgenesis offers considerable opportunity for advances in medicine and livestock enterprise. The ability to insert new genes for economically important characteristics such as fecundity, resistance to or tolerance of other environmental stresses would represent a major breakthrough in the breeding of commercially superior stock. Another opportunity that transgenic technology could provide is in the production of medically important proteins such as insulin and clotting factors in the milk of domestic livestock. The genes coding for these proteins have been identified and the human factor IX construct has been successfully introduced into sheep and expression achieved in sheep milk (Clark *et al.*, 1990). Transgenic sheep have recently been generated which express the visna virus envelope gene (Clements *et al.*, 1994). Table 1 summarized genes of the different economic important traits used for the production of transgenic animals.

Table 1 : Genes of some economic traits, which have been used for producing transgenic farm animals.

Genes	Traits
Growth hormone and their releasing factors	Growth
Insulin growth factor	Growth
Post weaning growth factor	Growth
Growth hormone	Milk quantity
GH releasing factor	Milk quantity
Casein genes	Milk quality
β-lactoglobulin	Milk quality
Booroola (F) genes	Ovulation rate
Cystein genes	Wool Production
MHC-ped	Preimplantation embryo development
HSV-thymidine kinase	Enzymatic activity

Source: Sharma and Balakrishnan (1992)

In the field of animal breeding, there were diverse opinions on how the technology might affect livestock genetic improvement programme. However, Smith *et al.* (1987) presented a comprehensive evaluation of strategies for developing, testing, breeding and disseminating transgenic livestock in the context of quantitative improvement of economic traits.

Recent advances in *in-vitro* technology (*in vitro* fertilization and maturation) will increase the number of zygotes available for gene transfer purposes. This, plus the utilization of embryonic stem cell and primordial germ cell technologies should enhance the efficiency of gene transfer in cattle and sheep considerably.

Molecular characterization of animal genetic resources

Developing countries are endowed with the majority of the domestic animal diversity. Some indigenous livestock breeds in these countries are in immediate danger of loss through indiscriminate crossbreeding with exotic breeds. The importance of indigenous livestock breeds lies in their adaptation to local biotic and abiotic stresses and to traditional husbandry systems. However, most of these animal genetic resources are still not characterized and boundaries between distinct populations are unclear. Lack of information on the characteristics and extent of indigenous breeds has led their under utilization, replacement and dilution through crossbreeding. So, characterization of indigenous breeds at molecular level unravels molecular differentiation between breeds, which lead to animal improvement programmes. It is in this sphere that biotechnology has an important role.

Genetic uniqueness of population is measured by the relative genetic distances of such populations from each other. Polymorphism in gene products such as enzymes, blood group systems and leukocyte antigens which have traditionally been used for measuring genetic distance are being rapidly replaced by polymorphism at the level of DNA, both nuclear (Jeffreys and Morton, 1987) and mitochondrial (Loftus *et al.*, 1994) as a source of information for the estimation of genetic distances. The first DNA polymorphism to be used widely for genome characterization and analysis were the restriction fragment length polymorphism (RFLP) (Southern, 1975) which detect variations ranging from gross rearrangements to single base changes. Minisatellites sequences of 60 or so bases repeated many hundreds or thousands of times at one unique locus within the genome have been used to generate DNA fingerprints typical of individuals within species (Jeffreys and Morton, 1987). Microsatellites (Weber and May, 1989) repeats of simple sequences, the commonest being dinucleotide repeats are abundant in genomes of all higher organisms, including livestock. Ease of identification and of sequence determination and need for only small amounts of DNA, are some of the advantages of microsatellites.

Randomly amplified polymorphic DNA (RAPD) (Williams *et al.*, 1990) has been extensively used for genetic characterization of a wide range of organisms. The technique uses short (up to 10 bases) primers to amplify nuclear DNA in the PCR. The procedure does not require knowledge of the sequence of DNA under study; primers are designed randomly. The basis of the polymorphism detected by this method is that products are either generated in PCR or not.

Complete sequencing of the genome is the ultimate form of genetic characterization. Sequencing has traditionally been expensive and laborious, but with the advent of automated sequencing this is changing rapidly. However, sequencing is unlikely to be used as a technique of choice for genetic characterization.

Conservation of animal genetic resources

Conservation of indigenous animal genetic resources should be one of the priority livestock development activities for developing countries. There are several ways to conserve animal genetic resources i.e. *ex-situ* and *in-situ* conservation as per FAO (1992). *Ex-situ* approaches include: maintenance of small populations in domestic animal zoos; cryopreservation of semen (and ova); cryopreservation of embryos; and some combinations of these. Whereas, *in-situ* approaches include maintenance of live animal population of a breed in its native environment or as close to it as practically possible. Developing and utilizing a genetic resource is considered the most rational conservation strategy.

Cryopreservation of gametes, embryos or DNA segments can be quite an effective and safe approach for breeds or strains whose populations are too small to be conserved by any other means. Cryopreserved bovine oocytes have been successfully matured and fertilized *in vitro* and zygotes developed to blastocyst stage (Lim *et al.*, 1991). In cattle, the transfer of frozen-thawed embryos is now a commercial practice and embryo survival rate after thawing can be as high as 80% with a pregnancy rate of about 50%. Respective pregnancy rates of 58 and 50% for fresh and frozen-thawed *in vitro* produced embryos have been reported (Lu *et al.*, 1990). Also, calves have been produced from transfer of both split and frozen-thawed *in vitro* produced embryos. Economic aspects of genetic conservation in farm animals have been assessed by Brem *et al.*

(1984). The study concluded that costs of *ex-situ* live animal conservation was moderate to high while costs of long-term cryopreservation of gametes were low.

Development in genetic engineering, cryobiology, cell biology and embryology will provide techniques that may enhance our ability to preserve germplasm *in vitro*. Techniques such as transfer of DNA within and between species and the production of viable transgenic animals are far from practical application. However, biotechnology will certainly contribute newer and cheaper methods for preservation such as storage of catalogued DNA. At present, other than live animal and embryo preservation, the other techniques do not allow preservation of genomes in a form which can be reactivated as such at a later stage, but they permit the preservation of individual genes or gene combinations for possible future regeneration. The use of microsatellite in genetic distancing of breeds is gaining momentum. Its application to concerted animal genetic resource utilization and conservation programme is a major objective of the Global Strategy for the Management of Farm Animal Genetic Resources, which FAO has recommended to implement.

Animal nutrition

Nutrition is one of the most important limiting factors to livestock production in developing countries. The availability of feed resources is inadequate in both quality and quantity, particularly during the dry seasons. Fibrous feeds, including crop residues, of low digestibility constitute the major proportion of feeds available to most ruminants in developing countries. The associated low productivity of animals due to low quality forages can be overcome to some extent by several means, like balancing of nutrients for the growth of rumen microflora, which facilitates efficient fermentative digestion and provides small quantities of by-pass nutrients to balance the products of fermentative digestion, enhancing digestibility of fibrous feeds through treatment with ammonia or alkali or by manipulating the balance of organisms in the rumen and genetic manipulation of rumen micro-organisms. Rumen micro-organisms can also be manipulated by adding probiotics, protein degradation protectors, methane inhibitors, buffer substances, bacteria or rumen content and/or branched chain volatile fatty acids and by eliminating or reducing rumen ciliate protozoa (defaunation),. Biotechnological options are available for improving rumen fermentation and enhancing the nutritive value and utilization of agro-industrial by-products and other forages (Kundu and Kumar, 1987). Genetic manipulation of rumen microorganism has been proved as potentially the most powerful tool for enhancing the rate and extent of digestion of low quality feeds.

Increasing digestibility of low-quality forages : As the majority of livestock farmers in developing countries are not capable of proving optimum feeding regimes, low-quality forages are usually the major component of ruminant diets. The characteristic feature of forages is their slow rate of microbial breakdown in the rumen with the result that much of the nutrients of the feed are voided in the faeces. The slow rate of breakdown also results in reduced outflow rate of feed residues from the rumen, which consequently depresses feed intake. At present, the main treatment methods for forages such as cereal straws are either mechanical (e.g grinding), physical (e.g. temperature and pressure treatment) or a range of chemical treatments of which sodium hydroxide or ammonia is among the more successful.

The lignification of the cell walls prevents degradation by cellulase or hemicellulase enzymes. Fortunately, it is possible to use lignase enzyme produced by the soft-rot fungus (*Phanerochaete chrysosporium*) which causes a high degree of depolymerisation of lignin (Tien and Kirk, 1983). The enzyme acts like a peroxidase and causes cleavage of carbon-carbon bonds. At present the levels of the lignase enzyme produced by the basidiomycete fungi are insufficient for the treatment of straw on a commercial scale. However, it is conceivable that the use of recombinant DNA engineering techniques will allow the modification of the lignase genes and associate proteins to increase their efficiency and stability. The lignin gene has to date been cloned and sequenced from *P. chrysosporium* (Zhang *et al.*, 1986).

Utilization of agricultural by-products as animal feed : In certain Asian countries like Indonesia, Malaysia, Thailand and Sri Lanka where animals could be integrated with perennial crops, plantation (oil palm, rubber, coconut) undergrowth remains the major feed resource. In these countries, large quantities of byproducts from mills processing primary products are not only wastes but also become the source of environmental pollution. Therefore, biotechnological methods for waste recovery and recycling to convert them to animal feed are necessary.

Production of single cell proteins (SCP) using these agro-industrial wastes or other wastes as substrate is a distinct biotechnological possibility. SCP is a generic term for a crude or refined source of protein whose origin is from unicellular or simple multicellular organisms. Single cell proteins are mainly used to supply quality protein to the livestock for better growth, higher milk production and wool production. SCP contains higher level of essential amino acids. Following microorganisms, which have been most often used in studies of SCP production, are:

- Algae: *Chlorella, Scenedesmus, Spirulina*
- Bacteria: *Bacillus, Hydrogenomonas, Methanomonas, Pseudomonas*
- Yeast: *Candida, Rhodotorula, Saccharomyces*
- Fungi: *Aspergillus, Rhizopus, Fusarium, Penicillium.*

Palm oil mill effluent (POME) has been used as a substrate for the growth of microorganisms in Malaysia. Production of a large quantity of SCP in a relatively short period of time is possible, but there is some doubt where feeding of SCP in ruminants will be cost effective or not. There are some biotechnological means through which small farmers can enhance the digestibility of animals in smallholdings. This includes supplies of molasses/urea blocks containing a little digestible fibre and the availability of small quantities of amino acids and peptides (slowly fermented protein) for the rumen microbes by providing some proteins in the blocks. Studies at National Dairy Research Institute, Karnal, India, suggested that urea-mineral-molasses block (UMMB) might improve nutrient utilization and production potential in ruminants under straw feeding. Additionally, mineral supplementation to grazing cattle encourages growth of animal (Prasad and Gowda, 2002).

Eliminating anti-nutritive factors from feeds : The different anti-nutritive factors in plant tissues include protease inhibitors, tannins, phytohaemagglutinins and cyanogens in legumes, and glucosinolates, tannins and sanapine in oilseed rape (Brassica napus) and other compounds in feeds belonging to the Brassica group. The adverse effects of these compounds are more marked in non-ruminants than in ruminants due to amino acid deficiency. Conventional plant breeding has been used to reduce and, in some cases, eliminate such anti-nutritive factors. An example is the introduction of cultivars of oilseed rape, which are low in, or free from glucosinolates. A combination of genetic engineering and conventional plant breeding should lead to substantial reduction or removal of the major anti-nutritive factors in plant species of importance as animal feeds. Transgenic rumen microbes could also play a role in the detoxification of plant poisons (Gregg, 1989) or inactivation of anti nutritional factors. Successful introduction of a caprine rumen inoculum obtained in Hawaii into the bovine rumen in Australia to detoxify 3-hydroxy 4(IH) pyridine (3,4 DHP), a breakdown product of the non-protein amino acid mimosine found in Leucaena forage (Jones and Megarrity, 1986) demonstrates the possibilities. Elimination of protozoa from rumen microbes was found to increase the digestibility of straw and cause increased wool growth in sheep (Bird and Leng, 1985).

Improving nutritive value of cereals : Moderate protein content and low amounts of specific amino acids limit the nutritive value of cereals and cereal by-products (e.g. barley is low in lysine and theonine). This is a major limitation in the ration formulation for non-ruminant livestock, which necessitates addition of expensive protein supplements. The study has been conducted to enhance the low level of lysine in barley by genetically engineering the grain genome (Shewry and Kreis 1987). Genetic modification through insertion of genes into rice protoplasts and generation of transformed plants has already been achieved.

Improving nutritive value of conserved fodder : Surplus fodder of kharif or winter (rabi) seasons may be conserved to supply essential nutrients to domestic animals for maintaining their production level in terms of milk or growth. The conservation of plant material as silage depends upon anaerobic fermentation of sugars in the material, which in turn is influenced by the ability of naturally occurring lactic acid bacteria to grow rapidly on the available nutrients under the existing physical environment. Unless the ensiled material is sterilized, lactic acid bacteria are always present. However, the ensiling conditions may not always be ideal for their development. In addition to the number and type of bacteria, other interrelated factors may affect quality of silage, including availability of water-soluble carbohydrates, the dry-matter content, the pH and extent of air exclusion. For example, lack of water-soluble carbohydrates may be overcome by wilting the material to raise the dry matter to a level at which less acid is required to stabilize the fermentation. The availability of sugars in the material and the rate at which the different microorganisms multiply also influences the ensilage process. Throughout this century, research workers have investigated ways through which the fermentation process in silage making can be controlled in order to improve the feeding quality of the resulting silage.

An ingredients of non-nutritive nature that accelerates growth or other type of performance, improves the efficiency of feed utilization or beneficial for health or metabolism of the animal considered as feed additive (Church and Pond, 1988). Use of additives, to restrict the activity of the pathogenic microorganisms or to stimulate the fermentation by the lactic acid bacteria or simply as nutrients has been one of the approaches. The additives hold great promise in augmenting nutrition, production and health of poultry and livestock sector includes enzymes, probiotics, antibiotics, mould inhibitors and herbal inhibitors (Verma and Mandal, 2000). Additives used in the early studies included chloroform, toluene and cresol (to inhibit bacterial growth) and sulphuric acid and hydrochloric acid (to reduce the pH). Other fermentation inhibitors, which have been studied, include organic acids, salts of acids, formaldehyde and other aldehydes, sodium hydroxide, and antibiotics. Of these, formic acid is probably the most widely used and studied and has been reported to have a beneficial effect on the fermentation process and on the nutritive value of silage. Use of acids is a hazard on the farm and can be particularly dangerous if recommended to farmers. Salts of acids are safer to handle but are less effective than the acids from which they are derived.

The hazardous nature of some of the chemical additives has necessitated a search for alternative compounds for improving the ensilage process. A group of compounds classified as fermentation stimulants have been widely studied. These include sugar sources (e.g. molasses and whey), enzymes and inocula of lactic acid bacteria. Molasses is of particular relevance to small hold farmers in developing countries in the tropics where sugarcane is produced and processed.

Enzymes can also improve the nutrient availability from feedstuffs, lower feed costs and reduce output of wastes into the environment. The primary objective of adding enzymes to animal feeds is to improve the utilization of nutrients in raw materials. This is achieved by one or more of the following mechanisms: 1) degradation of specific bonds in ingredients not usually degraded by endogenous digestive enzymes, 2) degradation of anti-nutritive factors that lower the availability of nutrients; 3) increased accessibility of nutrients to endogenous digestive enzymes and 4) supplementation of the enzyme capacity of young animals. Enzymes are essential for the breakdown of cell-wall carbohydrates to release the sugars necessary for the growth of the lactic acid bacteria. Although resident plant-enzymes and acid hydrolysis produce simple sugars from these carbohydrates, addition of enzymes derived from certain bacteria, e.g. *Aspergillus niger* or *Trichoderma viridi* (Henderson *et al.*, 1982) increases the amount of available sugars. Commercial hemicellulase and cellulase enzyme cocktails are now available and improve the fermentation process considerably (Hooper *et al.*, 1989). The enzymes that are used widely by the industry are the xylanases and beta-glucanases that cleave the non-starch polysaccharides in specific cereals, and, recently microbial phytases that target the phytate-complexes in plant-derived ingredients (Bedford and Schulze, 1998). The microbial phytase, the classic and most popular feed enzyme worldwide, not only causes the release of phytate-bound phosphorus and enables the reduction in the level of inorganic phosphates in animal diets, but also improves nutrient digestibility (Selle *et al.*, 2000) and lowers nutrient load in the manure. Several enzyme preparations, designed towards improving the utilization of protein, starch or lipids in specific ingredients, are also available. In particular, proteases are of interest because protein is the most expensive item in the cost of animal diets. Poorly digested proteins not only lower the efficiency of nitrogen utilization, but also contribute to odour problems and ground-water pollution. The use of proteases can be valuable to improve the nutritive value of protein meals by hydrolyzing certain protein types that are resistant to digestive enzymes by complementing the animals' own digestive system.

High protein yeast cell products are being used as a feed additive for cattle, pigs and poultry. Highly palatable and nutritious, these products also help to create a healthy balance of bacteria in the digestive tract, and prevent bacterial diarrhea.

The supplement of probiotics, a live microbial feed supplement, beneficially affects the host animals by improving its intestinal microbial balance (Fuller, 1989). For ruminants, the yeast (*Saccharomyces cerevisias*) has been considered as the promising probiotic culture for efficient nutrient utilization. Dutta (1998) reviewed that supplementation of probiotic culture to large and small ruminants increased feed intake, enhanced in weight gains and feed conversion efficiency in meat producing animals. The yeast, *S. cerevisias,* has potential to improve the milk production in dairy cows, increased milk, fat, milk protein and lactose content of milk.

Commercial bacterial inoculants (microbial silage additives) designed to add sufficient homofermentative lactic acid bacteria to dominate the fermentation are now available. The objective of using such additives is to ensure the rapid production of the required amount of lactic acid from the carbohydrates present to preserve the ensiled material. Most such inoculants contain *Lactobacillus plantarum* with or without other bacteria such as *L. acidophilus* and *Streptococcus thermophilus.* In general, the results with bacterial inoculants have been quite variable. However, with an effective product, it is possible to improve the fermentation of low dry-matter silages and to enhance the efficiency of their utilization. In order to improve the effectiveness of microbial inoculants in breaking down structural carbohydrates to glucose, detailed knowledge of the lactobacilli bacteria is essential. Work already undertaken on the molecular biology of *Lactobacillus plantarum* and other species (Armstrong and Gilbert, 1991) suggest that the rapid progress in this area will make it possible to construct novel genes encoding highly active fibre-degrading enzymes. Such genes could then be inserted into strains of *L. plantarum.* Successful silage making incorporating these technologies can only be achieved with strict adherence to recommended application procedures, including rates of additives, inoculants etc.

Improving rumen function : Genetic engineering play a crucial role to improve the functioning of rumen bacteria or of gut organisms, in general. The potential of application of molecular techniques in achieving the goals of rumen manipulation are enormous (Flint, 1994). Armstrong and Gilbert (1985) have reviewed the major areas of rumen function, which might benefit from transgenic technology. These include development of transgenic bacteria with enhanced cellulytic activity, capability to cleave lignohemicellulose complexes, reduced methane production capability decreased proteolytic and/or deaminase activities and increased ability for microbial production of specific amino acids. The first successful transfer of foreign genes into rumen bacteria (Bacteriodes ruminicola) was reported by Thomson and Flint (1989). For genetic rumen manipulation, the first approach would be to select the desirable genes and to express them in predominant rumen bacteria. Naturally present microorganisms in the rumen can be genetically modified to enhance their capacity of defined functions or to add new functions (Chang, 1996). Introduction of diverse genes into gut microorganisms have been extensively explored (McSweeney et al., 1999). The genetically modified microorganisms are either able to digest fibrous components and lignins of forage, or degrade toxins, synthesize essential amino acids, reduce ruminal methane production and tolerate acids (Forsberg et al., 1993). The second approach would be to introduce new species or strains of microorganisms into the gut (Stewart et al., 1988). Application of the said two approaches has a great potential to increase digestibility of feedstuffs and to improve animal production. However, the developed country is still a long way from commercial production of genetically engineered rumen bacteria.

Although several workers have isolated genes encoding plant structural carbohydrate-degrading enzymes from rumen bacteria, there are limited reports on the genetic engineering of these microorganisms. In contrast to conditions in which single species of organisms are grown in controlled environments and where the energy supply is usually in excess of demand, the rumen environment is very complex, competition between different microbial species is intense and energy is usually the limiting growth factor. This is probably the main reason why reintroduction of genetically modified rumen bacteria into their natural habitat has met with variable success (Flint *et al.*, 1989). Advances being made in transformation methods for obligate anaerobic bacteria will certainly result in successful genetic engineering of a range of rumen bacteria. However, it is not possible to predict if any of these bacteria will be capable of colonizing the rumen. It can be concluded that there are several potential opportunities for improving the efficiency of ruminant digestion and possibilities for utilizing a wider range of feeds than is currently possible. Modification of rumen microbial population (Flores, 1989) is one such opportunity. However, technical difficulties associated with making genetic modifications to individual species of rumen bacteria hinder progress in this area.

Designer feed ingredients : The term 'designer ingredients' which specifically refers to nutritionally enhanced animal feeds, as compared with the terms 'genetically modified' (GM) or 'transgenic ingredients' that are used widely. The designer ingredients are all products of genetic manipulation. The scope for genetic improvement of ingredients designed specifically for animal feeding has become possible with the advances in genetic engineering. A number of potential targets are being pursued to improve the value of seeds for animal feeding. Current product targets in plant breeding that are designed towards the feed industry include: 1) increased energy or protein content in maize; 2) enhanced protein quality – increased methionine/ lysine content in maize, oilseeds and grain legumes; 3) reduction of anti-nutritional factors and 4) reduced grain moulds and mycotoxins. Lysine supplementation through designer feed improves the nutrient balance of feed and feed conversion rates of livestock. However, the inclusion of genetically modified organisms in animal feeds raises issues of risk, e.g., of the transfer of foreign genes introduced into GM feed grain crops, especially antibiotic resistance marker genes, which might render common infectious diseases untreatable. The main conclusion of research to-date (Ryan and Mae-Wan, 2000) is that DNA is not degraded under most commercial processing condition, nor in silage, and that therefore further studies on GM use in animal feeding would be indicated. Potential feed safety problems related to the use of antibiotic marker genes may be available through the use of alternative transgenic systems, which are now become available.

Reproductive Physiology

Improving female reproductive performance is an important objective for increasing the profitability of livestock enterprise. There is much greater potential for increasing both biological and economic efficiency of animal production through genetic improvement in reproductive rate than through improvement in growth rate or body composition. Several reproduction techniques are available for increasing the reproduction rates of animals. The techniques, which are widely used, are artificial insemination (AI), embryo transfer and associated technologies. Measurement of progesterone in milk or blood, which is a widely used technique for monitoring ovarian function and for pregnancy test, is also an important technology for managing the reproductive function of the animal.

Artificial insemination

The term 'artificial insemination' implies the collection of semen from males and artificially introducing it into females to obtain pregnancy. The progress in semen collection and dilution, and cryopreservation techniques now enables a single bull to be used simultaneously in several countries. By this technique, a very small number of top bulls can be used to serve a large cattle population. In addition, each bull is able to produce a large number of daughters in a given time thus enhancing the efficiency of progeny testing of bulls. Van Vleck (1981) reported that optimum plans under natural within herd breeding systems could at best provide a gain of 0.5-0.6% per year. Whereas, optimum use of AI in large populations could increase this to 2.0 – 2.5% to, perhaps, 3% per year. The reasons are the increased accuracy of evaluation, together with more intense selection of sires of dams, sires of sires, and dams of sires. Generation intervals could be reduced somewhat through AI since progeny of sires can be obtained over a shorter period of time than with natural service.

The use of AI can accelerate the rate of genetic improvement, reduce the transmission of venereal diseases in a population and also lowers the costs of farmers to maintain their own breeding males. Hence, it facilitates more accurate recording of pedigree and minimize the cost of introducing improved stock. However, success of AI technology depends on accurate heat detection and timely insemination. Proper heat detection is inevitable for successful artificial insemination programme, which could be done by using monoclonal antibodies. One British Company developed a monoclonal antibody test that indicates the time of estrus of dairy cows. AI technology use is still more generally associated with dairy cattle than other domestic livestock species.

In sheep and goats, the failure to develop a simple, non-surgical insemination procedure has prevented extensive exploitation of the technology in sheep (Robinson and McEvoy, 1993). However, the technical success of laparoscopic intrauterine insemination has prompted research into less invasive trans-cervical procedures (Buckrell *et al.*, 1992). AI is credited for providing the impetus for many other developments, which have had a profound impact on reproductive biotechnology. Foote (1982) noted that studies of oestrus detection and ovulation control which evolved out of a need to correctly time inseminations, led to the development of embryo-transfer technology. A few developing countries have taken this technology to the field, mostly for programmes of "upgrading" indigenous stock and as a service to a limited number of commercial farmers keeping exotic dairy cattle breeds.

Embryo transfer (ET)

Embryo transfer has had its major use in obtaining more than the normal number of offspring from highly prized females or females of highly priced breeds. The goal was a high sale value (Cunningham, 1976). Although not economically feasible for commercial use on small farms at present, embryo technology can greatly contribute to research and genetic improvement in local breeds. The first successful embryo transfer in farm animals was done in 1934 in sheep and goats (Warwick *et al.*, 1934). The first offspring after transfer of embryos in cattle and swine were reported in 1951 (Willet *et al.*, 1951). There are two procedures presently available for production of embryos from donor females. One consists of super-ovulation, followed by AI and then flushing of the uterus to gather the embryos. The other, called *in vitro* fertilization (IVF), consists of recovery of eggs from the ovaries of the female then maturing and fertilizing them outside the body until they are ready for implantation into foster females. IVF facilitates recovery of a large number of embryos from a single female at a reduced cost thus making ET techniques economically feasible on a larger scale. Additionally, IVF makes available embryos suitable for cloning.

There are many possible uses of embryo transfer in animal production (Bhattacharya, 1984). The principal benefit of embryo transfer is the possibility to produce several progeny from a female, just as AI can produce many offspring from one male. For example the average lifetime production of a cow can be increased from 4 to 25 calves. Increasing the reproductive rate of selected females has the following benefits: genetically outstanding animals can contribute more to the breeding programme, particularly if their sons are being selected for use in AI; the rate of genetic change can be enhanced with specially designed breeding schemes which take advantage of increased intensity of female selection combined with increased generation turnover; transport of embryos is much cheaper than that of live animals; risk of importing diseases is avoided; facilitates rapid expansion of rare but economically important genetic stocks; and the stress to exotic genotypes can be avoided by having them born to dams of local breeds rather than importing them as live animals. Embryo transfer seems unlikely to lead to increased accuracy of predicted genetic value of animal (Cunningham, 1976).

Embryo transfer is still not widely used despite its potential benefits. In developing countries this is mainly due to absence of the necessary facilities and infrastructure. Even in developed countries, cost considerations still limit the use of commercial embryo transfer in specialized niches or for a small proportion of best cows in the herds. Commercial embryo transfer is more popular with cattle than other species. This is mainly because ET is relatively easier in cattle than the other species and also because it is more economical in cattle (i.e. cattle are worth more). Additionally, the low reproductive rate and the long generation interval of cattle make ET much more advantageous in cattle. But the general conclusion is that costs of embryo transfer would need to be reduced by a large multiple in order to be economically feasible if only genetic gain is considered (Van Vleck, 1981).

Production of several closely related, and hence genetically similar, individuals through ET techniques can make critical contributions to research. For example a project at the International Laboratory for Research on Animal Diseases (ILRAD) to locate the genes responsible for tolerance of some cattle populations to trypanosomiasis required large numbers of closely related crosses of trypanotolerant and trypanosusceptible cattle. Use of ET has made it possible to generate such families thereby facilitating the search for genetic markers of trypanotolerance. Additionally, ET could be useful in studying the extent to which a trait is influenced by the embryo (direct component) or the reproductive tract (maternal component).

Embryo sexing

Although embryo sexing may not have dramatic effects on rates of genetic gain (Colleau, 1991) it can considerably increase efficiency. Taylor *et al.*, (1985) concluded from a study that an all-female heifer system using ET was 50% more efficient than the highest achievable in a traditional system. In addition, to preventing freemartins with twinning, embryo sexing could reduce the number of embryos transferred and recipients needed. It has been suggested that, if multiple sexed-embryo transfer became as routine an operation as AI is, beef operations based on this system could become competitive with pig and poultry production in terms of efficiency of food utilization.

Embryo splitting and cloning

The techniques of embryo splitting in bovines has been available since 1981, but has not been used widely so far. Generally, embryos are split more than twice, producing four genetically identical embryos. The net result is a greater number of offspring per donor, despite the lower survival rates of split embryos as compared to whole embryos. Currently the applications of this technique in dairy cattle are limited. The intensity of selection is not increased, since the animals are genetically identical. However, it can be used to improve the accuracy of selection. Split embryos can also be used to estimate maternal effects and genetic trends of economically important traits.

Clones may be produced by embryo splitting and nuclear transfer (Macmillan and Tervit, 1990). Each clone is genetically an exact duplicate and naturally occurs only in pairs of identical twins. These offer the possibility for creating large clone families (Woolliams and Wilmut, 1989) from selected superior genotypes which, in turn, can be used to produce commercial clone lines. Some attention has been paid to the potential applications of clones produced by the development of embryo splitting (Nicholas and Smith, 1983) or cell culture techniques (Van Vleck, 1981). The potential gains described give a very substantial genetic lift. Using selected clones in the commercial herd could increase genetic merit by over 25% (Nicholas and Smith, 1983). However, some studies have concluded that cloning of embryos will not increase rates of genetic progress in the nucleus, but that it offers considerable advantages in increasing the rate of dissemination of tested superior genotypes in commercial populations. Other potential applications of cloning include efficient evaluation of genotype x environment interactions and testing and/or dissemination of transgenics. From a research standpoint, production of identical siblings should, by eliminating variability among animals, greatly reduce the size and hence the cost of experiments.

Hormonal treatment

Genetic improvement of fertility rate of animals would be possible by exogenous hormonal treatments. Use of hormonal assays to *monitor* the reproductive function can be rewarding for both research purposes and commercial livestock operations. The different hormones like progesterone and PMSG treatment and immunization against androstenedione increased ovulation rate in sheep.

One of the major challenges for increasing the reproductive rates in small ruminants is seasonal reproduction, which reduces the effectiveness the accelerated kidding/lambing program in every year. The reproductive seasonality of animals can be managed very effectively by use of the exogenous hormone. Seasonal anoestrus in ewes lowers reproductive efficiency of flocks, and affects the reproductive performance of rams (Haynes and Schanbacher, 1993). The treatment involving melangesterol acetate can provide effective manipulation of the breeding season in sheep and goats. Also, exogenous melatonin treatment of barren Rahmani ewes resulted in increased proportion of ovulating ewes and a higher ovulation rate (Aboul-Naga *et al.*, 1992). These responses, however, did not result in increased litter size because of increased ova wastage. Thus, in addition to the impracticability arising from prohibitive prices of hormonal preparations and the problems with hormonal administration at farm level, there are other technical problems associated with the use of these technologies. For example, technologies aimed at increasing litter size in traditional small ruminant production systems should not be applied unless management, including nutrition, can be improved in concert to ensure the survival of the additional progeny.

Molecular techniques

The seasonality of breeding in small ruminants effectively limits the expression of performance of females causing low fertility in females. Use of molecular techniques to aid in the identification of genetically superior individuals is appealing for traits involved in seasonal breeding. For genetic improvement of seasonality, two broad options exist. Firstly, the direct identification of genes responsible for seasonal breeding and development of DNA tests to identify favorable variants at these loci. Secondly, use of marker genes, bits of polymorphic DNA which do not themselves influence seasonality but which are known to be associated with genes that do, through proximity to these genes on a chromosome. This approach may be adopted when the location of gene influencing performance is not known. However, marker genes normally differ among breeds and families within breeds, so reliable use of marker genes requires relatively extensive genotyping of animals in each flock. Large numbers of mating with intensive evaluation of animals for traits of interest is required to identify useful genes or markers. Reports are available regarding association between expression of reproductive seasonality and alleles of the gene for Mel (1a) receptor in the ewe (Pelletier *et al.*, 2000).

It has been suggested that improvements in reproductive performance may be achieved by introducing changes to the regulatory elements of genes that govern the number of ovulations or the seasonal pattern of reproduction (Wilmut *et al.*, 1990). Incorporation of Booroola (F) genes by gene transfer techniques in native flock of sheep increases the ovulation rate in ewes (Bindon and Piper, 1986).

Physiology of Lactation and Growth

Administration of recombinant bovine somatotropin (BST) results in accelerated growth and leaner carcasses in meat animals and increased milk production in dairy cows. Recombinant bovine somatotropin (BST) is a genetically engineered synthetic analog of the natural growth hormone. Under good management and feeding, regular BST administration to lactating dairy cows increases milk yield by 15-30% and also increases efficiency of milk production. BST is now readily available and is already in commercial use in the developed countries like United States. The appropriateness of BST use for developing countries is still a matter of debate.

Phipps *et al.* (1990) have shown that feed conversion efficiency declines in subsequent lactations. The study by Phipps *et al.* (1990) also indicated a decline in the incremental amount of milk yield from BST application in subsequent lactations. Additionally, the effect of BST on reproductive performance is likely to be more adverse in the presence of higher biotic and abiotic stresses, including nutritional stresses (Burton *et al.*, 1990). There is also need to examine the economics of BST application in view of the known association of its use with mastitis (Burton *et al.*, 1990) and other infections. BST use is thus bound to be associated with increased use of antibiotics and other veterinary drugs. Thus, in evaluating the potential role of BST in developing countries, one needs to consider not only the possible response levels of the cattle in these countries to BST treatment, but also the cost of BST, the amount and cost of other incremental inputs required for effective use of BST, and the milk prices. Ultimately, the main technical constraint to BST use in developing countries will not only be its cost, but the absence of an efficient delivery system; current use of the technology requires regular injections.

Porcine somatotropin (PST) and recombinant growth hormone stimulatory peptides (e.g. growth hormone releasing factor, GRF) along with BST have been shown to increase growth rates by 8-38% in cattle, sheep and pigs. In almost all cases, administration of exogenous growth hormones have been associated with increased carcass protein and reduced carcass fat (Hart and Johnson, 1986). Other growth-promoting agents (e.g. anabolic steroids and beta agonists) have been shown to have even larger effects but public concerns over the possible residual effects in meats have led to their being banned in most developing countries.

Animal Health

One of the important benefits from biotechnology is the diagnosis of livestock diseases. Domestic animals are prone to wide range of infections, which damage heath and productivity. Biological techniques can produce cheaper and more efficient drugs. In cases where a natural source material is prohibitively expensive, genetic engineering (in microbial or tissue culture systems) can be used to produce drugs of high value for humans and animals. Examples are insulin, human growth hormone and tissue plasminogen activator etc. Use of DNA biotechnology in animal health may contribute significantly to improved animal disease control, thereby stimulating both food production and livestock trade.

Disease diagnosis

The accurate diagnosis of diseases is the most important step for successful control and prevention of diseases. This has been greatly improved in recent years through developments in biotechnology. Advanced biotechnology based diagnostic tests make it possible to identify the disease causing agent(s) and to monitor the impact of disease control programmes, to a degree of diagnostic precision (sub-species, strain, bio-type level) not previously possible.

Molecular epidemiology is a fast growing discipline that enables characterization of pathogen isolates (virus, bacteria, parasites) by nucleotide sequencing for the tracing of their origin. This is particularly important for epidemic diseases, where the possibility to pinpoint the source of infection can significantly contribute to improved disease control. Furthermore, the development of genetic probes, which allow the detection of pathogen DNA/RNA (rather than antibodies) in livestock, and the advances in accurate diagnostic kits considerably enhance animal health programmes.

The ability to generate highly specific antigens by recombinant DNA techniques has made it possible for an increasing number of enzyme-linked immunosorbent assays (ELISA) to have the capacity to differentiate between immune responses generated by vaccination from those due to infection (Robinson and McEvoy, 1993). It helps to overcome one of the major drawbacks of antibody detection tests: the fact that, because antibodies can persist in animals for long periods, their presence may not indicate current infection. The major development in the field of disease diagnosis is that it is now possible to immortalize individual antibody-producing cells by hybridization to produce antibodies of a given class, specificity and affinity (i.e. monoclonal antibodies) which has provided a tool that permits the analysis of virtually any antigenic molecule. The use of monoclonal antibodies has revealed that the failure of vaccines (e.g. of rabies) to provide protection in all parts of the world was due to the diversity in the antigenic composition of the causative virus. The (monoclonal antibody) technology is relatively simple and can readily be applied in developing countries. This test reveals current infections and facilitates differentiation between the major trypanosome species.

The advent of PCR has enhanced the sensitivity of DNA detection tests considerably. For example, PCR used in combination with hybridization analysis, has been shown (Brandon *et al.*, 1991) to provide a sensitive diagnostic assay to detect bovine leukosis virus. PCR and DNA probes are being increasing employed for the identification of genus and species of *Brucella* causing abortion in bovine animals. Identification of *Brucella* species is very important from the epidemiological standpoint (Mukherjee, 2000).

Other diagnostic techniques include nucleic acid hybridization (NAD) and restriction endonuclease mapping (REM). As has been indicated above, one of the most valuable features of these molecular techniques is their specificity and sensitivity. A good example of the specificity of NAD is its application in distinguishing infections caused by peste des petite ruminants (PPR) virus from rinderpest (Lefevre and Diallo, 1990), diseases whose symptoms are clinically identical and which cannot be distinguished antigenically with available serological reagents. This technique also allows comparison of virus isolates from different geographical locations. Nucleotide sequence analysis of the FMD virus protein 1D gene (VP1) allows direct evaluation of the degree of relationship amongst the strains. It helps in revealing the origin and course of epizootics and offers the possibility of preventing further outbreaks (Beck and Strohmaier, 1987). Detailed overview of biotechnological tools for diagnosis of livestock diseases has been provided by Robinson and McEvoy (1993).

Other available diagnostic techniques which may have application in small ruminants and/or cattle include: nucleic acid probes (NAP) for heartwater *(Cowdria ruminantium), chlamydia psitacci*, Paratuberculosis and Bluetongue (Blancou, 1990; Knowles and Gorham, 1990); restriction endonuclease reaction (RENR) for diagnosis of *Corynebacterium pseudotuberculosis* (Knowles and Gorham, 1990); PCR for characterising subtypes of Bluetongue from different geographical regions (Osburn,1991); monoclonal antibodies (MAB) for differentiating false positive anti-Brucella titres caused by *Yersinia enterocolitica* and true positive anti-Brucella titres in latent infected animals (Haas *et al.*, 1990); and MAB for diagnosis of Toxoplamosis, Pasteurellosis, *Mycoplasma* spp, PPR and Boder Disease (Blancou, 1990). Examples of diseases for which diagnostic tests are available are presented in Table 2.

Table 2 : Diseases for which probes and monoclonal antibodies (MAB) are available

Viral diseases	Bacterial diseases
Foot-and-mouth disease	Anaplasmosis (*A. centrale* and *A. marginale)*
Peste des petite ruminants	Contagious bovine pleuropneumonia
Rinderpest	Cowdriosis
African horse sickness	Contagious caprine pleuropneumonia
Bluetongue	Contagious agalactia
Vesicular stomatitis	Haemorrhagic septicemia (only MAB available)

Source: Lefevre (1992).

Vaccines

The conventional means of controlling major livestock diseases include chemotherapy, vector control, vaccination, slaughter of infected stock, and other management practices (including grazing management and controlled stock movements). These control measures are often unaffordable to farmers in developing country. Moreover, where these drugs or pesticides are used, resistance by parasites is often encountered and reinfection following administration of drugs against parasitic diseases usually occurs. Additionally, in many cases drugs either are not readily available locally or ineffective due to their improper preservation and lack of purity.

One of the most economical means of preventing specific diseases is immunization. An effective vaccine can produce long-lasting immunity. In some cases, vaccination can provide lifetime immunity. In general, vaccines offer a substantial benefit for comparatively low cost, a primary consideration for developing countries.

Vaccines have conventionally been produced by several methods some of which have become rather static with regard to efficacy, safety, stability and cost. Very effective vaccines against animal diseases such as rinderpest and pig cholera have been in use and it helped significantly to reduce the incidence of these diseases worldwide. Although vaccines developed using traditional approaches have had a major impact on the control of many epidemic and endemic viral, mycoplasmal and bacterial diseases affecting livestock, recombinant vaccines offer various advantages over conventional vaccines. There are safety (no risk of reversion to virulent form, reduced potential for contamination with other pathogens etc.) and specificity, better stability and importantly, such vaccines, coupled with the appropriate diagnostic test, allow the distinction between vaccinated and naturally infected animals. The latter characteristics is important in disease control programmes as it enables continued vaccination even when the shift from the control to the eradication stage is contemplated. Through the use of monoclonal antibodies and recombinant DNA technologies, it is now possible to define and produce immunogenic components much more rapidly. These technologies are increasingly being used to clarify the pathogenetic mechanisms and immune response to microbial diseases (McCullough, 1993). This should lead to the production of more effective vaccines in the future. Recombinant DNA technology also provides new opportunities for the development of vaccines against parasites (e.g., ticks, helminthes etc.) where conventional approaches have failed. To date many vaccines have been produced by these techniques. However, only few of these are being produced commercially.

Table 3 summarizes some vaccines developed by recombinant DNA technology. The following are of particular relevance to small ruminants: recombinant vaccines against Bluetongue and Rift Valley Fever (Osburn, 1991); a hybrid virus vaccine against Orf (Reid,

Table 3 : Examples of some novel animal vaccines

Species	Disease	Producer
Bovine	Papilloma virus	Molecular genetics
	Viral diarrhoea	California Biotechnology
	Brucellosis	Ribi ImmunoChem
	Rinderpest	USDA and University of California, Davis
Swine	Parvovirus	Applied Biotechnology
	Dysentery	Codon
Equine	Influenza, Herpes	California Biotechnology
		Applied Biotechnology
Avian	Coccidiosis,	Genex and A.H. Robins
	Newcastle virus	Codon and Salsbury labs
Companion	Canine parvovirus	Applied Biotechnology

Sources: Van Brunt (1987); Cunningham (1990).

1989), a thermo-stable recombinant vaccine against PPR (Lefevre and Diallo, 1990) and recombinant vaccines against *Taenia ovis* and Echinococcus (Blancou, 1990); tick gut antigens and *Haemonchus* gut antigens produced by molecular techniques offer ways for prophylaxis against these parasites. Progress is also being made towards the development of vaccines against Babesiosis and Theileriosis in cattle (Wright, 1990), which may indicate prospects for similar immune-prophylaxis becoming available for sheep and goats as well.

Resistance to diseases

One of the major areas of interest for improving livestock production is to increase their resistance to different diseases. Although it is clear that disease resistance, as confirmed by the immunological system, is really complex and involves cell mediated and humoral immunity. Certain aspects of the immune process can be targeted for potential attraction by genetic manipulation. Current disease control strategies include vaccination, medication, isolation of animals from pathogens, improved sanitation, improved management practices and eradication of certain diseases. These approaches, however, are not always effective. Lack of veterinary services, unavailability or high costs of drugs and vaccines, increasing occurrence of drug resistance by pathogens, limited scope of improved management and increased attempts by consumers to change drug-use regulations because of fears of contamination of human food underscore the need for more effective methods to combat disease. An attractive alternative solution is to breed for disease resistance. The presence of certain diseases may result from strictly genetic control, whereas others may be caused by a contamination of genetic predisposition and exposure to pathogens. There is ample evidence that genetic variation for disease resistance in domestic animals exists (Axford *et al.*, 2000).

Genetic resistance to the important infectious diseases of domestic livestock is usually found to follow polygenic inheritance, as do the production traits. Disease resistance research has included measurement of genetic control of disease losses, estimation of heritabilities and characterization of breed or strain. Numerous examples of this research exist for leucosis and Marek's disease in poultry (Freeman and Bumstead, 1987), mastitis and ketosis in cattle (Solbu, 1982), atrophic rhinitis and respiratory diseases in sheep (Kennedy and Moxley, 1980) and parasitic infestation and foot rot in sheep (Bulgin *et al.*, 1988). The biotechnological technique like cloning of animal offers the direct possibilities to challenge the clones against a specific disease or disease for selection of individuals for genetic resistance for a particular disease or diseases. Immune responsiveness has been suggested as an indirect indicator for disease resistance (Rothschild, 1989). In livestock, the most extensive research with genetic control of immune response has been in poultry (Lamont and Dietert, 1990). In cattle and sheep, experiments (Muggli *et al.*, 1987) have revealed genetic variation for immune response to a variety of antigens. These results suggest that genetic control of immune response exists for all livestock species. A second indirect approach would be to consider *in-vitro* methods as indicators of disease resistance. These methods include, for instance, phagocytic and bactericidal actions of pheripheral blood monocytes against disease agents. A third method would be to locate genetic markers associated with disease resistance or the actual genes themselves. This approach was generalized first by those suggesting use of RFLP analyses in genetic improvement programs (Soller and Beckman, 1983). Modern immunology, however, has revealed that a group of genes, called the major histocompatibility complex (MHC) genes, seem to be intimately associated with both disease resistance and immune responsiveness (Lamont and Dietert, 1990). More recent molecular genetic approaches have included using gene maps and quantitative trait loci (QTL) scans to find genes associated with disease resistance and immune response. Most of the studies to identify QTL for resistance to disease and ultimately the gene or genes underlying the QTL are being carried out in pigs and poultry (Bishop *et al.*, 2002). Marker assisted selection could be used to develop genetic markers for susceptibility or resistance and to use these to make breeding and treatment decisions, including introgression of desirable alleles into susceptible population. Identification of QTL for resistance to disease and selection based on detected QTL will lead to genetic improvement programme.

Animal products

Biotechnology can lead to new and improved animal products. For example, it can modify the composition of milk, or the fat content of meat. It may be possible to manipulate the genes, which encode particular animal products. In principle, the introduction of additional copies of a gene will change the amount of mRNA available for translation and this might be expected to change the rate of synthesis of the product.

Wilmut *et al.* (1990) reported the potential of gene transfer in modifying the milk composition aimed at producing human therapeutic proteins in the milk of farm animals. Genetically transformed cows can produce 'designer milk' with superior properties for use of in various milk products. Added caseins in milk, for instance, can enhance cheese making. The introduction of additional peptide bonds with increased sensitivity to proteolysis would increase the rate of ripening of cheese (Kang and Richardson, 1985). Similarly, an increase in the number of serine residues available for phosphorylation by casein kinase could be expected to increase the thermal stability of casein micelles and prevent undesired gelation of milk during thermal processing of dairy products. Increasing the phosphate group in casein can enhance the level of calcium. The other major milk proteins in milk are the whey proteins, alpha-lactalbumin and beta-lactoglobulin. Improvement of milk composition would require a decrease in the amounts of these proteins. Reduction of alpha-lactalbumin by transferring genes containing anti-sense sequences of alpha-lactalbumin mRNA could, in theory, reduce the level of the protein and hence, the level of milk lactose (Mercier, 1986). Removal of the source of lactose intolerance in milk can have a significant impact on the market for dairy products, especially among the 90% of people with an Asian or African background who are lactose-tolerant. A similar approach could be taken for the reduction of beta-lactoglobulin in milk, to which a number of individuals show an allergic response. Some new products made possible by biotechnology in the Asia region include: 1) microbial rennet which can be used in place of animal rennet to coagulate milk during the making of cheese; 2) new lactic acid bacterium produced by cell fusion technique; 3) safe anti-microbial agent, made from skim milk and glucose, which prolongs the shelf life of fresh milk; 4) biologically active peptides extracted from animal blood at slaughterhouse, which can be used as an additive in functional foods to improve the human health; and 5) colorant extracted from animal blood, to partially replace nitrate in meat products.

Besides this, a large number of biotechnological applications are coming on-stream in the area of modification of product quality; reduced contamination and perishability; mitigation of environmental impact of livestock production; and new products ('pharming'). It may also be possible to engineer dairy animals to produce entirely new proteins in the milk. Of particular interest are medically important human proteins which include insulin, plasminogen activator, alpha-antitrypsin and coagulation factors VIII and IX. Gene transfer opened up the possibility that dairy animals could be used as "bioreactors" for the production of these proteins (Lathe *et al.*, 1986). Thus, transgenic animals are increasingly designed for a number of non-food purposes, including medical research, the production of pharmaceuticals products in milk and the production of organs for transplantation to humans.

Biotechnology in Livestock Waste Management

In India, about 30 millions cattle and buffaloes and 70 million sheep and goat are slaughtered annually for meat production. Ruminant animals such as cattle, deer, goat and sheep are estimated to produce 106 metric tones or 15-20% of the estimated global methane production (Pundir *et al.*, 2003). Most (nearly 70%) of this is attributed to cattle of which majority is situated in developing countries. Ruminants produce carbon-dioxide and methane during the anaerobic fermentation of feed inside their rumen and contribute directly to the global warming. Ruminant's account for 18 per cent of the global production of methane in the atmosphere (Aneja, 1991). It is estimated that the amount of methane emission from livestock in India is 11000 metric tones each year (Pundir *et al.*, 2003). Livestock wastes from different sources have the potential to contribute excess nutrients, pathogens, organic matter, solids, and odorous compounds to the environment and posing a serous threat to environment. This pollution can cause eutrophication of surface waters, degradation of ground water quality and threats to human health. Hence, there is a need for the effective management of animal waste, which should be site-specific and non-polluting. Biotechnology can play a crucial role to convert the wastes of livestock origin into useful products.

Treatment of animal by-products and slaughter waste

Animal by-products and slaughterhouse wastes seem to become one of the residues with the problematic handling. Biotechnology offers a cheaper and cleaner alternative. Anaerobic digestion can play an important role in this respect by virtue of its ability to stabilize heterogeneous organic wastes and residues to biomethane.

A bacterium, *Rhodopseudomonas capsulata*, has the ability to grow rapidly in simple synthetic media. It is being used in advanced swine waste treatment plants in both Japan and Korea. Short chain fatty acids, one of the main sources of the bad odour of swine wastes, decreased dramatically after treatment. The residues after treatment can be used as a safe fertilizer.

Biogas technology

Biotechnology is already benefiting developing countries by providing cheap, clean biomass fuel by biogas technology. Biogas technology, by recycling and making scientific use of animal wastes, yields manifold benefits. It eliminates the age old and inefficient practice of burning cattle dung and simultaneously provides fuel and fertilizer and keeps the environment clean. Biogas is produced by means of a process known as anaerobic digestion. It is the process whereby the organic matter is broken down by microbial activity and, as the name suggests, it is the process, which takes place in the absence of air. The digestion of animal waste yields several benefits:

1. the production of methane for use as fuel.
2. the waste is reduced to slurry which has a high nutrient content which makes an ideal fertilizer
3. During the digestion process bacteria in the manure are killed, which is a great benefit to environmental health.

Benefits of biogas technology

Most confined livestock operations handle manure as liquids, slurries, semi-solids or solids that are stored in concrete basins, tanks and other containment structures. Biogas technology can be a cost effective, environment and neighborhood friendly addition to existing manure management strategies. Biogas technologies anaerobically digest manure, resulting in biogas and a liquefied, low-odor effluent. By managing the anaerobic digestion of manure, biogas technologies significantly reduce Biochemical Oxygen Demand (BOD), and pathogen levels; remove most noxious odors; and convert most of the organic nitrogen to inorganic N (e.g., ammonium). The principal reasons a farmer would consider installing a biogas system are:

On-site farm energy : By recovering biogas and producing on-farm energy, livestock producers can reduce monthly energy purchases from electric and gas suppliers. It can be used for running diesel and petrol engines. Electricity can also be generated from it. The cost involved is low which leads to greater acceptability (Table 4).

Table 4 : Quantity of diesel saved by running a 5hp dual fuel engine on biogas

Size of biogas unit (m^3)	Tie period for which a 5hp engine is run twice a day	Quantity of diesel oil saved (litres/day)
8	4 hr	3.6
15	6 hr 30 minutes	5.8
25	12 hours	10.8

Reduced odors : Biogas systems reduce offensive odors from overloaded or improperly managed manure storage facilities. These odors impair air quality and may be a nuisance to nearby communities. Biogas systems reduce these offensive odors because the volatile organic acids, the odor causing compounds, are consumed by biogas producing bacteria.

Manure for agriculture and aquaculture : In the process of anaerobic digestion, the organic nitrogen in the manure is largely converted to ammonium, the primary constituent of commercial fertilizer, which is readily available and utilized by plants. Farmers conceive the importance of biogas unit in terms of the availability of large quantity of better quality manure. Generally speaking one-third to half of the cattle dung is burnt as fuel and is thus loss to soil. A biogas plant doubles the availability of organic manure. The manure produced through a biogas unit has comparative advantage over ordinary manure in terms of both quantity and quality. About 70-75% of the original weight of cattle dung is conserved in a biogas unit, while in open compost pits 50% or more is lost. Biogas manure, known as digested slurry contains higher percentage of other plant nutrients (Table 5).

Table 5 : Comparison of plant nutrients in digested slurry (DS) and farm yard manure (FYM)

Plant nutrient	DS (%)	FYM (%)
Nitrogen (N)	1.5-2.0	0.5-1.0
Phosphorus (P_2O_5)	1.0	0.5-1.0
Potash (K_2O)	1.0	0.5-0.8

It is a good source of micronutrients like zinc, iron, manganese and copper, which have become a limited factor in many soils.

The digested slurry improves the soil fertility and increases the crop yield by 10-20%.

Digested slurry has been found useful for raising fish. Common crap fry and fingerlings fed on mixture rice bran and digested slurry (1:3) and mustard oil cake and rice bran (1:1) on equal nitrogen basis showed faster growth with the bran slurry mixture.

Domestic fuel : Biogas technology makes optimal utilization of the valuable natural resource of dung; it provides nearly three times more useful energy that dung directly burnt. As a cooking fuel, it is cheap and extremely convenient. Based on the effective heat produced, a 2cu m biogas plant could replace, in a month, fuel equivalent of 26 kg if LPG (nearly two standard cylinders), or 37 litres of kerosene, or 88kg of charcoal, or 210 kg of fuel wood, or 740 kg of animal dung.

25 kg Fresh Dung

Direct Burning		Biogas
5 kg dry dung	← Product →	1 cu m biogas
10460 Kcal	← Gross energy →	4713 Kcal
10%	← Device efficiency →	55%
1046 Kcal	← Useful energy →	2592 Kcal
None	← Manure →	10 Kg air dried manure

For domestic purpose, a biogas is easy to use and saves time in the kitchen; biogas stove has an efficiency of about 55%, which is comparable to that of an LPG stove. Cooking on biogas is free from smoke and soot, and can substantially reduce the health problems, which are otherwise quite common in most rural areas in India where biomass is the chief source of fuel. The use of biogas is helpful to improve the quality of life in household. However, the use of biogas is by no means confined to cooking alone. It can be used, through a specially designed mantle, for lightning, too.

Reduced Surface and Groundwater Contamination : Digester effluent is a more uniform and predictable product than untreated manure. The higher ammonium content allows better crop utilization and the physical properties allow easier land application. Properly applied, digester effluent reduces the likelihood of surface or groundwater pollution.

Sanitation and Health : Digesters reduce pathogen populations dramatically by isolating the pathogens and allow pathogen kill and die-off prior to entering storage for land application. Mosquitoes and flies do not breed in digested slurry. Thus biogas units improve sanitation. Biogas being a clean fuel does not cause air pollution. Apart from the

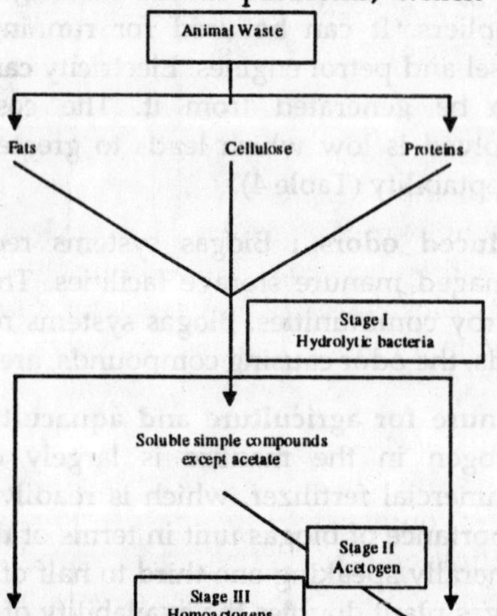

Fig. 1. Four stages of methane production from animal waste

direct benefits gleaned from biogas systems, there is other, perhaps less tangible benefits associated with this technology. By providing an alternative source of fuel, biogas can replace the traditional biomass based fuels, notably wood. Introduced on a significant scale, biogas may reduce the dependence on wood from forests, and create a vacuum in the market, at least for firewood (whether this might reduce pressure on forests however, is contestable).

Principle of Biogas Technology : Biogas production is a microbial process. All the microbes involved in biogas production grow in the absence of air (oxygen). The process involves the combined action of four groups of bacteria, in four stages, in a biogas plant (Fig. 1)

In the first stage of enzymatic hydrolysis, the extra cellular enzymes of microbes, such as cellulose, protease, amylase and lipase externally enzymolize organic material. Bacteria decompose the complex carbohydrates, lipids and proteins in cellulosic biomass into more simple compounds. During the second stage, acid-producing bacteria convert the simplified compounds into acetic acid (CH_3COOH), hydrogen (H_2), and carbon dioxide (CO_2). In the process of acidification, the facultatively anaerobic bacteria utilize oxygen and carbon, thereby creating the necessary anaerobic conditions necessary for methanogenesis. In order to produce more acetate, a third stage is involved in which organisms known as homoacetogens convert hydrogen and simple carbon compounds produced in first and second stages in to acetate. In the final stage, the obligatory anaerobes that are involved in methane formation decompose compounds with a low molecular weight, (CH_3COOH, H_2, CO_2), to form methane (CH_4) and CO_2 (Gate, 1999).

The resulting biogas, sometimes referred to as 'gobar' gas, consists of methane and carbon dioxide, and perhaps some traces of other gases, notably hydrogen sulphide (H_2S). Its exact composition will vary, according to the substrate used in the methanogenesis process, but as an approximate guide, when cattle dung is a major constituent of fermentation, the resulting gas will be between 55-60% CH_4, 40-45% CO_2, plus a negligible amount of H_2S and H_2 (KVIC, 1993). Biogas has the advantage of a potential thermal efficiency, given proper equipment and aeration, of 60%, compared to wood and dung that have a very low thermal efficiency of 17% and 11% respectively (KVIC, 1993).

Methanogenesis or more particularly, the bacteria involved in the fermentation process are sensitive to a range of variables that ultimately determine gas production, and it is worth briefly outlining these factors. Temperature is perhaps the most critical consideration. Gasification is found to be maximized at about 35°C, and below this temperature, the digestion process is slowed, until little gas is produced at 15°C and under. Therefore in areas of temperature changes, such as mountainous regions, or winter conditions that may be more accentuated inland, mitigating factors need to be taken into account, such as increased insulation (Kalia, 1988), or the addition of solar heaters to maintain temperatures (Lichtman, 1983).

Loading rate and retention period of material are also important considerations. In the KVIC model, retention ranges between 30-55 days, depending upon climatic conditions, and will decrease if loaded with more than its rated capacity (which may result in imperfectly digested slurry). KVIC state that maximum gas production occurs during the first four weeks, before tapering off, therefore a plant should be designed for a retention that exploits this feature.

Retention period is found to reduce if temperatures are raised, or more nutrients are added to the digester.

Other factors likely to affect methanogenesis are pH; gas production is found to decrease with increasing acidity, and can result from over-loading the plant, which may stimulate the more fecund acidophiles, at the expense of the more tardy methane-producing microbes. Improved nutrient content, also, as mentioned above will increase the digestion process, and can be manipulated by the addition of animal urine, while toxic substances, such as heavy metals may inhibit gas production (KVIC, 1983).

Understanding the process of methanogenesis allows manipulation, which can serve to maximize gas production in the field. Workers over the last twenty years have experimented with the digestion process, and have made strides in increasing gas yields, using techniques that can be similarly employed in a rural environment. Sharma *et al.* (1988), have shown that biogas generation is increased when the particle size of organic material is small, in this case, less than 1mm. The workers recommend that a physical pre-treatment, such as grinding would improve a system's performance, and could also reduce the size of digester needed. A manual machine for physical pre-treatment of material would be a viable piece of equipment in a rural environment; indeed, there may be a similar piece of equipment already in use. Other workers have found that biogas production is accelerated by the presence of metal ions in biomass (Geeta *et al.*, 1990).

Potential of Biogas

In India, the dissemination of large-scale biogas plants has began in the mid-seventies and the process has become consolidated with the advent of the National Project on Biogas Development (NPBD) in 1981, which has been continuing since. Against the estimated potential of 12 million biogas plants, 2.9 million family type and 2700 community and institutional plants have been set up till December 1999. This is estimated to have helped in a saving of 3 million tonnes of fuel wood per year and manure containing nitrogen equivalent to 0.7 million tonnes of urea. However, in terms of total dung that is available in the country, the potential is much more. The bovine population in India is 260 millions. As adult bovine produce an average of 10 kg of dung per day. Since grazing is a common practice in India, all the dung produced cannot be collected. If it is assumed that 75% of the dung is collected, nearly 2 millions tonnes of dung would be available everyday. At 25 kg per one cubic metre, this dung can feed as many as 40 millions biogas plants of 2 cubic metre capacity, which can be considered the ultimate potential for biogas technology. With such high potential, which can be routed to hitherto unemphasized applications of shaft power and electricity generation, biogas can make a significant contribution to the development of small industries and agriculture, and thus to the overall advancement of the rural areas.

Need for Biotechnological Research

The need for biotechnological research may be to increase the level of affluence for the developed countries, but for developing countries, it could reduce hunger and starvation. All available biotechnologies are not appropriate or relevant to all countries. Developing countries may need to adapt some of these technologies before they can use them. It is therefore important that developing countries should develop the capacity to maintain a strong base of applied and adaptive research and some level of training to keep abreast with new developments.

Some biotechnological techniques, which could have application in developing countries, are summarized in Table 6. While the applications of some of the available technologies are relatively simple, it is unlikely that developing countries will be able to retain those trained to support implementation of biotechnology programmes if adequate financing is not available to provide required equipment, chemicals etc to expand or even maintain such programmes.

The development of biotechnology capacity in developing countries is also justified in order to support research, which is only of interest to them or that, which because of restrictions in developed countries, can only be done in developing countries. For example, veterinary regulations in most developed countries generally prohibit the introduction of biological material (e.g. infectious agents required for vaccine development) or live animals for research purposes. Therefore developed countries must develop the capacity to undertake such research in the relevant countries.

Table 6 : Possible applications of biotechnology to the solution of problems of livestock production in developing countries

Problem	Possible solution	Scale of economic impact
Difficulty of implementing selection programmes	Selection in nucleus herds, using AI, ET, embryo sexing	Large
	Marker-assisted selection	Moderate
Difficulty of maintaining dairy cattle performance after F_1 cross	Use of IVF, ET and embryo sexing Selection among local breeds using AI, MOET	Large Medium/large
	Development of composite breeds	Large
Poor quality forages	Microbial treatment of forages	Moderate
	Modification of rumen microflora	Moderate/large
	Genetic improvement of forages and their symbionts	Moderate
Animal diseases	New vaccines and new diagnostics	Large
Cost and environmental challenge to imported cattle	Use of ET to import embryos	Small
Need for increased efficiency in intensive systems	Use of rBST and rPST in dairy and pig production	Large

Sources: Adapted (with additions) from Cunningham (1990); Doyle and Spradbrow (1990).

Biotechnology adoption and adaptation

Artificial insemination

The reason of non-implementation of technologies developed by developed countries with much success in the developing countries is the failure to recognize the importance of adapting technology to local conditions Artificial insemination is a good example. Where AI technology has been adopted in India, not much consideration has been given to adapting it to the circumstances in which it is to be applied to ensure sustainability. Government subsidies of the AI system are considered to be another main cause of failure of AI in developing countries. Privatization of the services may change the situation. For example, in some of these countries, the concept of farmer co-operatives is well developed and applied for specific cash crops or milk marketing. AI could easily be run by co-operatives organized in schemes in which farmers are "grouped" on the basis of such factors as transport requirements and similarity of systems of production. Semen can easily be delivered over short distances by motorbikes, bicycles, horses, donkeys etc. Under such circumstances, use of fresh semen collected from bulls belonging to the co-operative could be considered. This would eliminate the cost of freezing semen. Such schemes would make it easier to match genotypes to production systems within a country. Indeed, such a scheme could form the basis for the genetic improvement of localized indigenous breeds.

Embryo transfer

Embryo transfer (ET) could have a major impact on cattle breeding in developing countries (Cunningham 1990) especially as part of a nucleus breeding scheme. However, successful ET requires highly motivated, experienced staff and a high capital investment in facilities, equipment and drugs.

In general, the inappropriateness of ET for developing countries is ascribed to lack of infrastructure. However, in some instances, ET represents a solution to a lack of infrastructure. Thus, establishment of multiple ovulation embryo transfer (MOET) is considered an attractive means of genetic improvement where infrastructure for progeny testing is not available.

Most developing countries have limited financial resources. In addition, equipment and supplies tend to be more expensive than they are in developed countries due to transportation costs, import tariffs, lack of hard currency etc. These make ET technology prohibitive in these countries. However, it is possible to adapt ET techniques to local conditions thereby reducing the cost. Seidel and Seidel (1992) pointed out that a lot of the fancy equipment associated with ET is not essential for successful utilization of the technology. For example, fancy freezing equipment are no more effective than dry ice/alcohol baths for freezing embryos as these save labour and are more convenient. Similarly, filters are not essential for isolation of embryos, neither are disposable sterile syringes and fancy plastic dishes. Thus, by combining good imagination with knowledge of basic principles, the technology can be successfully adapted to local conditions. Therefore research, especially of an applied nature, on such technologies by institutions in developing countries is always justified and often essential. Moreover, researchers need to be exposed to new technologies or procedures to appreciate the power and limitations of such technologies.

Application of Biotechnology – Problems and Prospects

Biotechnology has become the most important new approach for the present and future of the high technology based countries. Major advances in biotechnology have made significant contributions to improved efficiency and productivity of farm animals and its role is exposed to increase in the future. Its application in fermentation technology, food processing, therapeutic and vaccination programmes has already given fruitful results. There is no doubt that biotechnology can bring the improvements in animal growth and production, nutrition, reproduction, health and animal products. It can do this while conserving natural resources and the environment. Gene transfer technology seems to be very promising for future genetic improvements of animals.

The research and development efforts in the area of biogas are aimed at developing new design of biogas plants and improved package of practices to increase the rate of biogas production, particularly at low and high temperatures, besides ensuring effective utilization of the designed slurry. Use of animal waste in the form of biogas can be a useful component in the rural economy in India, if well-coordinated dissemination system is adopted. Participation in biogas technology varies across socio-economic groups, and across regions. The goal to provide rural India with a viable and sustainable source of energy from the animal wastes can only be achieved by using biogas technology.

Problems in initiating biotechnology research are a lack of cooperative effort to work on goal-orientated projects and the absence of enough trained personnel in new genetic, reproductive, nutritional and veterinary techniques. Physical facilities and non-technical human resources are abundantly available but sometimes major equipment and costly chemicals are not easily available. These problems can be overcome with technical cooperation between developed and developing countries, and among developing countries themselves. One of greatest challenges facing the biotechnology industry is to convince the public of the safety of its products. This will require considerable communication skills and wide-ranging discussions to bring clarity and consensus into the moral and ethical issues relating to genetic engineering.

Priorities for biotechnology research and its application in developing countries including India should be in those areas where research has already been initiated, and where the developed countries have already been getting some benefit from the research investments. In those countries where biotechnology research is still to be undertaken, appropriate biotechnologies should be identified. In the higher learning institutes of the country, training facilities for modern biotechnological research should be given to young scientists of future generations. The different important areas like, identification of disease resistant gene for diseases of economic value and production of disease resistant flock for genetic improvement of farm animals, research on certain aspects of embryo transfer and associated technology, research on gene cloning and gene transfer, production of DNA vaccine for specific disease of economic importance, production of monoclonal antibodies, adoption of biotechnological approach to improve the low quality forages and agro-industrial wastes and genetic and non-genetic manipulation of rumen micro-flora for increasing the utilization of lingo-cellulose material. For better utilization of livestock wastes by using modern biotechnological techniques, the research and development projects are being sponsored in the following thrust areas: 1) studies in the field of microbiology, biochemistry and engineering for increasing the yield of biogas, especially at low and high temperatures; 2) development of cost effective design of biogas plants; 3) development of designs and methodologies for utilization of succulent biomass other than cattle dung for biogas production; 4) reducing the cost of biogas plants by using alternative building materials and construction methodology; and 5) diversified use of slurry for value added products.

Summary

India is a repository of domestic animal diversity. Among these, some local breeds of various species have been extinct or some are in the verge of extinction. Although the country is the largest milk producer in the world, but production potential of most of the animals is poor or marginal due to dilution of breeds and other factors. Improvement of production potential of animals in terms of milk, meat, products, skin, wool and other products is the burning issue to the scientific community. Techniques of modern biology such as molecular cloning of genes, gene transfer, genetic manipulation of animal and plant, embryo transfer, genetic manipulation of rumen microbes, chemical and biological treatment of low quality animal feeds for improved nutritive value, genetically engineered immunodiagnostic and immunoprophylactic agents as well as veterinary vaccines, *inter alia*, are a reality today and are finding their ways into research and development programmes of developing countries. Biotechnology is offering unprecedented opportunities for increasing agricultural productivity with special reference to animal productivity and for protecting the environment through converting the animal wastes into useful product. Conservation of our vast genetic diversity is also the key task to safe guard the social and bio-ecosystem of the country. In developing country like India, concerted and sequential efforts are required for research, development and implementation of technologies, which is best suited to animal farming community, for enhancing the animal productivity and to reduce the environmental pollution of animal origin. For achieving these objectives, advance training in the field of modern biological techniques is required for scientists, developmental workers, veterinarians and even advanced farmers. Proper infrastructure for conducting hi-tech research in the field of biotechnology should be given. The importance of biotechnology and its relevance is slowly being accepted by policy makers in developing countries. Government should make efforts to motivate/encourage the livestock entrepreneurs or farmers to use the latest technology for increasing the productivity of livestock. There is an urgent need of long term research strategy for identifying and quantifying the negative impact of livestock wastes on environment and the new research knowledge generated by biotechnology could be utilized for designing and implementing eco-friendly policies and programmes to reduce the pollution of livestock origin.

References

Aboul-Naga, A.M., Aboul-Ela, M.B. and Hassan, F. (1992) Manipulation of reproductive activity in subtropical sheep. Small Ruminant Research 7:151-160.

Aneja, R.P. (1991) Research and development priorities of the Indian dairy industry. Indian Dairyman, 1: 3-7.

Armstrong, D.G. and Gilbert, H.J. (1985) Biotechnology and the rumen: A mini review. Journal of the Science of Food and Agriculture 36:1039-1046.

Armstrong, D.G. and Gilbert, H.J. (1991) The application of biotechnology for future livestock production. In: Proceedings of the 7th International Symposium on Ruminant Physiology, Sendai, Japan. Academic Press, San Diego, USA. pp. 737-761.

Axford, R.F.E., Bishop, S.C., Nicholas, F.W. and Owen, J.B. (2000) Breeding for Disease Resistance in Farm Animals. 2nd Edition, CAB International, Wallinford, UK.

Baker, R.L. (1998) A review of genetic resistance to gastrointestinal nematodes in sheep and goats in the tropics and evidence for resistance in some sheep and goat breeds in sub-humid costal Kenya. Animal Genetics Resources Information Bulletin 24: 13-30.

Beck, E. and Strohmaier, K. (1987) Subtyping of European foot and mouth disease strains by nucleotide sequence determination. Journal of Virology 61: 1621-1626.

Bedford, M.R. and Schulze, H. (1998) Exogenous enzymes in pigs and poultry. Nutrition Research Reviews 11: 91-108.

Bhattacharya, N.K. (1984) Progress in embryo technology and its application in animal science. Paper presented at 1st Animal Physiology Research Workers' Conference, Karnal, India.

Bindon, B. M. and Piper, M.R. (1986) In: Genetic engineering of animals – An agricultural prospectives (Eds. J.W. Evans and A. Hollender), 637.

Bird, S.H. and Leng, R.A. (1985) Productivity responses to eliminating protozoa from rumen of sheep. In: Biotechnology and Recombinant DNA technology in the Animal Production

Industries - Reviews in Rural Science No. 6. (eds. R.A. Leng, J.S.F. Barker, D.B. Adams and K.J. Hutchinson, University of New England Publishing Unit, Armidale, Australia, pp. 109-117).

Bishop, S.C., Chesnais, J. and Stear, M.J. (2002) Breeding for disease resistance: Issues and opportunities. Proceedings of the 7th World Congress on Genetics Applied to Livestock Production, Montpellier, France 31: 597-604.

Blancou, J. 1990. Utilization and control of biotechnical procedures in veterinary science. Revue Scientifique et Technique de l'Office International des Epizooties 9:641-659.

Bodin L., Bibe B., Blance M.R. and Ricordean G. (1986) Parameters genetiques de la concentration plasmatique en FSH des agrelles de race Lacaune viande. Genetics, Selection, Evolution 18:55-62.

Bonneau, M. and Laarveld, B. (1999) Biotechnology in animal nutrition, physiology and health. Livestock Production Science 59: 223-241.

Brandon, R.B., Daniel, R.C.W. and Lavin, M.F. (1991) Early detection of bovine leukosis virus DNA in infected sheep using the polymerase chain reaction. Research in Veterinary Science 50:89-94.

Brem, G., Brenig, B., Goodman, H.M.Selden, R.C., Graf, E., Kruff, B., Springmann, K., Hondele, J., Meyer, J., Winnacker, E.L. and Krausslich, H. (1986) Gene transfer in rabbits and pigs. Proceedings of the 3rd World Congress on Genetics Applied to Livestock Production, Lincoln, Nebraska, XII. Biotechnology. Pp. 45-56.

Brem, G., Graf, F. and Krausslich, M. (1984) Genetic and economic differences among methods of gene conservation in farm animals. Livestock Production Science 1:65-68.

Buckrell, B.C., Buschbeck, C., Gartley, C.J., Kroetsch, T., McCutcheon, W., Martin, J., Penner, W.K. and Walton, J. S. (1992) A breeding trial using a transcervical technique for artificial insemination in sheep. In: Proceedings of the 12th International Congress on Animal Reproduction, The Hague, The Netherlands, 3:1531-1533.

Bulgin, M.S., Lincoln, S.D., Parker, C.F., Sowth, P.J., Dahmen, J.J. and Lane, V.M. (1988) Genetic-associated resistance to foot rot in selected Targhee sheep. Journal of the American Veterinary Medicine Association 192: 512-515.

Burton, J.H., MacLeod, G.K., McBride, B.W., Burton, J.L., Bateman, K., McMillan, I. and Eggert, R.G. (1990) Overall efficacy of chronically administered recombinant bovine somatotropin to lactating dairy cows. Journal of Dairy Science 73:2157-2167.

CBD (Convention on Biodiversity). (1992) The Convention on Biological Diversity. UNEP (United Nations Environment Programme), Nairobi, Kenya.

Chang, H. (1996) Genetic engineering to enhance microbial interference and related therapeutic applications. Nature Biotechnology 14: 423-431.

Church, D.C. and Pond, W.G. (1988) Basic Animal Nutrition and Feeding. 3rd Edition, John Wiley & Sons, Toranto, pp. 267-275.

Clark, A.J., Archibald, A.L., McClenghan, M., Simons, J.P., Whitelow, C.B.A. and Wilmut, I. (1990) The germ line manipulation of livestock: Progress during the last five years. Proceedings of the New Zealand Society for Animal Production 50:167-180.

Clements, J.E., Wall, R.J., Narayan, O., Haver, D., Sheffer, D., Powell, A.M., Zink, M. C. and Rexroad, C.E. (1994) Transgenic sheep that express the visna virus envelope gene. Theriogenology 41:180 (abstract).

Colleau, J.J. (1991) Using embryo sexing within closed mixed multiple ovulation and embryo transfer schemes for selection on dairy cattle. Journal of Dairy Science 74:3973-3984.

Constantini, F. and Lacy, E. (1981) Introduction of a rabbit beta-globin gene into the mouse germ line. Nature 294: 92-94.

Cunningham, E.P. (1976) The use of egg transfer techniques in genetic improvement. Proceedings of the EEC Seminar on egg transfer in cattle. (Edited by L.E.A. Rowson) pp. 345-353.

Cunningham, E.P. (1990) Animal production. In: Persley G.J. (ed), Agricultural Biotechnology: Opportunities for International Development. CAB (Commonwealth Agricultural Bureaux) International, Wallingford, UK. pp. 169-175.

Doyle, J.J. and Spradbrow, P.B. (1990) Animal health. In: Persley G.J. (ed), *Agricultural Biotechnology: Opportunities for International Development*. CAB (Commonwealth Agricultural Bureaux) International, Wallingford, UK. pp. 176-186.

Dutta, T.K. (1998) Studies on the potentials of probiotics on in vitro fermentation and productive performance of lactating dairy cows. Ph. D. Thesis submitted to National Dairy Research Institute, Karnal, Haryana, India.

FAO (2002) Year Book, Food and Agriculture Organization of the United Nations.

FAO (Food and Agriculture Organization of the United Nations). (1992) *The Management of Global Animal Genetic Resources. Proceedings of an FAO Expert Consultation*. FAO Animal Production and Health Paper 104. FAO, Rome, Italy. 309 pp.

Flint, H.J. (1994) Molecular genetics of obligate anaerobes from the rumen. FEMS Microbiological Letter 121: 259-267.

Flint, H.J., Bisset, J. and Webb, T. (1989) Use of antibiotic-resistance mutations to tract strains of obligately anaerobic bacteria introduced into the rumen of sheep. Journal of Applied Bacteriology 67:117-183.

Flores, D.A. (1989) Application of recombinant DNA to rumen microbes for the improvement of low-quality feed utilization. Journal of Biotechnology 10:95-112.

Foote, R.H. (1982) Cryopreservation of spermatozoa and artificial insemination: Past, present, and future. Journal of Andrology 3:85-100.

Forsberg, C.W., Cheng, K.J., Krell, P.J. and Phillips, J.P. (1993) Establishment of rumen microbial gene pools and their manipulation to benefit digestion by domestic animals. In: Proceedings of the Seventh World Conference on Animal Production, Edmonton, Alberta, Vol. 1. pp. 281-316.

Freeman, B.M. and Bumstead, N. (1987) Breeding for disease resistance-the prospective role of genetic manipulation. Avian Pathology 16: 353-365.

Fuller, R. (1989). Probiotics in man and animals. Applied Biotechnology 66: 365-378.

GATE (Deutsches Zentrum fur Entwicklungstechnologien)(1999): Biogas technology in Orissa, India. Eschborn, Germany.

Geeta, G.S, Jagadeesh, K.S, Reddy, T.K.R. (1990) Nickel as an accelerator of biogas production in water hyacinth (*Eichhornia crassipes*). Biomass 21: 157-161.

Gogolin-Ewens, K.J., Meeusen, E.N.T., Scott, P.C., Adams, T.E. and Brandon, M.R. (1990) Genetic selection for disease resistance and traits of economic importance in animal production. Revue Scientifique et Technique de l'Office International des Epizooties 9:865-896.

Gordon, I. and Lu, K.H. (1990). Production of embryos in vitro and its impact on livestock production. Theriogenology, 33: 77-88.

Gordon, J.W., Scango, G.A., Plotkin, D.J., Barbosa, J.A. and Ruddle, F.H. (1980) Genetic transformation of mouse embryos by microinjection of purified DNA. Proceedings of National Academic of Science, USA 77: 7380-7384.

Gregg, K. (1989) Enhancement of microbial detoxification by gene transfer. In: Proceedings of the 7th International Symposium on Ruminant Physiology, Sendai, Japan. Academic Press, San Diego, USA. p. 58.

Haas, L., Moenning, V. and Kaaden, O.R. (1990) Use of biotechnical methods in veterinary medicine. Revue Scientifique et Technique de l'Office International des Epizooties 9:245-251.

Hammer, R.E., Brinster, R.L., Rosenfeld, M.G., Evans, M.R. and Mayo, K.E. (1985) Expression of human growth hormone releasing factor in transgenic mice in increased somatic growth. Nature 315: 680-683.

Hart, I.C. and Johnson, I.D. (1986) Growth hormone and growth in meat producing animals. In: Buttery P.J., Haynes N.B. and Lindsay D.B. (eds), Control and Manipulation of Animal Growth. Butterworths, London, UK. pp. 135-159.

Haynes, N.B. and Schanbacher, B.D. (1983) The control of reproductive activity in the ram. In: W. Haresign (Ed.) Sheep Production. Butterworths, London, pp. 431.

Henderson, A.R., McDonald, P. and Anderson, D. (1982) The effect of cellulase preparation derived from *Trichoderma vivide* on the chemical changes during the ensilage of grass, lucerne and clover. Journal of the Science of Food and Agriculture 33:16-20.

Hooper, P.G., Sharp, R., Rowlinson, P. and Armstrong, D.G. (1989) An Assessment of the Potential of a Lactobacilli Inoculant as a Silage Additive. British Society of Animal Production, Winter Meeting, 13-15 March. Paper 63. Scarborough, UK.

Jasiorowski, H.A. (1990) Open nucleus breeding schemes - new challenge for the developing countries. In: Zurkowski M. (ed), Proceedings of a Conference on Open Nucleus Breeding Systems, Biatobrzegi, Poland, 1-19 January 1989. FAO (Food and Agriculture Organization of the United Nations), Rome, Italy. pp. 7-12.

Jeffreys, A.J. and Morton, D.B. (1987) DNA fingerprints of dogs and cats. Animal Genetics 18:1-15.

Jones, R.J. and Megarrity, R.G. (1986) Successful transfer of DHP degrading bacteria from Hawaiian goats to Australian ruminants to overcome the toxicity of *Leucaena*. Australian Veterinary Journal 63:259-262.

Kalia, A. K. (1988) Development and evaluation of a fixed-dome plug flow anaerobic digester. Biomass 16: 225-235.

Kang, Y. and Richardson, T. (1985). Genetic engineering of caseins. Food Technology, 39: 89-93.

Kennedy, B.W. and Moxley, J.E. (1980) Genetic factors influencing atrophic rhinitis in the pig. Animal Production 30: 277-283.

Kerr, R.J., Frisch, J.E. and Kinghorn, B.P. (1994) Evidence for a major gene for tick resistance in cattle. In: Proceedings of the 5th World Congress on Genetics Applied to Livestock Production, Guelph, Canada, 7-12 August 1994. Volume 20. International Committee for World Congress on Genetics Applied to Livestock Production, Guelph, Ontario, Canada. pp. 265-268.

Knowles, D.P. Jr. and Gorham, J.R. (1990) Diagnosis of viral and bacterial diseases. Revue Scientifique et Technique de l'Office International des Epizooties 9:733-757.

Kundu, S.S. and Kumar, V. (1987) Biotechnological processes to convert crop residues to quality feeds for livestock. Indian Dairyman 39:113-114.

KVIC (1993) Khadi and V.I Commission and its Non-Conventional Energy Programmes. KVIC, Bombay, India.

Lamont, S.J. and Dietert, R.R. (1990) Immunogenetics. In: Crawford, R.D. (Ed.), Poultry Breeding and Genetics. Elsevier, Amsterdam, UK.

Lande, R. and Thompson, R. (1990) Efficiency of marker-assisted selection in the improvement of quantitative traits. *Genetics* 124:743-756.

Lathe, R., Clark, A.J., Archibald, A.L., Bishop, J.O., Simons, P. and Wilmut, I. (1986) In: Exploiting new technologies in Animal Breeding (Smith, C., King, J. and McKay, J. eds.) pp. 91-102, Clarendon Press, Oxford.

Lefevre, P.C. (1992) Biotechnology and animal disease diagnosis in developing countries. In: Schwartz J.J. and Franzen H. (eds.), Potential and Limitations of Biotechnology in Livestock Production in Developing Countries. Proceedings of a Symposium, Humboldt, University of Berlin, October 1992. Part II. ATSAF (Council for Tropical and Subtropical Agricultural Research), Bonn, Germany. pp. 129-138.

Lefevre, P.C. and Diallo, A. (1990) Diagnosis of viral and bacterial diseases. Revue Scientifique et Technique de l'Office International des Epizooties 9:951-965.

Lichtman, R.J. (1983) Biogas Systems in India. VITA (Volunteers in Technical Assistance). Virginia, USA.

Lim, J.M., Fukui, Y. and Ono, H. (1991) The post-thaw developmental capacity of frozen bovine oocytes following *in vitro* maturation and fertilization. Theriogenology 35:1225-1235.

Loftus, R.T., MacHugh, D.E., Ngere, L.O., Balain, D.S., Badi, A.M., Bradley, D.G. and Cunningham, E.P. (1994) Mitochondrial genetic variation in European, African and Indian cattle populations. Animal Genetics 25:265-271.

Lu, K.H., Jiang, H.S., Wang, W.L. and Gordon, I. (1990) Pregnancies established in cattle by transfer of fresh and frozen embryos derived from *in vitro* maturation and fertilization of oocytes and their subsequent culture *in vitro*. Theriogenology 33:278 (abstract).

Macmillan, K.L. and Tervit, H.R. (1990) Available technologies and prospective developments in reproduction and embryology. Proceedings of the Australian Association of Animal Breeding and Genetics 8:9-18.

McCullough, K.C. (1993) The application of biotechnology to the diagnosis and control of animal diseases. Revue Scientifique et Technique de l'Office International des Epizooties 12:325-353.

McSweeney, C.S., Dalrymple, B.P., Gobius, K.S., Kennedy, P.M., Krause, D.O., Mackie, R.I. and Xue, G.P. (1999) The application of rumen biotechnology to improve the nutritive value of fibrous feed stuffs: pre and post ingestion. Livestock Production Science 59: 265-283.

Mercier, J.C. (1986) In: Exploiting new technologies in Animal Breeding (Smith, C., King, J. and McKay, J. eds.) pp. 122-132, Clarendon Press, Oxford.

Muggli, N.E., Hohenboken, W.D., Cundiff, L.V. and Mattson, D.E. (1987) Inheritance and interaction of immune traits in beef calves. Journal of Animal Science 64: 385-393.

Mukherjee, F. (2000) Molecular techniques for diagnosis of Brucellosis. Veterinary Research Abstracts Compendium. ISBN No. 81-90025-4-5. Indian Veterinary Science Congress, 18-19 February, Indian Veterinary Research Institute, Izatnagar, Bareilly , Uttar Pradesh, India.

Nicholas, F.W. (1979) The genetic implications of multiple ovulation and embryo transfer in small dairy herds. Proceedings of the 30th Annual Meeting of the EEAp, Harrogate, CG1, 11.

Nicholas, F.W. and Smith, C. (1983) Increase rate of genetic change in dairy cattle by embryo transfer and splitting. Animal Production 36:341-353.

Osburn, B.I. (1991) The Second International Symposium on Bluetongue, African Horse Sickness and Related Orbiviruses. Wiener Tierärztliche Monatsschrift 78:407.

Pelletier, J., Bodin, L. Hanocq, E., Malpaux, B. Teyssier, J., Thimonier, J. and Chemineau, P. (2000) Association between expression of reproductive seasonality and alleles of the gene for Mel1a receptor in the ewe. Biology of Reproduction 62: 1096-1101.

Phipps, R.H., Weller, R.F., Craven, N. and Peel, C.J. (1990) Use of prolonged-release bovine somatotropin for milk production in British Friesian dairy cows. I. Effect on intake, milk production and feed efficiency in two consecutive lactations of treatment. Journal of Agricultural Science (Cambridge) 115:95-104.

Prasad, C.S. and Gowda, N.K.S. (2002) Recent development in mineral supplementation for optimum production. In: The Proceedings of Globalization - Challenges to the Animal Nutritionists. IVth Biennial Conference and Exhibition, Nov. 20-22, Calcutta, pp. 97-111.

Pundir, R.S., Singh, K. and Saxena, R. (2003). Livestock-Environment interaction: Polices, Issues and options. Indian Dairyman 55(5): 31-39.

Reid, H.W. (1989) *Orf Virus Infection of Sheep*. Animal Diseases Research Association News Sheet 16.

Robinson, J.J. and McEvoy, T.G. (1993) Biotechnology - the possibilities. Animal Production 57:335-352.

Rothschild, M.F. (1989) Selective breeding for immune responsiveness and disease resistance in livestock. AgBiotech News and Information 3: 355-360.

Ruane, J. and Colleau, J.J. (1995) Marker assisted selection for genetic improvement of animal populations when a single QTL is marked. Genetical Research, Cambridge, 66: 71-83.

Ruane, J. and Colleau, J.J. (1996) Marker assisted selection for a sex-limited character in a nucleus breeding population. Journal of Dairy Science, 79: 1666-1678.

Ryan, A. and Ho, Mae-Wan (2000) Transgenic DNA in animal feed critique of MAFF report CS0116 "Effect of feed processing conditions on DNA fragmentation" Institute of Science in Society and Biology Department Open University Walton Hall, Milton Keynes MK 7 6 AA, UK. http://www.twnside.org.sg/title/feed-cn.html

Seidel G.E. and Seidel S.M. (1992) Analysis of applications of embryo transfer in developing countries. In: Niemann H., Franzen H. and Smidt D. (eds), Potential and Limitations of Biotechnology in Livestock Production in Developing Countries. Proceedings of a Symposium, Mariensee, Germany, May 1992. Part I. ATSAF (Council for Tropical and Subtropical Agricultural Research), Bonn, Germany. pp. 68-80.

Selle, P.H., Ravindran, V., Caldwel, R.A. and Vryden, W.L. (2000) Phytate and phytase: consequence for protein utilization. Nutrition Research Reviews 13: 255-278.

Sharma, A. and Balakrishnan, C.R. (1992) Prospects and retrospects of genetic engineering in farm animals. In: Perspectives in Cytology and Genetics (eds. G.K.Manna and S.C.Roy), 7: 271-287.

Sharma, S. K., Mishra, I. M., Sharma, M.P., Saini, J.S. (1988) Effect of particle size on biogas generation from biomass residues. Biomass 17: 251-263.

Shewry, P.R. and Kreis, M. (1987). Biotechnology and improvement of feeds. Proceedings of the Nutrition Society 46:379-385.

Smith, C. (1988a) Genetic improvement of livestock in developing countries using nucleus breeding units. World Animal Review 6:2-10.

Smith, C. (1988b) Applications of embryo transfer in animal breeding. Theriogenology 29: 203-212.

Smith, C., Meuwissen, T.H.E. and Gibson, J.P. (1987) On the use of transgenes in livestock improvement. Animal Breeding Abstracts 55:1-10.

Solbu, H. (1982) Heritability estimates and progeny testing for mastitis, ketosis and "all diseases". Zeitschrift Fur Tierzuchtung Zuchtungsbiologie 101: 210-219.

Soller, M. (1990) Genetic mapping of the bovine genome using deoxyribonucleic acid-level markers to identify loci affecting quantitative traits of economic importance. Journal of Dairy Science, 73:.2628-2646.

Soller, M. and Beckmann, J.S. (1983) Genetic polymorphism in varietal identification and genetic improvement. Theoretical and Applied Genetics 67: 25-33.

Southern, E.M. 1975. Detection of specific sequences among DNA fragments separated by gel electrophoresis. Journal of Molecular Biology 98:503-517.

Stewart, C.S., Fonty, G. and Gouet, Ph. (1988) The establishment of rumen microbial communities. Animal Feed Science Technology 21: 69-97.

Taylor, St. C.S., Moore, A.J., Thiessen, R.B. and Bailey, C.M. (1985) Efficiency of food utilization in traditional and sex-controlled systems of beef production. Animal Production 40:401-440.

Thomson, A.M. and Flint, H.J. (1989) Electroporation induced transformation of *Bacteroides ruminicola* and *Bacteroides uniformis* by plasmid DNA. FEMS Microbiology Letters 61:101-104.

Tien, M. and Kirk, K.T. (1983) Lignin degrading enzyme from the Hymenomycete *Phanerochaete chrysosporium* burds. Science 221:661-663.

Van Brunt, J. (1987) Bringing biotechnology to animal health care. Biotechnology 5:677-683.

Van Vleck, L.D. (1981) Potential genetic impact of artificial insemination, sex selection, embryo transfer, cloning and selfing in dairy cattle. In: Brackett B.G., Seidel, Jr. G.E. and Seidel, S.M. (eds.), New Technologies in Animal Breeding. Academic Press, New York, USA. pp. 221-242.

Verma, S.V.S. and Mandal, A.B. (2000) Chemical feed additives in poultry nutrition. Feed Microbiology (Karma et al., eds.) Centre for Advanced studies on Animal nutrition, Indian Veterinary Research Institute, pp. 153-161.

Warwick, B.L., Berry, R.O. and Horlacher, W.R. (1934) Resulting of mating rams to Angora female goats. Proceedings of 27th Annual Meeting of Animal Society of Animal Production, 225-227.

Weber, J.L. and May, P.E. (1989) Abundant class of DNA polymorphisms which can be typed using polymerase chain reaction. American Journal of Human Genetics 44:388-396.

Willet, E.L., Black, W.G., Casida,L.E., Stone, W.H. and Buckner, P.J. (1951) Successful transplantation of a fertilized bovine ovum. Science, 113: 247.

Williams, J.G., Kubelik, A.R., Livak, K.J.C., Rafalski, J.A. and Tingey, M.S.V. (1990) DNA polymorphisms amplified by arbitrary primers are useful as genetic markers. Nucleic Acids Research 18(22): 6531-6535.

Wilmut, I., Haley, C.S., Simons, J.P. and Webb, R. (1990) Prospects for molecular genetic manipulation to improve reproduction. In: W.G. Hiu, R. Thomsonand J.A. Woolliams (Eds.), Proceedings of 4th World Congress Genetics Applied to Livestock Production, Edinburgh, Vol. XVI, 4WCGALP Organising Committee, Edinburgh, pp. 281-290.

Woolliams, J.A. and Smith, C. (1988) The value of indicator traits in the genetic improvement of dairy cattle. Animal Production 46:333-345.

Woolliams, J.A. and Wilmut, I. 1989. Embryo manipulation in cattle breeding and production. Animal Production 48:3-30.

Wright, I.G. (1990) Immunodiagnosis of, and immunoprophylaxis against the haemoparasites *Babesia* spp and *Anaplasma* spp in domestic animals. Revue Scientifique et Technique de l'Office International des Epizooties 9:345-356.

Zhang, Y.Z., Zylstra, G.J., Olsen, R.H. and Reddy, C.A. (1986) Identification of cDNA clones for ligninase from *Phanerochaete chrysosporium* using synthetic oligonucleotide probes. Biochemical and Biophysical Research Communications 137:649-656.

Abbreviation

AI: Artificial insemination.
BST: Bovine somatotropin
DNA: Deoxyribonucleic Acid
DS: Digestive slurry
ELISA: Enzyme-linked immunosorbent assays
ET: Embryo transfer
FAO: Food and Agricultural Organization
FMD: Foot and mouth disease
FSH: Follicle stimulating hormone
FYM: Farm yard manure
GM: Genetically modified
GRF: Growth hormone releasing factor
NAD: Nucleic acid hybridization
REM: Restriction endonuclease mapping
ILRAD: Laboratory for Research on Animal Diseases
ILRI: International Livestock Research Institute
IVF: *In-vitro* fertilization
IVM: *In-vitro* maturation
MAB: Monoclonal antibody
BOD: Biochemical oxygen demand
MAI: Marker assisted introgression
MAS: Marker assisted selection
MHC: Major histocompatibility complex
MOET: Multiple ovulation and embryo transfer
mRNA: Messenger ribonucleic Acid
NAP: Nucleic acid probe
NPBD: National Project on Biogas Development
ONBS: Open nucleus breeding system
PCR: Polymerase chain reaction
PCV: Packed cell volume
PMSG: Pregnant mare serum gonadotropin
POME: Palm oil mill effluent
PPR: Peste des petite ruminants
QTL: Quantitative trait loci
RAPD: Randomly amplified polymorphic DNA
rBST: Recombinant bovine somatotropin
RCNR: Restriction endonuclease reaction
RFLP Restriction fragment length polymorphism
rPST: Recombinant porcine somatotropin
SCP: Single cell protein